Data
Transportation
and Protection

Applications of Communications Theory
Series Editor: R. W. Lucky, *AT & T Bell Laboratories*

A Continuation Order Plan is available for this series. A continuation order will bring delivery of
each new volume immediately upon publication. Volumes are billed only upon actual shipment.
For further information please contact the publisher.

Data Transportation and Protection

John E. Hershey
The BDM Corporation
Boulder, Colorado

and

R. K. Rao Yarlagadda
Oklahoma State University
Stillwater, Oklahoma

Plenum Press • New York and London

Library of Congress Cataloging in Publication Data

Hershey, J. E. (John E.)
  Data transportation and protection.

  (Applications of communications theory)
  Bibliography: p.
  Includes index.
  1. Data transmission systems. 2. Data transmission systems—Security measures. I.
Yarlagadda, R. K. Rao. II. Title. III. Series.
TK5105.H47   1986   005.7   86-15065
ISBN 0-306-42257-3

© 1986 Plenum Press, New York
A Division of Plenum Publishing Corporation
233 Spring Street, New York, N.Y. 10013

Printed in the United States of America

TO MY WIFE ANNA
AND
OUR CHILDREN
DAVID
JOHN
JIMMY

(JOHN E. HERSHEY)

TO MY WIFE MARCEIL
AND
OUR CHILDREN
TAMMY
RYAN
TRAVIS

(R. K. RAO YARLAGADDA)

Preface

A new breed of engineer is developing in our contemporary society. These engineers are concerned with communications and computers, economics and regulation. These new engineers apply themselves to data—to its packaging, transmission, and protection. They are *data engineers.*

Formal curricula do not yet exist for their dedicated development. Rather they learn most of their tools "on the job" and their roots are in computer engineering, communications engineering, and applied mathematics. There is a need to draw relevant material together and present it so that those who wish to become data engineers can do so, for the betterment of themselves, their employer, their country, and, ultimately, the world—for we share the belief that the most effective tool for world peace and stability is neither politics nor armaments, but rather the open and timely exchange of information. This book has been written with that goal in mind.

Today numerous signs encourage us to expect broader information exchange in the years to come. The movement toward a true Integrated Services Digital Network (ISDN) is perhaps the clearest of these. Also, the development of formal protocol layers reflects both a great deal of brilliance and compromise and also the desire for a common language among data engineers.

The first eight chapters of this book deal with some very basic mathematical "tools." The reader has undoubtedly been exposed to many of them, but it is our experience that many engineers are deficient in at least some of them. This is perhaps because many of the topics, when taught, are not properly motivated or integrated into the larger picture. There is also a wide assortment of useful tools that simply do not fit into any of the extant curricula for electrical or computer engineering. We have therefore assembled what, at first, may seem to be an unusual assortment of topics. These topics were not chosen haphazardly but rather as a collection of those concepts that have been of the greatest use to us as practicing data engineers. We do not treat the topics with rigor. Rather, what we hope to impart is a "visceral feel" for the material. We hope to inculcate in the reader a belief that real problems can be analyzed and solved by the application of very powerful, "user-friendly" tools that are already out there, just waiting to be used.

Chapter 1 deals with the issues of representing and manipulating data. Uniradix and mixed-radix number systems are reviewed. The specific topics chosen were selected to stimulate thought, to serve as the foundation for later chapters, or both. The latter part of the chapter touches on the vast and rich field of boolean functions. Most students never come to know and appreciate the beauty of the mathematics that intertwines with the boolean algebras—and it's a pity! But also of concern is the hazard that powerful results may remain locked in the literature, unsprung and unapplied to real problems.

Chapter 2 recounts some of the very powerful and straightforward tools of combinatorics, the mathematics of counting and choice. These topics themselves are preparatory for the chapter's latter half, which is a modest refresher on discrete probability theory. Only a few concepts and distributions are covered, but the sad fact remains that it is in probability issues where engineers stumble most often and fall the hardest. The topics were chosen so that the reader will be aware of some of the important issues, learn a few very useful procedures, and be able to discern where it is appropriate to apply the rudiments so learned.

Chapter 3 is an introduction to number theory. Number theory is "the queen of mathematics" and a topic not usually introduced to engineers. Recently, however, number theory has become an essential mathematical tool for understanding and applying contemporary ideas and techniques in signal processing and cryptography, and, once again, one need not be thoroughly immersed in the topic to avail oneself of much of its power. In this chapter, as in others, we have attempted to organize a useful subset of the collective knowledge and still preserve some of the "spice" that has proved to be so highly addictive to countless mathematicians throughout the centuries, consuming their lives and their talents.

The next four chapters (4–7) are intended to compose a unit on matrices. Almost a monograph in itself, the four chapters take the reader through this vast and powerful field. Matrix representations and matrix operations are probably the most valuable tools in the data engineering arts, yet, sadly, most students seem to have little working knowledge in these disciplines. These four chapters are designed to remedy this problem and not only prepare the student for the latter chapters of the book but also lay the groundwork for a variety of other courses of study.

Chapter 8, the final chapter on mathematical tools, is devoted to random and pseudorandom sequences, the study of which is traditionally extremely difficult for most engineers. And rightly so! For it has only been recently that theories have been unified and many of the related questions solved or even properly posed. The subject needs study because, time and time again, the same questions are asked, particularly in synchronization problems. The Markov chain is introduced and reviewed because of its lawful universality: It is easily and naturally employed quite often when all else fails. The chapter concludes with a rather lengthy review of m-sequences cast, for the most part, in a matrix formalism. The section is offered with no apology or further motivation save for the observation that the m-sequence is one of the most prevalent, persistent, and, indeed, useful of all of the semiarcane topics in communications theory. The reader will not learn much high-powered math, but will be able to understand and employ the m-sequence—or, if appropriate, reject a suggestion to use it.

The final six chapters deal with selected techniques of data engineering that rest on the tools which preceded them. It was very difficult to pick topics out of the many that could have been chosen. Rather than attempting to cover everything, we focused on six particular topics that have enjoyed much publicity in the past decade.

The first of these chapters, Chapter 9, is devoted to source encoding. As consumer sensitivity to the communications time-bandwidth product continues to increase, it is more and more common to find attention paid to source encoding. Certainly it is a logical place to start and, as we often tell our students, "Your very first priority in data engineering is to understand your source." The topics we picked for the chapter are traditional and extremely powerful. The student should be able to employ them directly or use them as a springboard for researching and devising more sophisticated schema.

Chapter 10 is an introduction to cryptography, which has become one of the most vital topics of the present decade. We divided the chapter into three parts. The first deals with classical cryptography, specifically the Data Encryption Standard (DES) promulgated by the National Bureau of Standards in 1977. The DES is undergoing integration into many different data architectures as public awareness and acceptance of a credible threat to privacy grows. The second part of the chapter is concerned with so-called two-key or public key cryptography (PKC). The study of PKC systems is hampered by the technique's relative newness and the lack of any widely accepted standards. It is important for the data engineer to be aware of PKC systems, however, and we have also striven to create an awareness of the importance of the interplay between cryptosystem and protocol—an interplay which can have a tremendous, but often difficult to recognize, impact on total system security. The final part of the chapter is devoted to the important new technique of secret sharing systems or "shadow" systems, as they are sometimes called. All of these topics are prime candidates as tools for computer security.

Chapter 11 is on synchronization, an extremely important part of data communications. The techniques and concepts presented in this chapter are probably most useful and immediately applicable to data links that use the RF spectrum.

Chapter 12 introduces the data frame or packet concept. This concept arose naturally as a response to the requirement to send data over a two-way channel that was not error-free but yet not plagued with errors to the same degree as, for example, a battlefield communications channel operating in the presence of electronic warfare jamming.

Chapter 13 is an introduction to space division switching or connecting networks. These networks are seeing great utilization not only in data communications but also in signal processing and parallel and data flow computing architectures. We believe that space division switching will continue to grow in importance as the division between communications and computation becomes increasingly blurred.

The final chapter provides a brief introduction to network architectures, which must be designed to enhance network reliability or survivability. These topics are of extreme interest to many national programs and commercial data services as well.

In writing these chapters, we continually asked ourselves two questions: "What do I, as a practicing data engineer, need to know about this particular topic?" and "What particular methods, algorithms, ways of looking at issues, and tricks have been most useful to me in my professional career?" We were guided by the answers. The book that resulted is intended to be suitable as a text for motivated upper-level undergraduate students and first-year graduate students, and as a reference work.

John E. Hershey
R. K. Rao Yarlagadda

Boulder and Stillwater

Acknowledgments

I express my thanks to the BDM Corporation and its farsighted management team for providing me with a working atmosphere conducive to professional development. I also thank my colleagues who, through the years, have shared their wisdom and insights with me and helped me to grow professionally. Explicit thanks go to W. J. Pomper, H. P. Kagey, C.-L. Wu, H. M. Gates, A. D. Spaulding, M. Nesenbergs, L. W. Pederson, M. Marcus, G. Simmons, R. Kubichek, W. Hartman, and P. McManamon.

J.E.H.

I express my thanks to the Oklahoma State University and to the students I have worked with. It has been a pleasure.

R.K.R.Y

Contents

Data—Its Representation and Manipulation

1.1. Introduction

In this chapter we are concerned with the preliminaries of representing information, or data, using binary units or bits. We start with a most basic concept—number systems. The number systems considered are those common ones of "normal binary representation," negabinary, and Gray coding. We also introduce a less well known "mixed-radix system," based on the factorials. This representation will be of use to us later on when we look at combinatorics.

In this chapter we are also concerned with the manipulation of binary data by various boolean functions. The study of boolean functions is most often encountered in courses on logic design, but over the years we have found that boolean functions are extremely useful vehicles for examining all sorts of processes and data related issues.

1.2. Number Systems

Most number systems are positional. A number, or quantity, is represented by a concatenation of symbols or a symbol "string." The amount that a particular symbol contributes to the number is a function of the particular symbol's position on the string. Most familiar number systems are uniradix. For these systems a number, N, has the form

$$N = s_n s_{n-1} \cdots s_2 s_1 s_0 \tag{1}$$

which is to be interpreted as meaning

$$N = s_n b^n + s_{n-1} b^{n-1} + \cdots + s_2 b^2 + s_1 b + s_0 \tag{2}$$

where b is the radix or number of different symbols that can be used to write a number.

In this book we deal primarily with the radix $b = 2$ or the binary number system. This particular system is important simply because the vast majority of high-speed electronic logic operates in "base 2." We refer to binary numbers expressed in the form (1) as *normal* binary numbers.

It is important to be able to convert base 2 numbers to their equivalent value representation in other bases. Two other important bases are octal (base 8) and hexadecimal (base 16). Base 8 uses the symbols 0, 1, 2, 3, 4, 5, 6, 7 for the quantities 0–7; base 16 traditionally uses 0, 1, 2, 3, 4, 5, 6, 7, 8, 9, A, B, C, D, E, F for the quantities 0–15. There is a particularly simple way to convert between base 2 and bases 8 and 16. We demonstrate by example. To convert a base 2 number to a base 16 number, we group our base 2 number into four-bit segments starting from the right with the *least* significant bit. Write the equivalent hexadecimal symbols for the four-bit segments. The concatenation of hexadecimal symbols is the required conversion. For example, assume we wish to convert

$$1100101001110010101110101 \tag{3}$$

into base 16. We proceed as follows:

$$((0)\ (0)\ 11)\ (0010)\ (1001)\ (1100)\ (1010)\ (1111)\ (0101)$$
$$3 \qquad 2 \qquad 9 \qquad C \qquad A \qquad F \qquad 5$$

We see that our equivalent hexadecimal number is 329CAF5. (How do you convert from hexadecimal to binary?) Octal conversion is performed similarly but instead of grouping the base 2 number into four-bit segments, the segments are only three bits long.

Problems

1. The conversion of base 2 numbers to other bases such as the decimal base ($b = 10$) is more difficult. Derive algorithms that perform conversions from base 2 to base 10 and vice versa. Do the same for base 16 to base 10 and vice versa.

2. One obvious way of converting a binary number to a decimal number is to express it as an octal number $s_n s_{n-1} \cdots s_1 s_0$ and then carry out the recursive computation portrayed in Figure 1.1.

(a) Expressing (3) as an octal number we have 312345365. Show that application of the algorithm in Figure 1.1 yields the decimal equivalent of 53070581.

There is an alternate method, recorded by Croy (1961), that is both interesting and useful for "paper-and-pencil" computations. The algorithm involves only decimal-based subtractions and multiplications by 2. The algorithm comprises three steps:

1. Convert the binary number to an octal number.
2. Assuming the octal number has more than one digit, in which case we would be finished, take the most significant digit, double the octal digit leaving the result in decimal; align this number so that its most significant digit falls under the second most significant digit of the octal number which is being converted to decimal; subtract the decimal number from the octal number—for purposes of the subtraction treat the octal number *as though it were a* decimal number.
3. Step 2 is the algorithm's heart and is repeated, with slight modification each iteration, until a subtraction has involved the least significant digit of the octal number which is being converted to decimal. The modification referred to involves picking one more

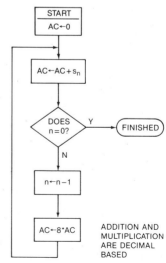

Figure 1.1. Procedure for converting an octal number to its decimal equivalent.

digit, starting with the most significant digit and moving right, for the number that is to be doubled.

Although the above algorithm is very easy to comprehend, it is a bit difficult to grasp just from words. An example seems warranted. We operate on the octal representation of (3): 312345365. The iterations are shown below. Readers are encouraged to plow through the computations and form their own descriptions of the algorithm.

312345365	$3 \cdot 2 = 6$
$\dfrac{6}{252345365}$	$25 \cdot 2 = 50$
$\dfrac{50}{202345365}$	$202 \cdot 2 = 404$
$\dfrac{404}{161945365}$	$1619 \cdot 2 = 3238$
$\dfrac{3238}{129565365}$	$12956 \cdot 2 = 25912$
$\dfrac{25912}{103653365}$	$103653 \cdot 2 = 207306$
$\dfrac{207306}{82922765}$	$829227 \cdot 2 = 1658454$
$\dfrac{1658454}{66338225}$	$6633822 \cdot 2 = 13267644$
$\dfrac{13267644}{53070581}$	$\leftarrow$ decimal conversion of the octal number

(b) Study the above algorithm and explain to someone else, in an informal way, the algorithm and why it works.

1.3. Negabinary Numbers

Most familiar number systems use a positive base, $b > 0$, in (2). The form (2) is thus incapable of representing negative integers. There is no reason, however, that b can not be allowed to assume negative or even imaginary and complex values. There is an important special case, the *negabinary* system, for which $b = -2$. Table 1 gives the first 16 negabinary numbers and their decimal equivalents.

Interest in data architectures using negabinary numbers ebbs and flows. Of all the attention, a considerable proportion has concerned applications

**Table 1. The First 16
Negabinary Numbers**

Negabinary	Decimal
0	0
1	1
10	−2
11	−1
100	4
101	5
110	2
111	3
1000	−8
1001	−7
1010	−10
1011	−9
1100	−4
1101	−3
1110	−6
1111	−5

to analog-to-digital conversion. Since data source representation and coding is one of our basic topics, we spend a little time reviewing some basic principles of negabinary arithmetic. We use work published by Wadel (1961).

The following are the important elementary relations between the binary and negabinary bases:

$$(2)^i = (-2)^i$$
$$-(2)^i = (-2)^{i+1} + (-2)^i \qquad (i, \text{even}) \qquad (4a)$$
$$(-2)^i = -(2^i)$$
$$(2)^i = (-2)^{i+1} + (-2)^i \qquad (i, \text{odd}) \qquad (4b)$$

To convert a *positive* binary number to negabinary, we write the binary number and underneath it we write a number that has ones in all the even positions which are one position to the left of a one in the number above it, and zeros elsewhere. We then add the two numbers using the negabinary addition rules:

$$0 + 0 = 0$$
$$0 + 1 = 1$$
$$1 + 0 = 1 \qquad (5)$$
$$1 + 1 = 110$$

For example, let us convert 1011, eleven, to negabinary:

$$\begin{array}{l} \downarrow\ \downarrow\ \downarrow \quad \text{(even positions)}\\ 1011\\ 10100 \end{array}$$

11111 ← (eleven in negabinary)

The above was an easy conversion. Now let us try to convert 1111, fifteen in normal binary:

$$\begin{array}{l} \quad\downarrow \qquad \text{(start of ``infinite carry'')}\\ 1111\\ 10100 \end{array}$$

. . . 00010011

Note that in this case an "infinite carry" is created. This happens when we attempt to add 1 to 11 according to the rules in (5). The carry produces zeros as it goes and as long as there are no significant bits to the left of the carry, the carry may be ignored. Thus we find that fifteen converts to 10011 in negabinary. The "infinite carry" is just one of the problems encountered in implementing a negabinary data architecture.

To convert a *negative* binary number to negabinary, we write down the absolute value of the binary number. Below it, we write a number that has ones in all odd positions which are one position to the left of a one in the number above it, and zeros elsewhere. We then add these numbers in accordance with (5) and appropriately dispense with infinite carries. For example, let us convert minus thirteen to negabinary:

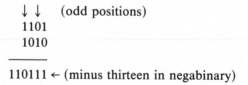

$$\begin{array}{l} \downarrow\ \downarrow \qquad \text{(odd positions)}\\ 1101\\ 1010 \end{array}$$

110111 ← (minus thirteen in negabinary)

Problems

1. Devise rules for adding and subtracting negabinary numbers.
2. Consider a number system with base $b = \sqrt{-1}$. Devise conversion rules from complex binary, $x + iy$, where x and y are normal binary and $i = b$. Derive rules for adding and subtracting these complex binary numbers.

What would be the significant problems in building computational hardware for this new base?

1.4. The Factorial Number System

As we said in the beginning of this chapter, most number systems are uniradix. There are, however, some important systems that are mixed-radix. One of the most important of these, for discrete mathematics, is the factorial number system. For our purpose, at this point, we consider the factorial as a function of the nonnegative integers. Let n be one of these integers. We write n factorial as $n!$ and define it as the product $n! = n(n-1)(n-2) \cdots (2)(1)$. A remarkable property of the factorials is that any nonnegative integer, m, can be *uniquely* written:

$$m = d_1 \cdot 1! + d_2 \cdot 2! + d_3 \cdot 3! + \cdots + d_n \cdot n! + \cdots = \sum_{i=1} d_i \cdot i! \qquad (6)$$

where the $\{d_i\}$ are the factorial digits of m and each factorial digit satisfies the constraint

$$0 \leq d_i \leq i \qquad (7)$$

The factorial digits for 99, for example, are $d_1 = 1$, $d_2 = 1$, $d_3 = 0$, and $d_4 = 4$ as $99 = 1 \cdot 1! + 1 \cdot 2! + 0 \cdot 3! + 4 \cdot 4!$ The factorial number system will serve as an important "aside" when we touch on combinatorics.

Problems

1. Prove that the representation given in (6) is unique for every non-negative m.
2. Develop an efficient algorithm for deriving the factorial digits of non-negative decimal numbers.

1.5. The Gray Code

Another, often useful, representation of data is the Gray code. The Gray code serves to represent the integers 0 to $2^n - 1$ by 2^n code words, each of n bits, that have the remarkable property that the code word representing integer i, $0 \leq i \leq 2^n - 2$, differs from the code work representing integer $i + 1$ by exactly (and only) one bit. A good way to visualize the Gray code for $n = 3$ is to first envision a three-dimensional, unit-volume cube whose eight vertices are naturally encoded as shown in Figure 1.2.

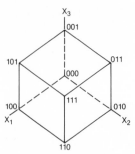

Figure 1.2. Three-dimensional unit-volume with coded vertices.

To generate a Gray code we must be able to visit each vertex exactly once and our path or "walk" on the cube must be confined to staying on the edges. This guarantees that our successive Gray code words will differ exactly and only by one bit. (Why?) One path that satisfies both of the above conditions is the path traversing the vertices as follows:

$$(0, 0, 0) \rightarrow (0, 0, 1) \rightarrow (0, 1, 1) \rightarrow (0, 1, 0) \rightarrow (1, 1, 0)$$
$$\rightarrow (1, 1, 1) \rightarrow (1, 0, 1) \rightarrow (1, 0, 0)$$

This particular path thus generates the following Gray code

Integer	Gray Code Words
0	0 0 0
1	0 0 1
2	0 1 1
3	0 1 0
4	1 1 0
5	1 1 1
6	1 0 1
7	1 0 0

There are many possible Gray codes for a general unit-volume hypercube ($n > 3$). We standardize on one specific form which we call the "normal" Gray code. This particular Gray code is often referred to as the "reflected" Gray code because its generation can be performed by a simple process involving iterated reflections as we will now demonstrate. To generate the normal Gray code for $n = 3$ (or any n) we proceed as follows. We write a zero and a one (the normal Gray code for $n = 1$) as follows:

0

1

We then write a "line of reflection" below the zero and copy the zero below the line of reflection. But, we start the copying by beginning with that entry which is just above the line of reflection and sequentially move to the top of the list:

$$
\begin{array}{c}
0 \\
1 \\
\hline
1 \\
0
\end{array}
\qquad \text{line of reflection}
$$

We now prefix the entries above the line with a zero and prefix those below with a one:

$$
\begin{array}{c}
00 \\
01 \\
11 \\
10
\end{array}
$$

Note that we have generated the normal Gray code for $n = 2$. To generate the normal Gray code for $n = 3$ we iterate the previous steps:

$$
\begin{array}{c}
00 \\
01 \\
11 \\
10 \\
\hline
10 \\
11 \\
01 \\
00
\end{array}
$$

and

$$
\begin{array}{c}
000 \\
001 \\
011 \\
010 \\
110 \\
111 \\
101 \\
100
\end{array}
$$

which is the eight-member set of code words for $n = 3$.

Problem

Generate the normal Gray code for $n = 4$, $n = 5$, and $n = 6$. Roughly, how far down the list of code words does the all-ones code word seem to fall? What would you expect "as a rule of thumb"?

We can easily convert from normal binary to normal Gray. If the n-bit normal binary word is denoted by $b_{n-1}b_{n-2} \cdots b_1b_0$ and the n-bit normal Gray word by $g_{n-1}g_{n-2} \cdots g_1g_0$ then

$$g_i = \begin{cases} b_i, & i = n - 1 \\ b_i + b_{i+1}, & i \neq n - 1 \end{cases} \tag{8}$$

where the addition in (8) is modulo 2 (i.e., $0 + 0 = 1 + 1 = 0$; $0 + 1 = 1 + 0 = 1$).

Conversion from normal Gray to normal binary is not as straightforward because we must maintain a "running sum," that is, we can not easily solve explicitly for a single bit as we do in (8) above. The conversion is given by

$$b_{n-1} = g_{n-1} \qquad \text{(addition is modulo 2)}$$

$$b_k = g_k + b_{k+1}, \qquad k \neq n - 1$$

Problems

1. Derive equations in the form of (8) for the ith binary bit in the normal binary to normal Gray conversion.

2. (Wang, 1966) There is a relatively efficient algorithm for normal Gray to normal binary conversion that takes advantage of the parallelism afforded by hardware registers in automatic data processing (ADP) equipment. Consider the algorithm shown in Figure 1.3. Registers B, G, and T are all n bits long. Register G contains a normal Gray code word, $g_{n-1}g_{n-2} \cdots g_1g_0$. At conclusion, Register B will contain the equivalent normal binary code word. Work through a few test cases for Wang's algorithm. Why does the algorithm work? Show that the algorithm of Figure 1.4 also works. Compare the execution times for the two algorithms.

If we encode the integers 0 through $2^n - 1$ using n-bit normal binary code words, it is simple to assess the magnitude of the errors in the decoded values. A single-bit error in code word transmission can yield at most an absolute difference of 2^{n-1}. If exactly two errors are made, the absolute error difference can be as high as $2^{n-1} + 2^{n-2} = 3 \cdot 2^{n-2}$. For exactly e errors, the difference can be as high as $2^{n-1} + 2^{n-2} + \cdots + 2^{n-e}$. If we encode with

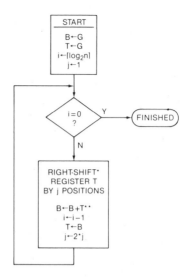

Figure 1.3. Parallel algorithm for converting normal Gray code to normal binary. (* Bits shifted beyond T_0 are lost. Zeros are shifted into T from the left. ** Addition here is parallel or bit-by-bit modulo 2 addition.)

normal Gray we have a different behavior. Cavior (1975) has derived an interesting upper bound for the maximum possible error. The bound states that if exactly e errors are made in transmitting an n-bit normal Gray code word representing the integers 0 through $2^n - 1$, then the magnitude of the error in the decoded values will be *less than* $2^n - (2^e/3)$, $e > 0$; thus, the *more* errors in transmission of the code word, the *smaller* the upper bound on the maximum possible error.

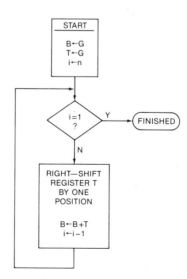

Figure 1.4. Another algorithm for converting normal Gray code to normal binary.

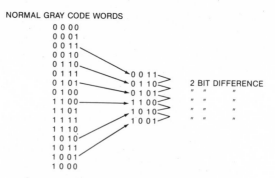

Figure 1.5. Six 4-bit normal code words of weight 2.

Problem

Experiment with Cavior's upper bound. For what error patterns is the bound "tight"?

It is often required to generate all n-bit words of weight $k \leq n$, that is, all words of n bits that have exactly k ones and $n - k$ zeros. The normal Gray code has a property that naturally commends its use to this task. Bitner et al. (1976) took this property and developed it into an efficient algorithm which we shall discuss below.

The relevant property exhibited by the normal Gray code is simply that successive n-bit code words of weight k differ in only and exactly two positions. To see this, look at Figure 1.5 where we generate all members of the $n = 4$ bit normal Gray code and then extract all $\binom{n}{k} = \binom{4}{2} = 6$ words of weight $k = 2$.

Figure 1.6 is a flow chart of the Bitner et al. algorithm for generating all the $\binom{n}{k}$ n-bit words of weight k.

As an example of the operation of the algorithm, the following is the output generated for the parameter values $n = 8$, $k = 6$:

$$
\begin{array}{cccccccc}
0 & 0 & 1 & 1 & 1 & 1 & 1 & 1 \\
0 & 1 & 1 & 0 & 1 & 1 & 1 & 1 \\
0 & 1 & 1 & 1 & 1 & 0 & 1 & 1 \\
0 & 1 & 1 & 1 & 1 & 1 & 1 & 0 \\
0 & 1 & 1 & 1 & 1 & 1 & 0 & 1 \\
0 & 1 & 1 & 1 & 0 & 1 & 1 & 1 \\
0 & 1 & 0 & 1 & 1 & 1 & 1 & 1 \\
1 & 1 & 0 & 0 & 1 & 1 & 1 & 1 \\
1 & 1 & 0 & 1 & 1 & 0 & 1 & 1 \\
\end{array}
$$

```
1 1 0 1 1 1 1 0
1 1 0 1 1 1 0 1
1 1 0 1 0 1 1 1
1 1 1 1 0 0 1 1
1 1 1 1 0 1 1 0
1 1 1 1 0 1 0 1
1 1 1 1 1 1 0 0
1 1 1 1 1 0 1 0
1 1 1 1 1 0 0 1
1 1 1 0 1 0 1 1
1 1 1 0 1 1 1 0
1 1 1 0 1 1 0 1
1 1 1 0 0 1 1 1
1 0 1 0 1 1 1 1
1 0 1 1 1 0 1 1
1 0 1 1 1 1 1 0
1 0 1 1 1 1 0 1
1 0 1 1 0 1 1 1
1 0 0 1 1 1 1 1
```

1.6. A Look at Boolean Functions

The study of boolean functions is a vast field. Unfortunately, most students who encounter boolean functions in the course of their studies are exposed to only a part of the theory—generally techniques to minimize particular constraints in functional synthesis. While these are necessary and profitable techniques, the student often becomes disenchanted with what seems laborious bookkeeping. In this section we touch very briefly on aspects of boolean function theory that will be new to most readers. Our coverage is neither thorough nor elegant; rather, our purpose is to stimulate thought and introduce concepts, in particular the Walsh-Hadamard transform, which will be needed later on. It is our hope that the interested reader will be inspired to view other problems in terms of boolean functions and be sensitized to the existence of powerful techniques that can be brought to bear on their analyses.

A boolean function of n variables is an assignment of a zero or one to each of the 2^n possible inputs. The most common artifice for considering a boolean function is the table of assignments known as a truth table. The general truth table for n inputs is shown in Figure 1.7.

$$\text{FOR } j=1 \text{ TO } k \text{ DO } \begin{cases} v_j \leftarrow 1 \\ p_j \leftarrow j+1 \end{cases}$$

$$\text{FOR } j=k+1 \text{ TO } n+1 \text{ DO } \begin{cases} v_j \leftarrow 0 \\ p_j \leftarrow j+1 \end{cases}$$

$t \leftarrow k$
$p_1 \leftarrow k+1$
$i \leftarrow 0$

WHILE $i < n+1$ DO $\left\{ \begin{array}{l} \text{OUTPUT } (v_n \cdots v_1) \\ i \leftarrow p_1 \\ p_1 \leftarrow p_i \\ p_i \leftarrow i+1 \\ \text{IF } v_i = 1 \text{ THEN} \left\{ \begin{array}{l} \text{IF } t \neq 0 \quad \text{THEN } v_t \leftarrow \overline{v_t} \\ \qquad\qquad\ \text{ELSE } v_{i-1} \leftarrow \overline{v_{i-1}} \\ t \leftarrow t+1 \end{array} \right. \\ \qquad\quad\ \text{ELSE} \left\{ \begin{array}{l} \text{IF } t \neq 1 \quad \text{THEN } v_{t-1} \leftarrow \overline{v_{t-1}} \\ \qquad\qquad\ \text{ELSE } v_{i-1} \leftarrow \overline{v_{i-1}} \\ t \leftarrow t-1 \end{array} \right. \\ v_i \leftarrow \overline{v_i} \\ \text{IF } t = i-1 \text{ OR } t = 0 \\ \qquad\qquad \text{THEN } t \leftarrow t+1 \\ \qquad\qquad \text{ELSE} \left\{ \begin{array}{l} t \leftarrow t - v_{i-1} \\[4pt] p_{i-1} \leftarrow p_1 \\[4pt] \text{IF } t = 0 \text{ THEN } p_1 \leftarrow i-1 \\ \qquad\qquad \text{ELSE } p_1 \leftarrow t+1 \end{array} \right. \end{array} \right.$

Figure 1.6. Algorithm for generating all n-bit words of weight $= k$ (the overbar indicates complementation, i.e., a zero becomes a one and a one becomes a zero).

Problem

Why are there 2^{2^n} different boolean functions of n variables?

There are 16 boolean functions of two variables. They are listed by their truth tables in Figure 1.8.

x_n	x_{n-1}	$\cdots$	x_2	x_1	$f(x_n, x_{n-1}, \ldots, x_2, x_1)$
0	0	$\cdots$	0	0	y_0
0	0	$\cdots$	0	1	y_1
0	0	$\cdots$	1	0	y_2
0	0	$\cdots$	1	1	y_3
					$\vdots$
1	1	$\cdots$	0	0	y_{2^n-4}
1	1	$\cdots$	0	1	y_{2^n-3}
1	1	$\cdots$	1	0	y_{2^n-2}
1	1	$\cdots$	1	1	y_{2^n-1}

Figure 1.7. General truth table of a boolean function of n variables.

		Function														
x_2	x_1	1	2	3	4	5	6	7	8	9	10	11	12	13	14	15	16
0	0	0	0	0	0	0	0	0	0	1	1	1	1	1	1	1	1
0	1	0	0	0	0	1	1	1	1	0	0	0	0	1	1	1	1
1	0	0	0	1	1	0	0	1	1	0	0	1	1	0	0	1	1
1	1	0	1	0	1	0	1	0	1	0	1	0	1	0	1	0	1

Figure 1.8. Sixteen Boolean functions of two variables.

Problem

Most of the 16 functions in Figure 1.8 are functions of x_1 and x_2. Which are really functions only of x_1? Of x_2? Of neither x_1 nor x_2?

Many of the functions of Figure 1.8 have acquired common names due to their utility. Function 2 is often called the AND function; function 7 the EXCLUSIVE-OR; function 8 the INCLUSIVE-OR; function 9 the NOR; and function 15 the NAND.

Problem

The NOR and NAND functions are the only two-input boolean functions that are "functionally complete," that is, all of the two-input boolean functions can be synthesized from *either* of them alone. Show that this is so by direct construction.

The following forms are very important, commonly encountered boolean functions:

1. *Self-dual functions.* A function $f(x_1, \ldots, x_n)$ is said to be self-dual if

$$f(x_1, \ldots, x_n) = \overline{f(\overline{x_1}, \ldots, \overline{x_n})} \qquad (9)$$

where the overbar indicates complementation, that is, 0 is replaced by 1 and 1 is replaced by 0.

2. *Symmetric functions.* A function f is symmetric if and only if

$$f(x_1, x_2, x_3, \ldots, x_{n-1}, x_n) = f(x_{i_1}, x_{i_2}, x_{i_3}, \ldots, x_{i_{n-1}}, x_{i_n})$$

where $(i_1, i_2, i_3, \ldots, i_{n-1}, i_n)$ is any ordering of $(1, 2, 3, \ldots, n-1, n)$. Some essential symmetric functions are the following:

(a) *Parity functions.* A function $f(x_1, \ldots, x_n)$ is said to be the parity function if

$$f(x_1, \ldots, x_n) = x_1 + x_2 + \cdots + x_n \qquad (10)$$

This function is useful in error detection coding.

(b) *Threshold functions.* A function $f(x_1, \ldots, x_n)$ is said to be a threshold function if

$$f(x_1, \ldots, x_n) = \begin{cases} 1, & \text{if and only if } \sum_{i=1}^{n} x_i \geqslant k \\ 0, & \text{otherwise} \end{cases} \tag{11}$$

These functions will be used when we discuss rapid acquisition sequences.

(c) *Majority functions.* A function $f(x_1, \ldots, x_n)$ where n is odd, is said to be a majority function if $f(x_1, \ldots, x_n)$ is a threshold function and $k = (n + 1)/2$. These functions will be used when we study component or JPL codes.

Problem

Convince yourself that a symmetric function of n uncomplemented variables, $f(x_1, \ldots, x_n)$, is completely described by listing those $\{k_i\}$, $0 \leqslant k_i \leqslant n$, for which $f(x_1, \ldots, x_n) = 1$ when *precisely* k_i of its arguments are unity. In fact, Shannon's notation for such a function is $S_{k_1, k_2, \ldots, k_m}(x_1, \ldots, x_n)$. What would Shannon's notation be for (1) a five-input parity function and (2) a five-input majority function?

As we have begun to see, the 2^{2^n} boolean functions of n variables may be broken into classes that share something in common. If we are given a function's truth table it is often not at all obvious if the function fits into a general class. Fortunately, there are a number of tools that can be used to aid us. Two of the most powerful of these tools are the truth table autocorrelation function and the Walsh–Hadamard transform.

To start, we define the autocorrelation of $f(x_1, \ldots, x_n)$, denoted by R_f, as

$$R_f(\delta) = \sum_{i=0}^{2^n - 1} f(i) f(i + \delta) \tag{12}$$

The argument $i + \delta$ in (12) needs special interpretation. It is the component-wise modulo 2 addition of the normal binary representations of i and δ. To make this clear we calculate $R_f(3)$ for a three-variable function:

$$R_f(3) = f(0)f(0 + 3) + f(1)f(1 + 3) + f(2)f(2 + 3) + f(3)f(3 + 3)$$

$$+ f(4)f(4 + 3) + f(5)f(5 + 3) + f(6)f(6 + 3) + f(7)f(7 + 3) \tag{13}$$

We now write the arguments of the second f's of (13) in component binary form:

$$R_f(3) = f(0)f((0, 0, 0) + (0, 1, 1)) + f(1)f((0, 0, 1) + (0, 1, 1))$$
$$+ f(2)f((0, 1, 0) + (0, 1, 1)) + f(3)f((0, 1, 1) + (0, 1, 1))$$
$$+ f(4)f((1, 0, 0) + (0, 1, 1)) + f(5)f((1, 0, 1) + (0, 1, 1))$$
$$+ f(6)f((1, 1, 0) + (0, 1, 1)) + f(7)f((1, 1, 1) + (0, 1, 1)) \qquad (14)$$

We now perform the component-wise modulo 2 addition to get

$$R_f(3) = f(0)f(0, 1, 1) + f(1)f(0, 1, 0) + f(2)f(0, 0, 1) + f(3)f(0, 0, 0)$$
$$+ f(4)f(1, 1, 1) + f(5)f(1, 1, 0)$$
$$+ f(6)f(1, 0, 1) + f(7)f(1, 0, 0) \qquad (15)$$

Converting back to decimal notation we have

$$R_f(3) = f(0)f(3) + f(1)f(2) + f(2)f(1) + f(3)f(0) + f(4)f(7)$$
$$+ f(5)f(6) + f(6)f(5) + f(7)f(4) \qquad (16)$$

Problem

(a) Verify the expressions below for $R_f(0)$ to $R_f(7)$ for an arbitrary three-variable boolean function:

$$R_f(0) = f(0) + f(1) + f(2) + f(3) + f(4) + f(5) + f(6) + f(7)$$
$$R_f(1) = 2(f(0)f(1) + f(2)f(3) + f(4)f(5) + f(6)f(7))$$
$$R_f(2) = 2(f(0)f(2) + f(1)f(3) + f(4)f(6) + f(5)f(7))$$
$$R_f(3) = 2(f(0)f(3) + f(1)f(2) + f(4)f(7) + f(5)f(6))$$
$$R_f(4) = 2(f(0)f(4) + f(1)f(5) + f(2)f(6) + f(3)f(7))$$
$$R_f(5) = 2(f(0)f(5) + f(1)f(4) + f(2)f(7) + f(3)f(6))$$
$$R_f(6) = 2(f(0)f(6) + f(1)f(7) + f(2)f(4) + f(3)f(5))$$
$$R_f(7) = 2(f(0)f(7) + f(1)f(6) + f(2)f(5) + f(3)f(4))$$

(b) Flow chart a computer program to calculate $\{R_f(i), i = 0, \ldots, 2^n - 1\}$ for an n-variable boolean function.

We are now ready to make use of the autocorrelation function in an example of identifying function type. Our example concerns self-dual functions. Consider the autocorrelation of the three-variable self-dual function shown in Figure 1.9. Karpovsky (1976) has shown that a function of n variables is self-dual if and only if

$$R_f(2^n - 1) = \sum_{i=0}^{2^n-1} f(i) - 2^{n-1} \tag{17}$$

For our example of Figure 1.9, we see that we have 4 ones in the truth table. Subtracting $2^{3-1} = 4$ from the number of ones in the truth table we get 0 which checks with the entry for $R_f(7)$.

Those who are familiar with Fourier analysis will recall that performing an autocorrelation operation on a time series is irreversible because phase information is destroyed. A similar phenomenon obtains for the autocorrelation function of a boolean function. A boolean function is uniquely determined by its autocorrelation function up to complementation of its arguments. Consider, for example, that we have the two boolean functions:

$$f_1 = x_3 + x_1 x_2 \tag{18a}$$

and

$$f_2 = \overline{x_3} + \overline{x_1} x_2 \tag{18b}$$

Figure 1.10 illustrates the invariance of the autocorrelation function under the variable complementations introduced in (18b). This invariance property is extremely useful for recognizing basically equivalent boolean functions.

Problem

How does the autocorrelation behave if the entries in the boolean function's truth table are complemented?

Even more useful than the autocorrelation function is the Walsh–Hadamard transform. This transform is introduced and used by a number of authors such as Titsworth (1964) and Rothaus (1976). A fast algorithm, similar to the FFT, has been adapted from work done by Yates (1937). In

x_3	x_2	x_1	δ	$f(\delta)$	$R_f(\delta)$
0	0	0	0	0	4
0	0	1	1	1	2
0	1	0	2	0	2
0	1	1	3	0	2
1	0	0	4	1	2
1	0	1	5	1	2
1	1	0	6	0	2
1	1	1	7	1	0

Figure 1.9. Autocorrelation of a self-dual function.

x_3	x_2	x_1	δ	$f_1(\delta)$	$R_{f_1}(\delta)$	$f_2(\delta)$	$R_{f_2}(\delta)$
0	0	0	0	0	4	1	4
0	0	1	1	0	2	1	2
0	1	0	2	0	2	0	2
0	1	1	3	1	2	1	2
1	0	0	4	1	0	0	0
1	0	1	5	1	2	0	2
1	1	0	6	1	2	1	2
1	1	1	7	0	2	0	2

Figure 1.10. Demonstration of autocorrelation invariance under complementation of variables.

cookbook fashion, the fast algorithm for calculating the Walsh–Hadamard transform is as follows:

1. Write the entire truth table for $f(x_1, \ldots, x_n)$.
2. Change the zeros to ones; change the ones to minus ones.
3. Form n new columns of 2^n entries each. The entries in the top half of each column are the pairwise sums of the elements in the preceding column. The entries in the bottom half are the pairwise differences of the elements in the preceding column.

The nth new column contains the Walsh–Hadamard coefficients.† If these coefficients are each divided (normalized) by 2^n then they will be the cross-correlation between $f(\delta)$ and the linear sum $\delta_1 x_1 + \delta_2 x_2 + \cdots + \delta_n x_n$, where $(\delta_n, \ldots, \delta_2, \delta_1)$ is the normal binary representation of δ. This remarkable property will be of great use when we study phase synchronization in Chapter 11.

Let us try an example. Consider the function

$$f(x_1, x_2, x_3) = x_1 + \overline{x_1} \overline{x_2} \overline{x_3} \tag{19}$$

In Figure 1.11 we calculate the Walsh–Hadamard transform. The table $f^*(\delta)$ is the result of applying rule 2 above. Because $n = 3$, we perform the pairwise sums and differences three times. These columns are headed by 1, 2, and 3, respectively.

Forming the cross-correlations (or normalized spectral coefficients), $\{S(i)\}$, by dividing the entries in column 3 by $2^3 = 8$ we obtain $S(0) = -0.25$, $S(1) = 0.75$, $S(2) = 0.25$, $S(3) = 0.25$, $S(4) = -0.25$, $S(5) = -0.25$, $S(6) = 0.25$, $S(7) = 0.25$. [For clarity, consider the computation of $S(6)$. We represent 6 as $1, 1, 0$. The linear sum $0 \cdot x_1 + 1 \cdot x_2 + 1 \cdot x_3 = x_2 + x_3$ is compared to $f(\delta)$. We find five agreements and three disagreements. The cross-correlation is thus $5 - 3 = 2$ which is the coefficient shown in column 3. Dividing (normalizing) by 8 we obtain 0.25.]

† In Chapter 5, we use matrix concepts to compute these coefficients.

Problem

The Walsh–Hadamard transform is uniquely related to its boolean function. Devise an inverse transform and recover $f(\delta)$ from the $\{S(i)\}$ above.

The Walsh–Hadamard transform, or "spectrum," of a boolean function is extremely useful as an identification tool. As an introduction, consider that we transpose x_1 and x_3 and complement x_2 of (19). The spectral coefficients of this new function are $S(0) = -0.25$, $S(1) = -0.25$, $S(2) = -0.25$, $S(3) = -0.25$, $S(4) = 0.75$, $S(5) = -0.25$, $S(6) = -0.25$, $S(7) = -0.25$.

If we compare the new spectral coefficients with the older set we see that the magnitudes have remained the same but some signs have changed and there have been some shufflings. There are three simple operations that are very useful when dealing with Walsh–Hadamard spectral coefficients:

Operation I: If $f(\delta) \rightarrow \overline{f(\delta)}$, that is, if each of the truth table entries is complemented, then the spectral coefficient will have its sign reversed, that is, $S(i) \rightarrow -S(i)$.

Operation II: If a variable is complemented, that is, $f(x_n, \ldots, x_i, \ldots, x_1) \rightarrow f(x_n, \ldots, \overline{x_i}, \ldots, x_1)$, then the sign is changed for each $S(k)$ where i is nonzero in the binary representation of k.

Operation III: If two variables are transposed, that is, $f(x_n, \ldots, x_i, \ldots, x_j, \ldots, x_1) \rightarrow f(x_n, \ldots, x_j, \ldots, x_i, \ldots, x_1)$, then the $\{S(k)\}$ are shuffled so that $S(x_n, \ldots, x_i, \ldots, x_j, \ldots, x_1) \rightarrow S(x_n, \ldots, x_j, \ldots, x_i, \ldots, x_1)$.

Why are we bothering with all this? This is really a two-part question. First, why are we interested in Operations I–III and, second, assuming that we are interested in these operations, why bother to pursue their effects in the spectral domain rather than simply dealing with the truth table itself.

x_3	x_2	x_1	δ	$f(\delta)$	$f^*(\delta)$	I	II	III
0	0	0	0	0	1	0	−2	−2
0	0	1	1	1	−1	−2	0	6
0	1	0	2	1	−1	0	2	2
0	1	1	3	1	−1	0	4	2
1	0	0	4	0	1	2	2	−2
1	0	1	5	1	−1	0	0	−2
1	1	0	6	0	1	2	2	2
1	1	1	7	1	−1	2	0	2

Figure 1.11. Fast computation of the Walsh–Hadamard transform.

The first question is answerable as follows. There are a number of classes of boolean functions that retain their important properties under Operations I–III or a subset of these operations. As to the second question, the answer often lies strictly with convenience. If we work only with the truth table, then we encounter only the values 0 and 1. In the spectral domain, the diversity of values is generally much wider. We have positive and negative numbers of different magnitudes so it is easier to see what we are doing and where to head with our attempts.

Problems

1. Use Operations I–III to turn the new spectral coefficients of our example above into the old coefficients and thus unravel our operations on the boolean function.
2. Rothaus (1976) discovered that there is a class of functions, called "bent functions," whose spectral coefficients are all of the same magnitude. Show that the function $f(x_1, x_2, x_3, x_4) = x_1x_2 + x_3x_4$ is bent.

Problem (The Boolean Difference)

A very important, contemporary circuit theory discipline is fault detection and analysis. A very basic tool in fault analysis is a construct called the "boolean difference." Following Sellers et al. (1968), we denote a boolean function, f, of n variables in the usual way as $f(x_1, x_2, \ldots, x_n)$. We desire to know how a fault on an input line, say x_i, will affect the function f. A fault is defined as a bit inversion, $x_i \leftarrow \overline{x_i}$. For example, if f is an AND gate of two variables, $f(x_1, x_2) = x_1x_2$, then an error in x_1 will affect the f output if and only if $x_2 = 1$. The boolean difference is defined with respect to the variable of interest. We denote it as δf_{x_i} and define it as

$$\delta f_{x_i} = f(x_1, \ldots, x_i, \ldots, x_n) + f(x_1, \ldots, \overline{x_i}, \ldots, x_n)$$

where the plus sign denotes modulo 2 addition and the overbar indicates complementation. The boolean function is then also a function of the input variables and we can write $\delta f_{x_i} = g(x_1, \ldots, x_n)$. For the AND gate example above $g(x_1, x_2) = x_2$.

1. If $\delta f_{x_i} = 0$, is it then true that $f = f(x_1, \ldots, x_{i-1}, x_{i+1}, \ldots, x_n)$?
2. If $\delta f_{x_i} = 1$, is it then true that $f = f(x_1, \ldots, x_{i-1}, x_{i+1}, \ldots, x_n) + x_i$?
3. What do the Walsh–Hadamard transform coefficients of f tell you about δf_{x_i}?

References

Bitner, J., G. Ehrlich, and E. Reingold (1976), Efficient Generation of the Binary, Reflected Gray Code and Its Applications, *Communications of the ACM*, Vol. 19, pp. 517–521.

Cavior, S. (1975), An Upper Bound Associated with Errors in Gray Code, *IEEE Transactions on Information Theory*, Vol. 21, p. 596.

Croy, J. (1961), Rapid Technique of Manual or Machine Binary-to-Decimal Integer Conversion Using Decimal Radix Arithmetic, *IRE Transactions on Electronic Computers*, Vol. 10, p. 777.

Karpovsky, M. (1976), *Finite Orthogonal Series in the Design of Digital Devices*, Wiley, New York.

Rothaus, O. (1976), On "Bent" Functions, *Journal of Combinatorial Theory, Series A*, Vol. 20, No. 3, May.

Sellers, F., M-Y. Hsiao, and L. Bearnson (1968), *Error Detecting Logic for Digital Computers*, McGraw-Hill, New York.

Titsworth, R. (1964), Optimal Ranging Codes, *IEEE Transactions on Space Electronics and Telemetry*, pp. 19–30, March.

Wadel, L. (1961), Conversion from Conventional to Negative-Base Number Representation, *IRE Transactions on Electronic Computers*, p. 779.

Wang, M. (1966), An Algorithm for Gray-to-Binary Conversion, *IEEE Transactions on Electronic Computers*, pp. 659–660.

Yates, F. (1937), The Design and Analysis of Factorial Experiments, Imperial Bureau of Soil Science, Harpenden, England.

Counting and Probability

2.1. Counting

Counting, or more formally the mathematics of combinatorics, comprises some of the most difficult and intriguing problems in all of mathematics. It is the essential tool of discrete probability theory as well as serving in other fields of interest to data engineers such as fault isolation, cryptography, and network reliability and survivability.

Combinatorics is a vast and complex field; however, one need master only a small subset of techniques to deal with an impressive array of important, "real world" problems. In this chapter we sample a few elementaty techniques and problems. They were chosen partly because of their simplicity but also because they have been of use to us in our professional work.

2.2. Generating Functions

Generating functions are one of the most popular and powerful tools of combinatorial analysis or counting. They are very simple to grasp conceptually and apply. In essence, a generating function is usually represented as a polynomial in a dummy variable t, $f(t)$. The polynomial may contain either an infinite or finite number of terms. The powers of t in $f(t)$ may be either positive, negative, or both (mixed). For purposes of developmental discussion we consider only those cases for which

$$f(t) = \sum_i a_i t^i \tag{1}$$

where the upper limit may be finite or infinite. The function is thus defined through its coefficients. By definition, a_i is the number of ways of selecting, grouping, or otherwise achieving i items. For example, consider that we have four items, A, B, C, and D. Table 1 depicts the number of ways we can pick or group exactly i items from our set of four. Thus, $f(t)$ for this case is

$$f(t) = 1 + 4t + 6t^2 + 4t^3 + t^4 \tag{2}$$

Now we are ready to use a little trick. Recall the basic property of exponents, $t^m \cdot t^n = t^{m+n}$ and $t^0 = 1$. If we associated the binomial $t + 1$ with each of the four members of our set of items, we can use the exponent properties to perform our counting. The polynomial

$$f(t) = (t + 1)(t + 1)(t + 1)(t + 1) = (t + 1)^4 = 1 + 4t + 6t^2 + 4t^3 + t^4$$
$$ A\text{'s} \quad B\text{'s} \quad C\text{'s} \quad D\text{'s} \quad \text{(binomials)}$$

$$\tag{3}$$

is precisely our generating function. The reason for this is easily grasped. We recall that what we are trying to count is the number of ways of selecting exactly i items from our set. The coefficient of t^i which is a_i is this number. If A is present in a selection then A will contribute a count of one; otherwise A's absence will "contribute" a count of zero. The i in t^i is the sum of all contributions and thus the binomial $t + 1$, which can also be thought of as $t^1 + t^0$, represents the two possible cases: A is in the selection and contributes a one to the count, or A is absent from the selection and contributes nothing to the count.

The next case we consider is that of rolling a normal, perfect pair of dice and computing the sum of the spots showing on their tops. The outcome of each die can be represented by the polynomial $t + t^2 + t^3 + t^4 + t^5 + t^6$. Note that in this case, 1, or t^0, is not present since a normal die has no side

Table 1. Ways of Selecting Exactly i Items

$i = 0$	$i = 1$	$i = 2$	$i = 3$	$i = 4$	
	A	AB	ABC	ABCD	
	B	AC	ABD		
	C	AD	ACD		
	D	BC	BCD		
		BD			
		CD			
1	4	6	4	1	Total ways

devoid of spots. The generating polynomial for the sum of spots is then

$$f(t) = (t + t^2 + t^3 + t^4 + t^5 + t^6)^2$$

$$= t^2 + 2t^3 + 3t^4 + 4t^5 + 5t^6 + 6t^7 + 5t^8 + 4t^9 + 3t^{10} + 2t^{11} + t^{12}$$

$$(4)$$

Note that (4) shows us that there are $\sum_{i=2}^{12} a_i = 1 + 2 + 3 + 4 + 5 + 6 + 5 + 4 + 3 + 2 + 1 = 36$ different ways two dice can land and, for example, 6 ways out of the 36 that they will show a sum of 7 spots.

Our third case concerns a generating function for different elements. Consider that we wish to change a quarter and that we have an unlimited supply of dimes, nickels, and cents. We can easily determine the number of ways in which we can change a quarter by immediately writing down the generating polynomial

$$f(t) = \underbrace{(1 + t^{10} + t^{20} + t^{30} + \cdots)}_{\text{dimes polynomial}} \underbrace{(1 + t^5 + t^{10} + t^{15} + t^{20} + t^{25}}_{\text{nickels polynomial}}$$

$$+ t^{30} + \cdots)\underbrace{(1 + t + t^2 + t^3 + t^4 + \cdots)}_{\text{cents polynomial}}$$

$$(5)$$

The coefficient of t^{25}, which is a_{25}, will be the number of ways we can change the quarter. Let us look a little further at the implementation of calculating a_{25}. First, we need not consider or retain, in our computations, terms beyond t^{25}. We can thus set the "dimes polynomial" to $(1 + t^{10} + t^{20})$ and the "nickels polynomial" to $(1 + t^5 + t^{10} + t^{15} + t^{20} + t^{25})$ and multiply them to obtain

$$1 + t^5 + 2t^{10} + 2t^{15} + 3t^{20} + 3t^{25} \qquad (6)$$

We are now ready to multiply (6) by the "cents polynomial." Our second point to be made is that the cents polynomial need not be formally multiplied. Recall that at this final juncture, all we desire is a_{25} and not the full generating polynomial. Therefore, let us reason that for every term in (6) there will be a term in the cents polynomial which will bring the exponent sum to 25. We can thus merely sum the coefficients of (6) to determine $a_{25} = 1 + 1 + 2 + 2 + 3 + 3 = 12$. There are thus 12 ways to change a quarter using an unlimited number of dimes, nickels, and cents. Table 2 enumerates the ways as a check.

Our fourth case concerns developing a generating function for different elements but with a restriction on the number of elements available. For example, suppose that in the previous problem we did not have an unlimited supply of nickels but instead had only three of them. Our nickels polynomial

Table 2. Ways of changing quarter

Dimes	2	2	1	1	1	1	0	0	0	0	0	0
Nickels	1	0	3	2	1	0	5	4	3	2	1	0
Cents	0	5	0	5	10	15	0	5	10	15	20	25

would then become $(1 + t^5 + t^{10} + t^{15})$ versus $(1 + t^5 + t^{10} + t^{15} + \cdots)$. Multiplying our new nickels polynomial by our dimes polynomial and again discarding terms beyond t^{25} we have the new nickels–dimes polynomial

$$(1 + t^{10} + t^{20})(1 + t^5 + t^{10} + t^{15}) = 1 + t^5 + 2t^{10} + 2t^{15} + 2t^{20} + 2t^{25}$$

$$(7)$$

If we have an unlimited supply of cents, we can change the quarter in only $1 + 1 + 2 + 2 + 2 + 2 = 10$ ways. Again, Table 3 is our check.

Problems

1. (Niven, 1965) Show that there are 292 ways to change a dollar bill using half dollars, quarters, dimes, nickels, and cents.
2. The output of four random number generators, A, B, C, and D, are summed. Generator A produces either a 2 or a 3. Generator B produces either a -5 or a 0 or a 2. Generator C produces a 1 or a 3 or a 4. Generator D is broken and always produces a 1. Show that there are four ways that the sum will be either 0 or 7.

2.3. Permutations

Assume we have n distinct items and n numbered slots. Each slot is assigned exactly one item. We say that each particular assignment of items to slots constitutes a permutation of the items. For example, if we consider that our items are the first n natural numbers and the slots are spaces on a line then we have the following possible permutations for $n = 3$:

123
132
213
231
312
321

Table 3. Ways of changing a quarter using no more than three nickels

Dimes	2	2	1	1	1	1	0	0	0	0
Nickels	1	0	3	2	1	0	3	2	1	0
Cents	0	5	0	5	10	15	10	15	20	25

We note that there are six permutations possible if $n = 3$. The number of possible permutations of n distinct items is easily calculated by counting with a sequential assignment argument: namely, there are n choices or slots available for the first item to be assigned; $n - 1$ choices remain for the next assigned item, $n - 2$ for the next; and finally only one slot remains for the final or nth item. The total number of permutations is just the product of these choices or $n(n - 1)(n - 2) \cdots 3 \cdot 2 \cdot 1$ which is n factorial which is usually written $n!$

Problem

We defined $n!$ above as the product $n! = n(n - 1) \cdots 2 \cdot 1$. Another method of defining some functions is through recursion. Convince yourself that the following definition is equivalent

$$n! = \begin{cases} 1, & \text{if } n = 1 \\ n(n - 1)!, & \text{if } n \neq 1 \end{cases}$$

The quantity $n!$ grows rapidly and it is worthwhile to have an approximation in hand. An excellent one is Stirling's approximation which is

$$n! \simeq \sqrt{2\pi n} \, n^n e^{-n} \tag{8a}$$

We can make (8a) an exact equality by incorporating one more term:

$$n! = \sqrt{2\pi n} \, n^n e^{-n} e^{r_n} \tag{8b}$$

The term e^{r_n} was introduced by Robins (1925) who showed that

$$\frac{1}{12n + 1} < r_n < \frac{1}{12n} \tag{9}$$

Looking at (9) we see that the approximation (8a) is an excellent one for moderate to large n.

Table 4 is an exact table for $n!$ for $1 \leq n \leq 32$. The values are given both in base 10 and base 16 (hexadecimal).

Table 4. Exact values of the first 32 factorials

N	N Factorial (Decimal)	N Factorial (Hexadecimal)
1	1	1
2	2	2
3	6	6
4	24	18
5	120	78
6	720	2D0
7	5040	13B0
8	40320	9D80
9	362880	58980
10	3628800	375F00
11	39916800	2611500
12	479001600	1C8CFC00
13	6227020800	17328CC00
14	87178291200	144C3B2800
15	1307674368000	13077775800
16	20922789888000	130777758000
17	355687428096000	1437EEECD8000
18	6402373705728000	16BEECCA730000
19	121645100408832000	1B02B9306890000
20	2432902008176640000	21C3677C82B40000
21	51090942171709440000	2C5077D36B8C40000
22	1124000727777607680000	3CEEA4C2B3E0D80000
23	25852016738884976640000	57970CD7E2933680000
24	620448401733239439360000	83629343D3DCD1C00000
25	15511210043330985984000000	CD4A0619FB0907BC00000
26	403291461126605635584000000	14D9849EA37EEAC91800000
27	10888869450418352160768000000	232F0FCBB3E62C3358800000
28	304888344611713860501504000000	3D925BA47AD2CD59DAE000000
29	8841761993739701954543616000000	6F99461A1E9E1432DCB6000000
30	265252859812191058636308480000000	D13F6370F96865DF5DD54000000
31	8222838654177922817725562880000000	1956AD0AAE33A4560C5CD2C000000
32	263130836933693530167218012160000000	32AD5A155C6748AC18B9A580000000

Problem

Devise a simple algorithm for computing the number of ending zeros in $n!$ both for base 10 and base 16.

2.3.1. Generation of Permutations

There are $n!$ different permutations of n distinct items. We could label each permutation with a number and our numbering scheme could be lexicographical. To see where we are heading, consider Table 5 which lists all 24 permutations of the four digits 0, 1, 2, 3.

**Table 5. Lexicographical listing
of the 24 permutations of 0, 1, 2, 3**

Permutation number	Permutation
0	0 1 2 3
1	0 1 3 2
2	0 2 1 3
3	0 2 3 1
4	0 3 1 2
5	0 3 2 1
6	1 0 2 3
7	1 0 3 2
8	1 2 0 3
9	1 2 3 0
10	1 3 0 2
11	1 3 2 0
12	2 0 1 3
13	2 0 3 1
14	2 1 0 3
15	2 1 3 0
16	2 3 0 1
17	2 3 1 0
18	3 0 1 2
19	3 0 2 1
20	3 1 0 2
21	3 1 2 0
22	3 2 0 1
23	3 2 1 0

Is there a way whereby given a number, i, in the range $0 \leq i \leq n! - 1$, we can derive the corresponding lexicographically listed permutation? Indeed there is! It was discovered by Lehmer (1964) and is known as Lehmer's Lexicographical Method. It is a beautiful procedure and we will first define it and then illustrate it with the aid of Table 5. The method has six steps:

1. Pick, i, $0 \leq i \leq n! - 1$.
2. Calculate the factorial digits of i: $d_1, d_2, \ldots, d_{n-1}$.
3. Create a lattice of points (i, j). Write a zero at $i = j = 0$.
4. Write the factorial digit d_i at (i, i). (You should now have a 45° diagonal line of n numbers. We will proceed to fill in the triangle array below this line.)
5. The following procedure is to be done for each i, $0 \leq i \leq n - 2$: Starting with $j = i$ and sequentially proceeding to $j = n - 2$, do the following: Compare $d_{i,j}$ to $d_{i+1,i+1}$. If and only if $d_{i,j} \geq d_{i+1,i+1}$, write $d_{i,j} + 1$ at $(i + 1, j)$; otherwise, write $d_{i,j}$ at $(i + 1, j)$.

6. The permutation corresponding to i will be $d_{n-1,n-1}$, $d_{n-1,n-2}$, $d_{n-1,n-3}, \ldots, d_{n-1,0}$.

Let us try an example. Suppose we want to determine the permutation corresponding to $i = 15$. We first form the factorial digits of 15: $15 = 2 \cdot 3! + 1 \cdot 2! + 1 \cdot 1!$ Thus, $d_1 = 1$, $d_2 = 1$, and $d_3 = 2$. We now form our diagonal as per steps 3 and 4.

$$2$$
$$1$$
$$1$$
$$0$$

Let us fill out the bottom row. We take the zero and compare it to the factorial digit heading the column to its immediate right. The factorial digit is a one. Since the zero is not equal to or greater than the factorial digit, the zero is simply copied under the one. The same is true as we progress further right and the filled-in row is seen to be

$$2$$
$$1$$
$$1$$
$$0 \quad 0 \quad 0 \quad 0$$

Now we proceed to fill in the next row. We take the one and compare it to the factorial digit leading the column to the right. This time we find that the column header is not greater, so we increment the one by unity and write it in. Now, moving to the final column, we find that the two heading the column is not greater than the new value to be compared and so we again increment by unity and the second row is now complete. We complete the third row in a similar fashion as shown in Figure 2.1. The desired permutation is derived by extracting, from top to bottom, the entries of the rightmost column. We see then that for Lehmer's Lexicon, permutation number 15 is 2 1 3 0, which agrees with Table 5.

We thus have a guaranteed method of randomly deriving permutations. All we need do is generate a random number between 0 and $n! - 1$ and employ Lehmer's method.

Figure 2.1. Lehmer's triangle for transforming factorial digits into a permutation.

Problem

Attempting the random generation of permutations without carefully understanding what you are doing can be disastrous. For many years, the professional gamblers held to the notion that "luck runs in cycles." Gardner (1968) published a clever application of sequence theory due to Gilbreath that might shed some light on the visceral origins of this notion. The first thing to do is obtain a deck of playing cards. There are $52! \simeq 10^{68}$ different permutations of the deck. How can we achieve a random permutation? Well, the most common approximation is by the riffle shuffle. We will prearrange the deck and modify this shuffle so as to highlight a wonderful effect. First, we order the cards through the entire deck so that the same suit sequence, for example, Hearts, Diamonds, Clubs, Spades, is faithfully repeated throughout. Second, we break the deck into two parts as follows. We take cards, *one at a time*, off the top of the deck to build another stack. We stop after the new stack contains c cards where c may be any number in the range $0 \le c \le 52$. If c is chosen close to 26, however, the effect will be more dramatic. The two parts of the deck are now riffle shuffled together. Two elements of randomness are thus injected. The first is the choice of c; the second is the riffling. Now take the cards off the resulting deck four at a time. *Every grouping of four cards will have one card of each suit.* Why does this work?

2.4. Combinations

We have seen that there are $n!$ different or distinct linear orderings of n distinct items. Suppose now that some, say r, of the n items are indistinguishable. How does this reduce the number of distinguishable orderings of the n items? The question is easily answered by a simple consideration. Let us assume that all n items are initially distinct. We write down one of the $n!$ distinct permutations of the n items which we labeled $i_1, i_2, \ldots, i_n$:

$$i_{j_1} i_{j_2} i_{j_3} \cdots i_{j_n} \tag{10}$$

Now suppose that we wish to permute a specific subset of r of the items in (10). There are $r!$ possible permutations of the subset. Each of these $r!$ permutations is, in itself, a distinct permutation of the n items, that is, each of the $r!$ permutations is imbedded in the $n!$ possible permutations. But now suppose that the r items suddenly become indistinguishable. Well, if they are indistinguishable, all the $r!$ permutations will appear the same and thus the total number of distinct permutations of the n elements is reduced to

$$\frac{n!}{r!} \tag{11}$$

Problem

Show that the number of distinct permutations of n items in which r_1 are of one type and indistinguishable from each other, r_2 are of another type and indistinguishable from each other, ..., r_k still another type and indistinguishable from each other, such that $r_1 + r_2 + \cdots + r_k = n$ is

$$\frac{n!}{r_1! r_2! \cdots r_k!} \tag{12}$$

We are now in a position to study a useful dichotomy. We select k items from n distinct items, $k \le n$. In how many ways can we do this? What we are doing is electing an implicit mapping. We are dividing the n distinct items into two classes—the class of items that are selected and the class of items that are not selected. Under this mapping all selected items are indistinguishable from each other and all nonselected items are indistinguishable from each other. The number of such selections is then

$$\frac{n!}{k!(n-k)!} \tag{13}$$

It is customary to introduce a special symbol for expression (13):

$$\binom{n}{k} = \frac{n!}{k!(n-k)!} \tag{14}$$

The form $\binom{n}{k}$ is sometimes read as "the number of *combinations* of n (distinct) items taken k at a time." It is also the (binomial) coefficient of $a^{n-k}b^k$ in the expansion of $(a + b)^n$.

2.4.1. Combinatorial Identities

The development of combinatorial identities is an extremely rich subset of combinatorics. Many of the simple identities are easily derived from the binomial theorem, which, as the reader will recall, is

$$(x + y)^n = \sum_{i=0}^{n} \binom{n}{i} x^{n-i} y^i \tag{15}$$

Problem

Prove the identity (Liu, 1968)

$$\sum_{i=1}^{n} i \binom{n}{i} = n2^{n-1} \tag{16}$$

[Hint: Set $x = 1$ in (15) and appropriately differentiate and evaluate.]

Many other identities are derivable from simple combinatorial arguments. Again, drawing from Liu's excellent work (1968), we derive an extremely useful identity in this manner. The identity is developed by first noting that the number of ways we can select n distinct items from $2n$ distinct items is, as we have seen,

$$\binom{2n}{n} \tag{17}$$

Now let us count the ways in a slightly different manner. Suppose we first divide our pile of $2n$ distinct items into two piles of n items each. Now we proceed to create two new piles of n items each by selecting i items from one pile and $n - i$ items from the other. We can do this in a total of

$$\binom{n}{0}\binom{n}{n} + \binom{n}{1}\binom{n}{n-1} + \binom{n}{2}\binom{n}{n-2} + \cdots + \binom{n}{n-1}\binom{n}{1} + \binom{n}{n}\binom{n}{0}$$

different ways. This is clearly just another way of counting the number of ways of selecting n distinct items from $2n$ distinct items and thus

$$\sum_{i=0}^{n} \binom{n}{i}\binom{n}{n-i} = \binom{2n}{n} \tag{18}$$

Problem

Show that $\binom{n}{r} = \binom{n}{n-r}$ and then prove the identities

$$\sum_{i=0}^{n} \binom{n}{i}^{j} = \begin{cases} n+1, & j = 0 \\ 2^{n}, & j = 1 \\ \binom{2n}{n}, & j = 2 \end{cases}$$

(There is, incidentally, no closed-form expression for $j > 2$.)

Perhaps the single most important identity is

$$\binom{n}{m} = \binom{n-1}{m} + \binom{n-1}{m-1} \tag{19}$$

Using this identity on $\binom{n-1}{m}$ and $\binom{n-1}{m-1}$ we obtain

$$\binom{n}{m} = \binom{n-2}{m} + 2\binom{n-2}{m-1} + \binom{n-2}{m-2} \tag{20}$$

Using the identity on the new right-hand terms we obtain

$$\binom{n}{m} = \binom{n-3}{m} + 3\binom{n-3}{m-1} + 3\binom{n-3}{m-2} + \binom{n-3}{m-3} \qquad (21)$$

Replicating once again

$$\binom{n}{m} = \binom{n-4}{m} + 4\binom{n-4}{m-1} + 6\binom{n-4}{m-2} + 4\binom{n-4}{m-3} + \binom{n-4}{m-4} \qquad (22)$$

This process can be continued as long as the coefficients are defined, that is, as long as i and j of $\binom{i}{j}$ are proper values; j may not exceed i, neither may i nor j be negative.

Note that the coefficients of the right-hand terms appear to be the binomial coefficients. This suggests that

$$\binom{n}{m} = \sum_{i=0} \binom{d}{i}\binom{n-d}{m-i} \qquad (23)$$

The suggested identity (23) is indeed correct and is known as the Vandermonde Convolution. It is one of the most widely used combinatorial identities. There are many ways to prove (23). The following is suggested as intuitive and instructive in its own right.

Consider that we are on a corner in a large city and that we are p blocks west and q blocks south of where we wish to go (see Figure 2.2) We wish to be efficient in our travel to our destination and therefore we impose the simple rule that at any cross-street we may go north or east providing, or course, we go no further north or east than our ultimate destination. How many distinct paths are available to us?

The problem is easily answered once we realize that each path from (p, q) to $(0, 0)$ must consist of exactly $n = p + q$ steps composed of p steps east and q steps north. Let a step east be denoted by E and a step north by N. Each path is specified then by a $p + q$ long string of E's and N's. How many such strings are there? The answer is immediate. There are $\binom{n}{p}$ distinct paths.

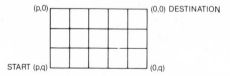

Figure 2.2. The intra city path problem.

Now let us break our travel from (p, q) to $(0, 0)$ into two paths. Consider the locus of points r steps from the destination at $(0, 0)$. We assume $r < n$. The locus is plotted as a dashed line in Figure 2.3. It is clear that each path from (p, q) to $(0, 0)$ must pass through the locus of points r steps away from $(0, 0)$. We should therefore be able to count the number of distinct paths from (p, q) to $(0, 0)$ by calculating the number of distinct paths from (p, q) to the locus and then to the origin. Note that in Figure 2.3 we have drawn the case for which $r < p$. If $r \geq p$, the locus will intersect the leftmost edge of the grid. It is this and other similar considerations that lead to our leaving off limits on $\sum$, similar to what was done in (23).

The first observation we make is that the locus of points r steps from $(0, 0)$ is also the locus of points $p + q - r$ steps from (p, q). The locus of points can be represented by the general point (x, y) where $x + y = r$. The number of distinct paths from (p, q) to (x, y) is $\binom{p+q-x-y}{p-x} = \binom{n-r}{p-x}$ and the number of distinct paths from (x, y) to $(0, 0)$ is $\binom{r}{x}$. Thus, the total number of paths, $\binom{n}{p}$, is also

$$\binom{n}{p} = \sum \binom{r}{x}\binom{n-r}{p-x} \tag{24}$$

2.4.2. Combinatorial Tables

It is useful to have a short table of $\binom{n}{i}$ and $\sum_{j=0}^{i}\binom{n}{j}$. Table 6 is such a table to three significant digits. The following shorthand has been used:

Special Character	Implied Power of 10 Multiplying Entry
=	−2
−	−1
(blank)	0
A	1
B	2
C	3
D	4
E	5
F	6
G	7

Figure 2.3. The locus of lattice points r steps away from $(0, 0)$.

Table 6. Combinatorial Tables

n	i	( )/Σ				
1						
	0–4	100 = /100 =				
2						
	0–4	100 = /100 =	200 = /300 =			
3						
	0–4	100 = /100 =	300 = /400 =			
4						
	0–4	100 = /100 =	400 = /500 =	600 = /110 −		
5						
	0–4	100 = /100 =	500 = /600 =	100 − /160 −		
6						
	0–4	100 = /100 =	600 = /700 =	150 − /220 −	200 − /420 −	
7						
	0–4	100 = /100 =	700 = /800 =	210 − /290 −	350 − /640 −	
8						
	0–4	100 = /100 =	800 = /900 =	280 − /370 −	560 − /930 −	700 − /163
9						
	0–4	100 = /100 =	900 = /100 −	360 − /460 −	840 − /130	126/256
10						
	0–4	100 = /100 =	100 − /110 −	450 − /560 −	120/176	210/386
	5–9	252/638				
11						
	0–4	100 = /100 =	110 − /120 −	550 − /670 −	165/232	330/562
	5–9	462/102A				
12						
	0–4	100 = /100 =	120 − /130 −	660 − /790 −	220/299	495/794
	5–9	792/159A	924/251A			
13						
	0–4	100 = /100 =	130 − /140 −	780 − /920 −	286/378	715/109A
	5–9	129A/238A	172A/410A			
14						
	0–4	100 = /100 =	140 − /150 −	910 − /106	364/470	100A/147A
	5–9	200A/347A	300A/648A	343A/991A		
15						
	0–4	100 = /100 =	150 − /160 −	105/121	455/576	137A/194A
	5–9	300A/494A	501A/995A	644A/164B		
16						
	0–4	100 = /100 =	160 − /170 −	120/137	560/697	182A/252A
	5–9	437A/689A	801A/149B	114B/263B	129B/392B	
17						
	0–4	100 = /100 =	170 − /180 −	136/154	680/834	238A/321A
	5–9	619A/940A	124B/218B	194B/412B	243B/655B	
18						
	0–4	100 = /100 =	180 − /190 −	153/172	816/988	306A/405A
	5–9	857A/126B	186B/312B	318B/630B	438B/107C	486B/155C
19						
	0–4	100 = /100 =	190 − /200 −	171/191	969/116A	388A/504A
	5–9	116B/167B	271B/438B	504B/942B	756B/170C	924B/262C

Table 6 (*continued*)

n	i	( )/Σ				
20						
	0–4	100 = /100 =	200 − /210 −	190/211	114A/135A	485A/620A
	5–9	155B/217B	388B/605B	775B/138C	126C/264C	168C/432C
	10–14	185C/617C				
21						
	0–4	100 = /100 =	210 − /220 −	210/232	133A/156A	599A/755A
	5–9	203B/279B	543B/822B	116C/198C	203C/402C	294C/696C
	10–14	353C/105D				
22						
	0–4	100 = /100 =	220 − /230 −	231/254	154A/179A	732A/911A
	5–9	263B/354B	746B/110C	171C/281C	320C/600C	497C/110D
	10–14	647C/174D	705C/245D			
23						
	0–4	100 = /100 =	230 − /240 −	253/277	177A/205A	886A/109B
	5–9	336B/446B	101C/145C	245C/391C	490C/881C	817C/170D
	10–14	114D/284D	135D/419D			
24						
	0–4	100 = /100 =	240 − /250 −	276/301	202A/233A	106B/130B
	5–9	425B/555B	135C/190C	346C/536C	735C/127D	131D/258D
	10–14	196D/454D	250D/704D	270D/974D		
25						
	0–4	100 = /100 =	250 − /260 −	300/326	230A/263A	127B/153B
	5–9	531B/684B	177C/246C	481C/726C	108D/181D	204D/385D
	10–14	327D/712D	446D/116E	520D/168E		
26						
	0–4	100 = /100 =	260 − /270 −	325/352	260A/295A	150B/179B
	5–9	658B/837B	230C/314C	658C/972C	156D/253D	312D/566D
	10–14	531D/110E	773D/187E	966D/284E	104E/388E	
27						
	0–4	100 = /100 =	270 − /280 −	351/379	293A/330A	176B/209B
	5–9	807B/102C	296C/398C	888C/129D	222D/351D	469D/819D
	10–14	844D/166E	130E/297E	174E/471E	201E/671E	
28						
	0–4	100 = /100 =	280 − /290 −	378/407	328A/368A	205B/242B
	5–9	983B/122C	377C/499C	118D/168D	311D/479D	691D/117E
	10–14	131E/248E	215E/463E	304E/767E	374E/114F	401E/154F
29						
	0–4	100 = /100 =	290 − /300 −	406/436	365A/409A	238B/278B
	5–9	119C/147C	475C/622C	156D/218D	429D/647D	100E/165E
	10–14	200E/365E	346E/711E	519E/123F	679E/191F	776E/268F
30						
	0–4	100 = /100 =	300 − /310 −	435/466	406A/453A	274B/319B
	5–9	143C/174C	594C/768C	204D/280D	585D/866D	143E/230E
	10–14	300E/530E	546E/108F	865E/194F	120F/314F	145F/459F
	15–19	155F/614F				
31						
	0–4	100 = /100 =	310 − /320 −	465/497	450A/499A	315B/365B
	5–9	170C/206C	736C/943C	263D/357D	789D/115E	202E/316E

(*continued*)

Table 6 (*continued*)

n	i	( )/Σ				
	10-14	444E/760E	847E/161F	141F/302F	206F/508F	265F/773F
	15-19	301F/107G				
32						
	0-4	100 = /100 =	320 - /330 -	496/529	496A/549A	360B/414B
	5-9	201C/243C	906C/115D	337D/451D	105E/150E	280E/431E
	10-14	645E/108F	129F/237F	226F/462F	347F/810F	471F/128G
	15-19	566F/185G	601F/245G			

The tables present $\binom{n}{i}/\sum_{j=0}^{i}\binom{n}{j}$. The entries under $i = 7$ for $n = 28$ is $118D/168D$. According to the above shorthand then, $\binom{28}{7} \simeq 1180000$ and $\sum_{j=0}^{7}\binom{28}{j} \simeq 1680000$.

2.5. Recurrence Relations/Difference Equations

A very powerful tool in counting theory is the recurrence relation or difference equation. Although we will be considering these topics in the matrix chapters, we felt that a "thumbnail sketch" at this point would be appropriate.

Following Tucker (1980), we say that a recurrence relation or difference equation is simply the expressing of a quantity's value at a particular time or other discrete index, $q(n)$, in deterministic relation to its value at other, usually earlier, instants along with a "forcing function" or external input, $f(n)$. In other words,

$$q(n) = c_1 q(n - 1) + c_2 q(n - 2) + \cdots + c_m q(n - m) + f(n) \qquad (25)$$

where the $\{c_i\}$ are constants.

If $f(n) = 0$, we see that $q(s)$ for all $s \geq n$ is precisely determined by the m values $q(n - 1)$, $q(n - 2), \ldots, q(n - m)$. This particular form, wherein a forcing function is absent, is termed a linear, homogeneous form. The linear, homogeneous equation is easily solved. All we need do is to form and solve what is known as the characteristic equation. This equation is formed by writing $r^n = c_1 r^{n-1} + c_2 r^{n-2} + \cdots + c_m r^{n-m}$, where r stands for root (of the characteristic equation), and then dividing by r^{n-m} to get

$$r^m - c_1 r^{m-1} - c_2 r^{m-2} - \cdots - c_{m-1} r - c_m = 0 \qquad (26)$$

Upon solving (26), one of two things may happen: either all the roots, $\{r_i\}$, will be distinct or they will not. If the roots are distinct, the general solution is

$$q(n) = C_1 r_1^n + C_2 r_2^n + \cdots + C_m r_m^n \qquad (27)$$

where the $\{C_i\}$ are determined by m initial conditions.

As an example, consider that we have a counter whose upper range is R. The counter is started at 0 and every second it is incremented by either one or two steps. In how many ways can it count from 0 to R? If R is unity, there is only one way. If R is two, there are two ways. (What are they?) Let $q(n)$ be the number of ways the counter can count from 0 to n. How can we write a recurrence relation? Probably the easiest way to see the answer is to consider that there are only two results possible at the first time instant: either the counter is incremented by one or it is incremented by two. If the counter increments by one, then we can consider that there are $q(n-1)$ ways to count from 1 to R. If the counter increments by two, there are then $q(n-2)$ ways to count from 2 to R. Thus, we can write the recurrence relation

$$q(n) = q(n-1) + q(n-2) \qquad (28)$$

The characteristic equation belonging to (28) is

$$r^2 - r - 1 = 0 \qquad (29)$$

Solving (29) we find that

$$r_1 = \frac{+1+\sqrt{5}}{2} \qquad (30)$$

and

$$r_2 = \frac{+1-\sqrt{5}}{2}$$

These roots are distinct so we can at once write

$$q(n) = C_1 \left(\frac{+1+\sqrt{5}}{2}\right)^n + C_2 \left(\frac{+1-\sqrt{5}}{2}\right)^n \qquad (31)$$

Now we have already noted that

$$q(1) = 1 \quad \text{and} \quad q(2) = 2 \qquad (32)$$

Using these two boundary conditions to solve for C_1 and C_2 we find that

$$q(n) = \frac{1}{2^{n+1}\sqrt{5}}[(1+\sqrt{5})^{n+1} - (1-\sqrt{5})^{n+1}] \qquad (33)$$

Equation (28) is a famous recurrence relation and coupled with the initial or boundary conditions (32) it gives rise to the famous sequence of numbers known as the Fibonacci numbers. It is noteworthy that (33) yields an integer for every integer $n \geq 0$. The first 68 Fibonacci numbers are shown in Table 7. The largest perfect square in the infinite Fibonacci series is, incidentally, known to be $F(12) = 144$ (Wyler 1964)

Problem

Show that the nth Fibonacci number, $n \gg 1$, can be closely approximated by $0.4472(1.618)^n$.

Returning to Equation (26), suppose that the roots are not distinct. In this case the solution takes a very different form. If $r = \rho$ is a root repeated s times, then s of the solution terms of (27) will be $C_1\rho^n + C_2 n\rho^n + C_3 n^2\rho^n + \cdots + C_s n^{s-1}\rho^n$.

If we now allow a forcing function, $f(n)$, to be present, we have what is termed an inhomogeneous equation. The solution to an inhomogeneous equation depends (perhaps risking redundancy) on a bit of insight and a bit of luck. What we do is to first solve the associated homogeneous equation, that is, we ignore $f(n)$. We then obtain, by some means, any solution to the full inhomogeneous equation. The homogeneous solution is then added to the particular solution found and initial conditions are used to resolve

Table 7. First 68 Fibonacci numbers

N	$F(N)$	N	$F(N)$	N	$F(N)$	N	$F(N)$
0	0	1	1	2	1	3	2
4	3	5	5	6	8	7	13
8	21	9	34	10	55	11	89
12	144	13	233	14	377	15	610
16	987	17	1597	18	2584	19	4181
20	6765	21	10946	22	17711	23	28657
24	46368	25	75025	26	121393	27	196418
28	317811	29	514229	30	832040	31	1346269
32	2178309	33	3524578	34	5702887	35	9227465
36	14930352	37	24157817	38	39088169	39	63245986
40	102334155	41	165580141	42	267914296	43	433494437
44	701408733	45	1134903170	46	1836311903	47	2971215073
48	4807526976	49	7778742049	50	12586269025	51	20365011074
52	32951280099	53	53316291173	54	86267571272	55	139583862445
56	225851433717	57	365435296162	58	591286729879	59	956722026041
60	1548008755920	61	2504730781961	62	4052739537881	63	6557470319842
64	10610209857723	65	17167680177565	66	27777890035288	67	44945570212853

the constants associated with the homogeneous solution; *but* this must be done using the *entire* solution.

As an example, consider that we have normal n-bit Gray code words which represent the numbers 0 through $2^n - 1$. One of the Gray code words will consist of all ones. Let $p(n)$ be the integer represented by the all ones n-bit code word. By examining the Gray code we quickly determine that $p(1) = 1, p(2) = 2, p(3) = 5, p(4) = 10$, and so on. Recalling that the normal Gray code can be generated by reflection, we can write a recurrence relation for $p(n)$.

Problem

Show that the recurrence for $p(n)$ can be written

$$p(n) + p(n-1) = 2^n - 1 \tag{34}$$

How do we solve (34)? Well, we could plunge right in or, as is sometimes very useful, we could manipulate it into a "friendlier" form. Let us try the latter. We see the recursion can be "backed up" if we wish, that is, we can write

$$p(n-1) + p(n-2) = 2^{n-1} - 1 \tag{35}$$

If we subtract (35) from (34) we obtain

$$p(n) - p(n-2) = 2^{n-1} \tag{36}$$

The associated homogeneous solution to (36) is $C_1(1)^n + C_2(-1)^n$. In this case, a particular solution to the inhomogeneous equation is obtained by looking for a series for $p(n)$ that contains the series for $p(n-2)$ so that the overlapped terms are subtracted out. Such a series should have all even-power terms of the same sign and similar behavior for all terms of odd powers. We note that (36) yields a term of degree $n - 1$. Noting that $2^{n-1} = 2^n - 2^{n-1}$, we can posit the following as a particular solution for the inhomogeneous equation: $2^n - 2^{n-1} + 2^{n-2} - 2^{n-3} + \cdots - 2^k$, where $k = 1$ for n even and $k = 0$ for n odd. We can now solve for C_1 and C_2. Remembering that we must use the full solution, we write

$$\begin{aligned} p(1) &= C_1 - C_2 + 2^1 - 2^0 = 1 \\ p(2) &= C_1 + C_2 + 2^2 - 2^1 = 2 \end{aligned} \tag{37}$$

Clearly, the solution to (37) sets $C_1 = C_2 = 0$ and thus (34) has no homogeneous component to its solution. In Chapter 7 we will discuss the use of matrices in solving difference equations.

Problem

Solve the preceding problem, the location of the all ones code word, by inspection using the normal Gray to normal binary conversion rule.

2.6. Probability

It is hard to define probability simply without becoming recursive and "begging the question." Rather than become mired in formalism, we approach the subject with perhaps the attitude of an engineer reaching for a tool.

First, we are interested in discrete probabilistic situations only. We touch on only a couple of important probability distribution functions but ones that are essential to many of the topics we will be considering. Second, to caution the reader against believing that our rather brute force approach has removed all its charming subtleties, we have included a few problems that should be disquieting as they introduce apparent paradoxes.

We introduce first the notion of an experiment or process sampling. Basically, we "do something" and observe one and only one of n possible outcomes. Before we perform the experiment we can attempt to predict the outcome by assigning what are called *a priori probabilities* of occurrence to the n possible outcomes $\{P_1, P_2, \ldots, P_n\}$. The a priori probability of occurrence of the ith event, p_i, can be interpreted, for our purposes, as meaning a ratio. If the experiment were to be run N times, under exactly the same conditions each time, and the ith event occurred N_i times, then

$$p_i = \lim_{N \to \infty} \frac{N_i}{N}$$

The calculation of discrete probabilities often devolves, to a counting problem. An event can occur in N different ways. All the ways are mutually exclusive and exhaustive; that is, they are all different and, taken together, constitute all possible outcomes. They "partition" the outcome space. Keynes' principle of indifference (see Gardner, 1970) states simply, "If you have no ground whatever for believing that any one of N mutually exclusive events is more likely to occur than any other, a probability of $1/N$ is assigned to each." Often, a set of outcomes are grouped together and the grouping is labeled a "favorable event." Famous probability problems are built on the difficulty of calculating the probability of a favorable event. Invariably, the problem devolves to enumerating all constituent micro-outcomes grouped under the heading "favorable event." The rule for determining the probability of a favorable event is very simple: simply sum the probabilities of all the micro-outcomes which constitute a favorable event.

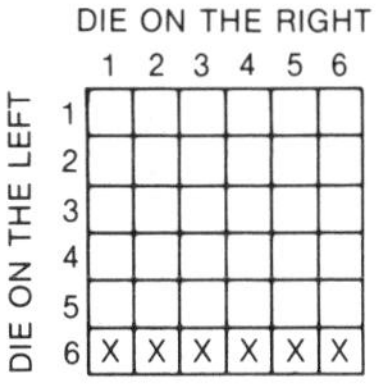

DIE ON THE RIGHT

Figure 2.4. Possibilities admitted by the latter statement. Because each of the six possible situations is equally likely, the probability of there being two 6's is $\frac{1}{6}$.

As an example, suppose we have a six-faced die whose faces bear the natural numbers 1, 2, 3, 4, 5, 6. The die is fair and therefore the probability that any particular one of the six faces shows after a roll of the die is, by the principle of indifference, $\frac{1}{6}$. Let us define a favorable event as the showing of a face whose number is divisible by two or three. If we simply counted those faces whose numbers were divisible by two we would get the faces showing 2, 4, or 6. Similarly, for three we would get the fces showing 3 and 6. Note that face six falls into both groups. We must be careful to avoid counting six twice so a favorable event is the showing of the faces 2, 3, 4, or 6 and the probability of a favorable event is $\frac{1}{6} + \frac{1}{6} + \frac{1}{6} + \frac{1}{6} = \frac{2}{3}$. It is the definition of a favorable event that often makes the counting difficult.

Now suppose we roll two normal and fair dice and report, "There is at least one 6 showing." What is the probability there are two 6's showing? If you answered, "One-sixth because the dice are independent," you would be wrong. Your answer would have been correct, however, if I had reported, "The die on my left is a 6." What then is the difference? Are the two reports inequivalent?

They are, and the reason is information content or, inversely, degree of ambiguity. The latter statement identifies a die and the only outcomes possible are limited to those marked with a X in the diagram shown in Figure 2.4.

The former report admits the possibilities shown in Figure 2.5.

An alternate way of looking at the former case is through the use of the Rev. Thomas Bayes' celebrated theorem. The most commonly stated form of Bayes' theorem, which is an essential component of probability theory, can be introduced as follows. Given that there are two events, A and B, and given that there are three observers, consider the following.

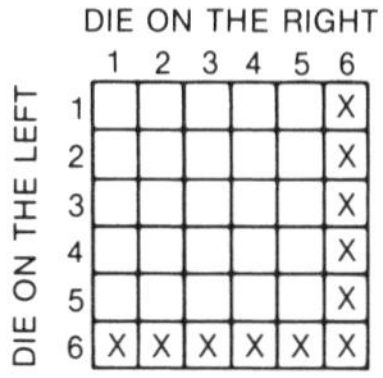

Figure 2.5. Possibilities admitted by the former statement. Here there are 11 possible, equally likely, situations and the probability of there being two 6's is $\frac{1}{11}$.

Observer 1 observes and compiles statistics only on Event A. Observer 2 similarly observes and compiles statistics only on Event B. Observer 1 determines that the a priori probability, the probability assumed before a particular experiment is undertaken, that event A occurs is P_A and Observer 2 similarly determines the a priori probability of Event B as P_B. Observer 3 is a more catholic observer, interested in a possible dependence between events A and B. For example, Observer 1 could be discerning the probability of rain on April 15 in Minneapolis. Observer 2 could be discerning the probability of rain on April 15 in St. Paul. Observer 3 will very likely find that there is a degree of dependency. Observer 3 determines the probability that both events A and B occur in the sampling period; we denote this probability as P_{AB}, and the conditional probability as $P_{A|B}$, which is the probability that A occurs given that B has occurred.

Now consider that any event involves a binary proposition: that is, either it occurs or it does not. The two cases E (event occurs) and $\bar{E}$ (event does not occur) partition the possibilities. We now claim that although even A may not be related in any way to event B, we can still write

$$P_A = P_{A|B}P_B + P_{A|\bar{B}}P_{\bar{B}} \tag{38}$$

We can also write

$$P_{AB} = P_{A|B}P_B = P_{B|A}P_A \tag{39}$$

From (39) we immediately see that

$$P_{B|A} = \frac{P_{A|B}P_B}{P_A} = \frac{P_{A|B}P_B}{P_{A|B}P_B + P_{A|\bar{B}}P_{\bar{B}}} \tag{40}$$

This is Bayes' theorem. The probability $P_{B|A}$ is the *a posteriori probability*; the probability that obtains after a particular experiment has yielded a particular outcome—here, after Event A has occurred.

We can now look at our two-dice problem; the "former statement" of Figure 2.5.

$$P_{\text{two6's}|\text{one6}} = \frac{P_{\text{one6}|\text{two6's}}P_{\text{two6's}}}{P_{\text{one6}}} = \frac{1 \cdot (1/36)}{1 - (5/6)^2} = \frac{1}{11} \tag{41}$$

2.7. Generating Functions in Probability Theory

Just as in combinatorics, generating functions are also extremely useful and important in discrete probability theory. The generating function couples powers of a dummy variable, t, to probabilities of the discrete outcomes in a natural way. Thus, for example, Feller (1950) demonstrates that the generating function for the probability distribution associated with the outcome of a roll of a normal and perfect die is

$$\tfrac{1}{6}(t + t^2 + t^3 + t^4 + t^5 + t^6) \tag{42}$$

We assume, without loss of generality, that our discrete variate will assume nonnegative integral values $X(X \geq 0)$ only, and that the corresponding probabilities are p_X. The generating function is then

$$G(t) = \sum_{X=0}^{\infty} p_X t^X \tag{43}$$

The nice thing about having the generating function for a distribution is that the moments of the distribution, if they exist, are easily obtained. For example, consider that we differentiate $G(t)$ with respect to t:

$$G'(t) = \sum_{X=0}^{\infty} X p_X t^{X-1} \tag{44}$$

The form (44) is seen to yield the mean value of X, $E(X)$, if we set $t = 1$. Similarly, the variance, the expected value of the square of the difference between X and $E(X)$, can be derived straightforwardly.

Problem

Show that the variance is gotten from the generating function by $G''(1) + G'(1) - [G'(1)]^2$.

2.8. The Bernoulli Source

The Bernoulli source presents, upon sampling, either a zero or a one. The outcome at sampling time t is independent of all previous outcomes and is governed by the following probability distribution:

$$\begin{aligned}
\text{probability of a one} &= p, \\
\text{probability of a zero} &= 1 - p.
\end{aligned} \tag{45}$$

When $p = \tfrac{1}{2}$, the Bernoulli source is said to be *balanced*.

The "integral" of the Bernoulli source is the *binomial distribution*. This distribution is concerned with the *total* number of ones produced in n outcomes from a Bernoulli source, not the *order* of the production of the ones. The probability that there will be exactly k ones in n outcomes ($k \leqslant n$) is

$$\binom{n}{k} p^k (1-p)^{n-k} \tag{46}$$

It is easy to show that the generating function for the binomial distribution is

$$(q + pt)^n \tag{47}$$

where $q = 1 - p$.

Problem

Show that the mean value of (46) is np and that the variance is npq.

It is often necessary to find the probability that a binomial variate will fall in one of the distribution's "tails," that is k in (46) will be in a region either near zero or near n. There are a number of approximations to the tail region probabilities. One that leads to a nice manageable form and has been of use to the authors is an approximation derived by Brockwell (1964). Brockwell considers the sum

$$S_r(m) = \sum_{l=m}^{mr} \binom{mr}{l} p^l q^{mr-l} \tag{48}$$

where m and mr but not necessarily r are integers and $rp < 1$. Brockwell's approximation is

$$S_r(m) = \binom{mr}{m} p^m q^{mr-m} \left(\rho_0(p^{-1}) + \frac{\rho_1(p^{-1})}{m} + \cdots + \frac{\rho_k(p^{-1})}{m^k} + R_k \right) \tag{49}$$

where

$$\rho_0(y) = \frac{y-1}{y-r} \tag{50}$$

$$\rho_1(y) = \frac{y(y-1)}{y-r} \frac{d}{dy} \rho_0(y) \tag{51}$$

and

$$\rho_k(y) = \left(\frac{y(y-1)}{y-r}\frac{d}{dy}\right)^k \rho_0(y) \qquad (52)$$

in general. The remainder term, R_k, in Equation (49) satisfies the following bound:

$$0 \leqslant (-1)^{k+1} R_k < \left|\frac{\rho_k(p^{-1})}{m^k}\right| \qquad (53)$$

Problems

1. Carry out Brockwell's approximation to order $k = 1$ for $p = q = \frac{1}{2}$ and show that

$$S_r(m) \simeq 2^{-mr}\binom{mr}{m}\left(\frac{1}{2-r} - \frac{(2/m)(r-1)}{(2-r)^3}\right)$$

and show that the maximum error from (53) is

$$\text{Max } R_1 = \frac{2(1-r)}{m(2-r)^3}$$

2. In classical probability theory, we find extensive use of "urn" problems. An urn problem is often a convenient vehicle for setting down probabilistic processes and problems so that they can be visualized. Basically, what we have is an opaque container, an urn that is filled with a finite number of balls. Usually, each ball is painted with one of a number of colors. Balls are then drawn at random from the urn, that is, the selection probability of a particular ball is the same for all balls. (The "indifference principle" obtains.) The drawn ball is examined, its color noted and recorded. The ball is then either "replaced" in the urn and reacquires equal probability with all the other balls of being drawn at the next pick, or, the ball is "discarded." The former case is known as *drawing with replacement*; the latter as *drawing without replacement.*

We have in our possession an urn with $N > 1$ unpainted balls. We wish to paint W of the balls white and the remainder black so that the probability of getting two differently colored balls on two drawings *without* replacement is $\frac{1}{2}$. Show that this is possible if and only if N is a perfect square.

3. A three-input boolean OR gate is connected to three balanced Bernoulli sources. The output of the OR gate is a one. What is the probability that all three sources have a one as their output?

4. (Efron's Dice; Gardner, 1970) Consider that we have four dice each with six sides. Die A shows 2, 3, 3, 9, 10, 11 on its sides. Die B shows 0, 1, 7, 8, 8, 8 on its sides. Die C shows 5, 5, 6, 6, 6, 6 on its sides. Die D shows 4, 4, 4, 4, 12, 12 on its sides. We play the following games. You pick a die and then I pick a die. We roll them. The die showing the higher number wins. Why is this game unfair to you and what is my best strategy for winning? This is a very subtle question. It can be rephrased somewhat by asking: "Is there a best die to pick and, if so, which one is it?" Calculate the following probabilities: (1) the probability that Die A beats Die B, (2) the probability that Die B beats Die C, (3) the probability that Die C beats Die D, and (4) the probability that Die D beats Die A. Is "probability that x beats y" a transitive relation?

5. (Gardner, 1976) There are three tables: Table 1, Table 2, and Table 3. Each table has a Red urn and a Blue urn sitting on it:

> The Red urn on Table 1 has 5 black balls and 6 white balls.
> The Blue urn on Table 1 has 3 black balls and 4 white balls.

> The Red urn on Table 2 has 6 black balls and 3 white balls.
> The Blue urn on Table 2 has 9 black balls and 5 white balls.

> The Red and Blue urns on Table 3 are both empty.

(a) You approach Table 1. You select one of the urns and draw and replace one ball. You wish to maximize your probability of drawing a black ball. Which urn should you choose and what is the probability that you will draw a black ball?

(b) You approach Table 2. You select one of the urns and draw and replace one ball. You wish to maximize your probability of drawing a black ball. Which urn should you choose and what is the probability that you will draw a black ball?

(c) Your friend comes along and dumps the contents of the two Red urns on Tables 1 and 2 into the previously empty Red urn on Table 3. Your friend also dumps the contents of the two Blue urns on Tables 1 and 2 into the previously empty Blue urn on Table 3. You now approach Table 3. You select one of the urns and draw one ball. You wish to maximize your probability of drawing a black ball. Which urn should you choose and what is the probability that you will draw a black ball?

What are the rather profound implications of this problem? What does it tell you about combining statistics derived from sampling populations of different sizes?

2.9. Some Important and Famous Problems

We will often be interested in problems that require us to assemble various collections of items. Two of the most famous and important problems representing this subject are (1) the so-called "birthday problem" and (2) coupon collecting. We study the birthday problem first.

Assume that we have N initially empty urns labeled $1, 2, \ldots, N$ and suppose that we repeatedly reach into a box containing balls also numbered $1, 2, \ldots, N$ and extract one ball at random. (The box contains an infinite number of balls of all types and the probability of drawing a particular number is and remains $1/N$.) We note the number on the ball we have extracted and we place it in the urn bearing the same number. We are interested in determining the probability that no urn contains more than one ball after k replications of the experiment, $k \leq N$. It is clear, by the pigeonhole principle, that if we repeat the experiment $N + 1$ times, at least one urn will contain more than one ball.

The solution to this problem is quite straightforward. Let E_i be the event that we pick the ith ball and that its urn is empty, $i \leq N$. The probability we seek is then the joint probability,

$$P(k) = P_{E_1, E_2, \ldots, E_k} \tag{54}$$

The joint probability (54) can be immediately expanded in terms of the conditional probability

$$P_{E_1, E_2, \ldots, E_k} = P_{E_k | E_1, E_2, \ldots, E_{k-1}} P_{E_1, E_2, \ldots, E_{k-1}} \tag{55}$$

but $P_{E_1, E_2, \ldots, E_{k-1}}$ is just $P(k - 1)$. In general then,

$$P(i) = P_{E_i | E_1, E_2, \ldots, E_{i-1}} P(i - 1) \tag{56}$$

The term

$$P_{E_i | E_1, E_2, \ldots, E_{i-1}} \tag{57}$$

is easily evaluated. All it asks for is the probability that a random ball will be assigned to an empty urn. Since $i - 1$ urns are occupied, the probability of being assigned to an empty urn is simply

$$P_{E_i | E_1, E_2, \ldots, E_{i-1}} = \frac{N - i + 1}{N} \tag{58}$$

We can thus write $P(k)$ directly as

$$P(k) = \prod_{i=1}^{k} \left(\frac{N - i + 1}{N} \right) \tag{59}$$

Problems

1. Show that, for N and $N - k$ both sufficiently large, (59) can be approximated as

$$P(k) \simeq \left(\frac{N}{N - k} \right)^{N-k+1/2} e^{-k} \tag{60}$$

2. (The Birthday Problem) Assume a person's birthday to be randomly and equally distributed over 365 days. Using (60) find the smallest grouping of people such that the probability is $\frac{1}{2}$ or greater that at least two people have the same birthday.

Coupon collection asks a series of slightly different questions. Let us assume we are again drawing from our box containing balls numbered 1, 2, ..., N. Again consider that the box contains an infinite number of balls of all types and that the probability of drawing a particular number is and remains $1/N$. We are often interested in knowing how many balls we must draw until we can expect to have a specified collection of numbered balls. Coupon collection or "occupancy" problems must be very carefully stated and understood. All ambiguity must be removed. This is often overlooked.

There are two coupon collection problems of great interest. Parzen (1960) considers these and others at length. We do not derive the expressions but merely quote them for use.

The first problem asks for the minimum number of balls that must be drawn, D_r, to collect r *different* numbers. For this variable,

$$E(D_r) = N \sum_{i=1}^{r} \frac{1}{N - r + i} \tag{61}$$

$$\text{Var}(D_r) = N \sum_{i=1}^{r-1} \frac{i}{(N - i)^2} \tag{62}$$

The second problem asks for the minimum number of balls that must be drawn, S_r, to collect r *specific* numbers. For this variable,

$$E(S_r) = N \sum_{i=1}^{r} \frac{1}{r - i + 1} \tag{63}$$

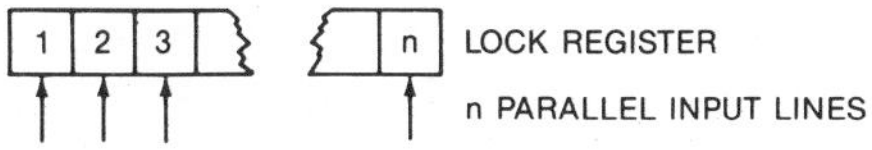

Figure 2.6. An electronic lock.

and

$$\text{Var}\,(S_r) = N \sum_{i=1}^{r} \frac{N - r + i - 1}{(r - i + 1)^2} \tag{64}$$

Problem

Use the approximation $\sum_{i=1}^{N} (1/i) \simeq \ln N$ (for large N) to show that (61) can be approximated as

$$E(D_r) \simeq N \ln \frac{N}{N - r + 1} \tag{65}$$

Now consider the special case of (65) for which $r = N$. What we are asking for is the expected number of drawings required to amass a collection containing at least one of each of the numbers. For this case

$$E(D_r) \simeq N \ln N \tag{66}$$

Equation (66) is very interesting. Suppose that we are trying to open the electronic lock of Figure 2.6 and that we do not know what n-bit vector unlocks the lock. If we connect an n-bit binary counter to the input lines and let it count, it will open the lock, on the average, after 2^{n-1} steps. If, on the other hand, we attach a balanced Bernoulli source to each input line, we will have to wait longer on the average for the lock to open. How much longer, on the average, do we have to wait? (Hint: Be careful.)

2.10. Random Mappings

Another classical problem of great importance relates to probabilistic mappings. There are two types of random mapping which are very important because they are often called upon to serve as models. We will see such application when we look at cryptography in a later chapter. The first mapping is a random mapping that is one-to-one, *a permutation*, on a set of N elements. We are interested in the average number of cycles induced by such a permutation. For example, suppose we have 16 elements labeled from 0 through 15; and suppose that our arbitrarily chosen random mapping, $f(X)$, is as shown in Table 8.

**Table 8. Random Permutation
Mapping of the 16 Elements**

x	$f(x)$	x	$f(x)$
0	3	8	13
1	8	9	4
2	2	10	14
3	15	11	9
4	7	12	0
5	10	13	1
6	5	14	12
7	11	15	6

In Figure 2.7 we graph the permutation specified by Table 8. Figure 6 shows us that the permutation of Table 8 induces four cycles. Note that each element has a unique successor *and* a unique predecessor. Of course, this is a required characteristic of a permutation.

Greenwood (1953) has shown that for N large, the mean number of cycles induced by a random permutation is

$$\ln N + 0.577 \qquad (67)$$

with a variance of

$$\ln N - 1.068 \qquad (68)$$

The second type of mapping of interest is a mapping function that is totally random. The probability that element i will be mapped to element j is $1/N$ for every i and j. An example of such a mapping, $f(X)$, is given in Table 9.

In Figure 2.8 we graph the mapping specified by Table 9. Note that every element has a unique successor *but* not every element has a unique predecessor and as a consequence not every element indeed has a predecessor.

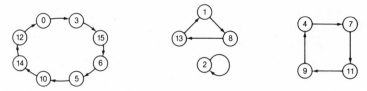

Figure 2.7. Graph of the permutation of Table 8.

**Table 9. Random Mapping of
the 16 Elements**

x	$f(x)$	x	$f(x)$
0	6	8	2
1	15	9	4
2	10	10	12
3	1	11	13
4	5	12	1
5	5	13	9
6	4	14	15
7	11	15	8

The mapping specified in Table 9 is said to induce "components" rather than cycles which would be the case for a permutation mapping. A specific component comprises all those states in a cycle along with the "trees" of states that lead into the cycle. From Figure 2.8 we see that the mapping of Table 9 has induced two components: one with a single-state cycle part, the other with a six-state cycle part.

Kruskal (1954) has shown that for N large, the expected number of components is

$$\tfrac{1}{2}\ln 2N + 0.289 \tag{69}$$

2.11. Redundancy and the Perfect Voter

A technique that is often used to increase reliability is to triplicate items that may fail and derive an output that is more likely correct by arbitrating the outputs with a perfect voter, that is, an arbitration module that is assumed flawless. The voter looks at an odd number of binary outputs and selects the majority of the ouputs as the correct one. The relevant mathematics of this approach have been worked out by Deo (1968) for the device of Figure 2.9.

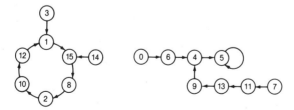

Figure 2.8. Graph of the mapping of Table 9.

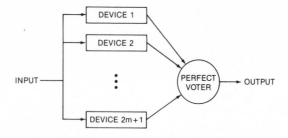

Figure 2.9. Perfect voter overseeing $2m + 1$ imperfect binary output devices.

The $2m + 1$ devices of Figure 2.9 can produce one of two outputs. The perfect voter outputs the majority of these device outputs. The probability of an *incorrect* output from the voter, P_{2m+1}, has been shown to be

$$P_{2m+1} = \frac{(2m + 1)!}{m!} \, p^{m+1} \sum_{k=0}^{m} \frac{1}{k!(m - k)!(m + k + 1)} (-p)^k \qquad (70)$$

where p is the independent probability of failure of any particular device.

Problems

1. Show that $P_3 = p^2(3 - 2p)$
2. Show that

$$P_{2m-1} - P_{2m+1} = \binom{2m - 1}{m} p^m (1 - p)^m (1 - 2p) \qquad (71)$$

2.12. Bias

Very often we wish to estimate a parameter. This is usually done by first conducting an ensemble of experiments, the outcome of which is influenced by the parameter's value. We then work backward from the ensemble outcomes to estimate the parameter of interest. The science of estimation is a well-developed, highly complex discipline. It deals with such questions as, "Is the ensemble of proposed experiments the most efficient set I can use to estimate the parameter?" and "How do I deal 'fairly,' in a statistical sense, with what I suspect is erroneous data?" One topic in estimation theory that is easily grasped and can be extremely important is *estimator bias*. It is important for two reasons. First, estimator bias can lead to very significant and costly errors if not recognized. Second, sometimes costly and ill-designed experiments have already been performed to estimate a parameter. Instead of discarding the costly data obtained, it is often possible to calculate the bias that has been introduced and remove it.

The general problem in bias estimation can be succinctly stated as follows. Given that a parameter θ is estimated by the estimator $\hat{\theta}$, what is the expected value of $\hat{\theta}$? If $E(\hat{\theta}) = \theta$, then $\hat{\theta}$ is said to be an *unbiased estimator* of θ. Some examples should help to make this clear. Suppose we can observe the output of a Bernoulli source and we wish to estimate p. Our first approach is to count the ones in n outputs and estimate p as the ratio of the number of ones divided by n. Let us investigate the bias of this estimator. The first step is to determine the probability, $P(x)$, that we will observe *exactly* x ones in the n outputs. Clearly,

$$P(x) = \binom{n}{x} (1 - p)^{n-x} p^x \tag{72}$$

The estimator for p is $\hat{p}$ and by definition of the estimator

$$\hat{p} = \frac{x}{n} \tag{73}$$

The expected value of $\hat{p}$ is

$$E(\hat{p}) = \sum_{x=0}^{n} \frac{x}{n} \binom{n}{x} (1 - p)^{n-x} p^x = \frac{1}{n} \sum_{x=0}^{n} x \binom{n}{x} (1 - p)^{n-x} p^x = \frac{1}{n} np = p \tag{74}$$

Thus, $E(\hat{p}) = p$ and our estimate is seen to be unbiased.

Our second approach to estimating p is the following experiment. We observe the outputs of the source until a one appears. We estimate p by the reciprocal of the number of bits observed up to and including the first one. (This method incidentally, is quite seductive for those cases for which $p \ll 1$.) Again, we are looking at strings of the form

$$000 \cdots 001$$
$$n - 1 \text{ zeros} \tag{75}$$

The probability, P_n, that the string (75) has exactly n bits is

$$P_n = (1 - p)^{n-1} p \tag{76}$$

The estimator for p is $\hat{p}$ and by definition is

$$\hat{p} = \frac{1}{n} \tag{77}$$

The expected value of $\hat{p}$ is

$$E(\hat{p}) = \sum_{n=1}^{\infty} \frac{1}{n}(1-p)^{n-1}p = \frac{p}{1-p} \sum_{n=1}^{\infty} \frac{(1-p)^n}{n} = \frac{-p \ln p}{1-p} \qquad (78)$$

Note that this second estimator is biased as $E(\hat{p}) \neq p$.

Problem

Graph $E(\hat{p})$ versus p for (78). Comment on the percentage of bias as $p \to 0$. Simulate a Bernoulli process by flipping a coin (heads = 1, tails = 0) and estimate p for the process using the above technique. Use an ensemble of at least 20 replications and average your estimators. The bias in estimator (77) is removable. How do you remove it? Apply the removal procedure to your coin flipping caper. Comment on your results.

2.13. Maximum Likelihood Estimation

Maximum likelihood estimation is a common and very useful technique for estimating parameters that are not directly observable. The maximum likelihood estimator (MLE) of a variable is that value which maximizes the probability of an observed, dependent event.

Let us plunge right into an example. Suppose we have a faulty counter. There are two possible inputs to the counter, zero and one. If a zero is sent to the counter, the counter does not increment. If a one is sent to the counter, the counter increments with probability $1 - p$, that is, it fails with probability p. (Our faulty counter is analogous to a "z channel" which we shall encounter in a later chapter.) Suppose we initialize our counter to zero and then send n ones to it and read the final count, c. If $c < n$, we know that the counter failed and we know it failed $n - c$ times. Intuitively, we would estimate the failure probability as

$$\hat{p} = \frac{n-c}{n} \qquad (79)$$

Let us approach the problem slightly differently. Let us work "backward" and estimate the probability of failure by finding that $\hat{p}$ which maximizes the probability of observing our final count, c. The probability, P_c, that we will observe a count of *exactly* c is

$$P_c = \binom{n}{c} p^{n-c}(1-p)^c \qquad (80)$$

We form the MLE by differentiating (80) with respect to p and solving for the $\hat{p} = p$ that forces the derivative to zero. Thus,

$$\frac{dP_c}{dp} = \binom{n}{c}(n-c)p^{n-c-1}(1-p)^c - \binom{n}{c}cp^{n-c}(1-p)^{c-1} = 0 \qquad (81)$$

resulting in

$$\hat{p} = \frac{n-c}{n} \qquad (82)$$

Note that our estimator (79) is the same, in this case, as our MLE (82).

Problems

1. Show that the particular MLE given in (82) is an unbiased estimator.

2. (a) Consider that our faulty counter was "fed" by n samples taken from a balanced Bernoulli source. Given that we observe a final count of c, $c < n$, derive the MLE for the counter failure probability p.
   (b) Investigate the bias of the MLE derived in (a).

References

Brockwell, P. (1964), An Asymptotic Expansion for the Tail of a Binomial Distribution and Its Application in Queueing Theory, *Journal of Applied Probability*, Vol. 1, pp. 163–169.

Deo, N. (1968), Generalized Parallel Redundancy in Digital Computers, *IEEE Transactions on Computers*, Vol. 17, p. 600.

Feller, W. (1950), *An Introduction to Probability Theory and Its Applications*, Vol. I, 2nd ed., Wiley, New York.

Gardner, M. (1968), Mathematical Games, *Scientific American*, Vol. 218, p. 112+, June.

Gardner, M. (1970), Mathematical Games, *Scientific American*, Vol. 223, p. 110+, December.

Gardner, M. (1976), Mathematical Games, *Scientific American*, Vol. 234, p. 119+, March.

Greenwood, R. (1953), The Number of Cycles Associated with the Elements of a Permutation Group, *American Mathematical Monthly*, Vol. 60, pp. 407–409.

Kruskal, M. (1954), The Expected Number of Components Under a Random Mapping Function, *American Mathematical Monthly*, Vol. 61, pp. 392–397.

Lehmer, D. (1964), The Machine Tools of Combinatorics, in *Applied Combinatorial Mathematics*, Wiley, New York, pp. 5–31.

Liu, C. (1968), *Introduction to Combinatorial Mathematics*, McGraw-Hill, New York.

Niven, I. (1965), *Mathematics of Choice*, Random House, New York.

Parzen, E. (1960), *Modern Probability Theory and Its Applications*, Wiley, New York.

Robins, A. (1925), A Remark on Stirling's Formula, *American Mathematical Monthly*, Vol. 32, pp. 26–29.

Tucker, A. (1980), *Applied Combinatorics*, Wiley, New York.

Wyler, O. (1964), Problem Solutions, *American Mathematical Monthly*, pp. 220–222.

The Natural Numbers and Their Primes

3.1. Introduction

The natural numbers are the positive integers $1, 2, 3, \ldots$. Perhaps the hoariest of mathematical jokes is that all the natural numbers are interesting! The proof is by induction and contradiction. Consider that the first natural number, unity, divides all the natural numbers. It is therefore "interesting." Now let k be the least natural number that is not "interesting." Is it then not "interesting" that k is the first noninteresting of the natural numbers? Ergo, all natural numbers are interesting.

All of the natural numbers are of interest to the mathematician but some are more interesting than others. These (natural) numbers of extra interest are the primes. A prime number, by definition, is a number that has exactly two divisors. Thus, 2, 3, 5, 7, 11, and 13 are all primes while 1, 4, 6, 8, 9, 10, and 12 are not. Table 1 lists all the 168 primes that are less than 1000. Let us select any nonprime number, excepting unity. We will find that it is expressible as a product of primes or powers of primes, for example, $12 = 2^2 \cdot 3$. We can always express a number, n, as

$$n = p_1^{\alpha_1} p_2^{\alpha_2} p_3^{\alpha_3} \cdots \tag{1}$$

where p_1 is the first prime (2), p_2 the second (3), p_3 the third (5) and so on, and the $\{\alpha_i\}$ are the powers of the primes. There are three classes of numbers and these classes form a partition of the natural numbers. The partition class is determined by

$$A = \sum_{i=1}^{\infty} \alpha_i \tag{2}$$

Table 1. The Primes Less Than 1000

N	p_N	N	p_N	N	p_N	N	p_N
1	2	43	191	85	439	127	709
2	3	44	193	86	443	128	719
3	5	45	197	87	449	129	727
4	7	46	199	88	457	130	733
5	11	47	211	89	461	131	739
6	13	48	223	90	463	132	743
7	17	49	227	91	467	133	751
8	19	50	229	92	479	134	757
9	23	51	233	93	487	135	761
10	29	52	239	94	491	136	769
11	31	53	241	95	499	137	773
12	37	54	251	96	503	138	787
13	41	55	257	97	509	139	797
14	43	56	263	98	521	140	809
15	47	57	269	99	523	141	811
16	53	58	271	100	541	142	821
17	59	59	277	101	547	143	823
18	61	60	281	102	557	144	827
19	67	61	283	103	563	145	829
20	71	62	293	104	569	146	839
21	73	63	307	105	571	147	853
22	79	64	311	106	577	148	857
23	83	65	313	107	587	149	859
24	89	66	317	108	593	150	863
25	97	67	331	109	599	151	877
26	101	68	337	110	601	152	881
27	103	69	347	111	607	153	883
28	107	70	349	112	613	154	887
29	109	71	353	113	617	155	907
30	113	72	359	114	619	156	911
31	127	73	367	115	631	157	919
32	131	74	373	116	641	158	929
33	137	75	379	117	643	159	937
34	139	76	383	118	647	160	941
35	149	77	389	119	653	161	947
36	151	78	397	120	659	162	953
37	157	79	401	121	661	163	967
38	163	80	409	122	673	164	971
39	167	81	419	123	677	165	977
40	173	82	421	124	683	166	983
41	179	83	431	125	691	167	991
42	181	84	433	126	701	168	997

where the $\{\alpha_i\}$ are as defined in (1). If

$A > 1,$ then n is termed a *composite* number

$A = 1,$ then n is a *prime* number

$A = 0,$ then n is unity

The latter case shows why number theorists often refer to unity as "the empty product of primes."

When n is written as in (1), it is said that n is "canonically decomposed" into its primes. This decomposition is unique and this fact is known as the Fundamental Theorem of Arithmetic.

Some primes are of a special form such as the Mersenne primes which are primes of the form $2^p - 1$, where p is also a prime. The following values of p yield Mersenne primes:

$p = 2, 3, 5, 7, 13, 17, 19, 31, 61, 89, 107, 127, 521, 607, 1279, 2203, 2281,$
3217, 4253, 4423, 9689, 9941, 11213, 19937, 21701, 23209, 44497, 86243, 216091.

3.2. Finding Primes: I

3.2.1. The Sieve of Eratosthenes

The Sieve of Eratosthenes allows us to find all primes in the interval, $[\sqrt{n}] + 1$ to n if we know all primes in the interval from unity to $\sqrt{n}$. The Sieve depends on the obvious fact that if n is composite, then one of the primes in its canonical decomposition must be less than or equal to n. The Sieve operates as follows:

1. Write down all the numbers in the interval $[\sqrt{n}] + 1$ to n.
2. Cross out all multiples of the primes in the interval 1 to $[\sqrt{n}]$ that fall in the interval $[\sqrt{n}] + 1$ to n.
3. The numbers remaining in the interval $[\sqrt{n}] + 1$ to n constitute all the primes in that interval.

As an example, we know that the primes in the interval 1–5 are 2, 3, 5. We apply the Sieve to the interval 6–25. Crossing out the multiples of 2 we have 6̸ 7 8̸ 9 1̸0̸ 11 1̸2̸ 13 1̸4̸ 15 1̸6̸ 17 1̸8̸ 19 2̸0̸ 21 2̸2̸ 23 2̸4̸ 25. Crossing out multiples of 3 we have 6̸ 7 8̸ 9̸ 1̸0̸ 11 1̸2̸ 13 1̸4̸ 1̸5̸ 1̸6̸ 17 1̸8̸ 19 2̸0̸ 2̸1̸ 2̸2̸ 23 2̸4̸ 25. Finally, crossing out multiples of 5 we obtain 6̸ 7 8̸ 9̸ 1̸0̸ 11 1̸2̸ 13 1̸4̸ 1̸5̸ 1̸6̸ 17 1̸8̸ 19 2̸0̸ 2̸1̸ 2̸2̸ 23 2̸4̸ 2̸5̸. The remaining numbers: 7, 11, 13, 17, 19, 23 are all of the primes in the interval 6–25. Now that we know all the primes in the interval 1–25 we can go on to find all the primes in the interval 1–125 and so on. The Sieve of Eratosthenes is a very simple tool and concept; the ideas behind it are very powerful and useful.

3.2.2. Number of Primes and Their Distribution

Suppose that we gather together all known primes, $p_1, p_2, \ldots, p_k$, multiply them together, and add 1 to the product. Let us denote this quantity as $P + 1$. None of the known primes will divide this product. Why? $P + 1$ will then either be a new prime or be composed of primes not previously known. Therefore, it follows that there are an infinite number of primes. This succinct proof is credited to Euclid.

A variation of Euclid's approach is the following. Consider that we partition all the known primes into two classes, A and B. We add the product of all members in class A to the product of all members in class B. This sum is not divisible by any known prime and is therefore a new prime or composed of unknown primes. Note further that the primes in each class may be taken to arbitrary positive powers without affecting the proof. If we choose the convention that if a class is empty, then we produce a one for the product of its members, the "empty product of primes," then this special case becomes essentially Euclid's argument.

Although there is an infinite number of primes, the density of primes becomes increasingly thin as n grows. The celebrated Prime Number Theorem (1896) shows that as n approaches infinity, the number of primes not exceeding n approaches $n/\ln(n)$. Indeed, there are intervals of arbitrary length that are prime-free. Consider, for example, the interval $n! + 2$ to $n! + n$. This interval is of length $n - 1$ and is prime-free. Why?

3.3. The Euclidean Algorithm

The Greatest Common Divisor (GCD) of two numbers n_1 and n_2, represented as (n_1, n_2), is the largest number d such that d divides both n_1 and n_2. The Euclidean Algorithm enables us to find this divisor, that is, $d = (n_1, n_2)$.

The algorithm is defined as follows. We assume, without loss of generality, that $n_1 > n_2$.

I.   Set $a = n_1$ and $b = n_2$.
II. Form $a = db + r$, where $0 \leqslant r < b$.
III. If $r = 0$, declare d the GCD of n_1 and n_2.
IV. Replace a by b, b by r, and then return to step II.

We define a replication of the Euclidean Algorithm as a pass through step

II. This step involves division which is the most time-consuming operation in the algorithm.

As an example, let us choose $n_1 = 581$ and $n_2 = 189$. The following table illustrates the flow of the algorithm.

Replication	Step	a	b	d	r
	I (Initialization)	581	189	—	—
1	II	581	189	3	14
1	IV	189	14	—	—
2	II	189	14	13	7
2	IV	14	7	—	—
3	II	14	7	2	0
3	III	—	$\boxed{7}$	—	0

Thus, we see that $(581, 189) = 7$ and that the Euclidean Algorithm required three replications to determine this. If n_1 and n_2 are independently derived from a uniform distribution over the interval $1-N$ (N large), this form of the Euclidean Algorithm requires about $0.843 \ln (N)$ replications on the average, and a maximum of about $2.078 \ln (N)$ replications.

Problem

Prove that the above algorithm (a) produces a common divisor of n_1 and n_2 and (b) produces the GCD.

The previous form of the Euclidean Algorithm required us to perform divisions. General division is a relatively time-consuming process compared to addition or subtraction. There have been a number of modern versions of the Euclidean Algorithm which dispose of general division. Following is one such version that requires only subtraction and division of even numbers by 2 to find the GCD of n_1 and n_2. Division of an even number by 2 is easily and efficiently performed, of course, on a binary digital computer because the only requirement is to shift the number one place to the right. This modified algorithm usually requires more replications than the classical one just presented; however, the absence of division generally allows the modified algorithm to finish first.

The algorithm is as follows. We assume that both n_1 and n_2 are odd. If they are not, they are to be made odd by dividing each of them by 2 until they are odd. Obviously, the factors of 2 that n_1 and n_2 have in common must be remembered and we must multiply the GCD by this factor.

   I. Set $a = n_1$ and $b = n_2$.
  II. Compute $d = |a - b|$.
 III. If $d = 0$, declare a the GCD of n_1 and n_2.
 IV. Divide d by 2.
  V. If d is even, return to step IV.
 VI. If $a \geq b$, replace a by d; otherwise replace b by d. Return to step II.

We define a replication of this algorithm as a pass through step VI.

As an example, let us again set $n_1 = 581$ and $n_2 = 189$. The following shows the flow.

Replication	Step	a	b	d
	I (Initialization)	581	189	—
0	II	581	189	392
0	IV	581	189	196
0	IV	581	189	98
0	IV	581	189	49
1	VI	49	189	—
1	II	49	189	140
1	IV	49	189	70
1	IV	49	189	35
2	VI	49	35	—
2	II	49	35	14
2	IV	49	35	7
3	VI	7	35	—
3	II	7	35	28
3	IV	7	35	14
3	IV	7	35	7
4	VI	7	7	—
4	II	7	7	0
	III	$\boxed{7}$		

We see that this modified version of the Euclidean Algorithm required four replications to determine that $(581, 189) = 7$.

If, as we previously assumed, n_1 and n_2 are independently derived in a random manner from a uniform distribution over the interval 1-N (N large), then this version of the algorithm will require about $1.019 \ln (N)$

replications on the average, and a maximum number of about $1.443 \ln (N)$ replications (Brent, 1976).

Problems

1. Find a pair of numbers, n_1 and n_2, for which the modified algorithm requires fewer replications than the first or classical version.
2. The Least Common Multiple (LCM) of two natural numbers is the smallest number divisible by both n_1 and n_2. The LCM of n_1 and n_2 is often written $\{n_1, n_2\}$. The LCM is related to the GCD. Show that $\{n_1, n_2\} = n_1 n_2 / (n_1, n_2)$.

3.4. Congruences

The number a is congruent to the number b modulo m, written $a \equiv b(m)$, if and only if, $m|(a - b)$, or, alternately, $a - b = km$, where k is some integer.

Problem

Show that the congruence relation, $\equiv$, is an equivalence relation.

We see immediately that a congruence relation mod m partitions the integers into m classes; the classes are indexed by the remainder, r ($0 \leqslant r < m$), for general n when n is divided by m. (Observe that if n is negative and the remainder is less than zero, it can be brought into the range $0 \leqslant r < m$ by adding a multiple of m to it.)
We now present some basic properties of congruences:

Multiplication and Exponentiation:

1. If $a \equiv b(m)$ and $c > 0$, then $ac \equiv bc(m)$ and $ac \equiv bc(mc)$ and $a^c \equiv b^c(m)$.
2. If $a \equiv b(m_1)$
   $a \equiv b(m_2)$
   $\vdots$
   $a \equiv b(m_r)$,
   then $a \equiv b(\{m_1, m_2, \ldots, m_r\})$, where $\{ \ \}$ is the LCM of all its arguments.

Division:

1. If $ac \equiv bc(m)$ and $(c, m) = 1$, then $a \equiv b(m)$.
2. If $ac \equiv bc(m)$ and $(c, m) = d$, then $a \equiv b(m/d)$.
3. If $a \equiv b(m)$ and n is a positive divisor of m, then $a \equiv b(n)$.

Problem

Prove all the above properties.

Whenever we write

$$f(x) \equiv 0(m) \tag{3}$$

we are asking for all solutions to (3) out of the m candidates $x = 0, 1, 2, \ldots, m - 1$. The study of the solutions to (3) for special classes of $f(\ )$ can be the subject of a series of academic courses. For our purposes, it is important to grasp that (3) may have (a) no solutions, (b) a unique solution, or (c) multiple solutions.

Problem

Find all the solutions to the following equations:

(a) $x^2 + 1 \equiv 0(3)$.
(b) $x^2 + x + 1 \equiv 0(3)$.
(c) $2x^2 + 1 \equiv 0(3)$.

3.5. Residue Sets

If a number k is divided by a number m, then the remainder, d, of the division will be one of the numbers in the set $R = \{0, 1, 2, \ldots, m - 2, m - 1\}$. The set R is called the principal complete residue set of the modulus m. Every number is congruent mod m to one of the numbers in R.

Two important theorems regarding principal complete residue sets are the following:

1. If the same number, b, is added to every member of R and the sums reduced mod m, then the new set of numbers, S, will be another representation of R. For example, if we choose $m = 3$, we have $R = \{0, 1, 2\}$. If we add $b = 2$ to each member of R and reduce the sums mod 3 we get $S = \{2, 0, 1\}$, which is merely a permutation of the elements in R.

2. If every element of R is multiplied by the same number a and if the products are reduced mod m and, furthermore, if $(a, m) = 1$, then the new set of numbers, S, will be another representation of R. For example, if we again choose $m = 3$ and set $a = 2$ we have $S = \{0, 1, 2\}$.

The above two theorems point to a more concise theorem:

If $(a, m) = 1$, then $\{ax + b\}$ generates or "spans" a complete residue set if x spans a complete residue set.

Note that we can generate all six permutations of $\{0, 1, 2\}$ by choosing different a's and b's:

$$a = 1, b = 0 \quad \{0, 1, 2\} \to \{0, 1, 2\}$$
$$a = 1, b = 1 \quad \{0, 1, 2\} \to \{1, 2, 0\}$$
$$a = 1, b = 2 \quad \{0, 1, 2\} \to \{2, 0, 1\}$$
$$a = 2, b = 0 \quad \{0, 1, 2\} \to \{0, 2, 1\}$$
$$a = 2, b = 1 \quad \{0, 1, 2\} \to \{1, 0, 2\}$$
$$a = 2, b = 2 \quad \{0, 1, 2\} \to \{2, 1, 0\}$$

Problems

1. Prove the above theorems. (Hint: prove by contradiction. Use the pigeon hole principle and what you have learned about congruences.)
2. Can the form $ax + b$ generate all permutations of the residue sets $\{0, 1, 2, 3\}$ and $\{0, 1, 2, 3, 4\}$? How can it—or why can it not? (Use a combinatorial argument.)

If you work on problem 2 above, you should come to the conclusion that the form $ax + b$ can not do either of the tasks. If we look at $ax + b$ as a "permutation polynomial", we can show that it can be made to effect all permutations on residue sets having only two or three members. In general, it can be shown that all permutations of a residue set can be accomplished with a permutation polynomial if the number of elements in the residue set is a prime, p, and if we allow polynomials of degree $p - 1$. To find these polynomials, $T_p(x)$, all we need do is write down the general permutation

$$\{0, 1, 2, \ldots, p - 1\} \xrightarrow{T_p(x)} \{e_0, e_1, e_2, \ldots, e_{p-1}\} \tag{4}$$

where $e_0, e_1, e_2, \ldots, e_{p-1}$ is any permutation of $0, 1, 2, \ldots, p - 1$, and solve, by brute force, for the coefficients of

$$T_p(x) = c_1 x + c_0, \qquad p = 2$$
$$T_p(x) = c_{p-2} x^{p-2} + c_{p-3} x^{p-3} + \cdots + c_1 x + c_0, \qquad p > 2 \tag{5}$$

from the p equations

$$\sum_{i=0}^{p-2} c_i x_j^i \equiv e_j \bmod (p)$$

where x_j is the jth element of the principal residue set, that is, $T_p(x_j) = e_j$. The permutation polynomials for the first four primes are given below.

For $p = 2$, we have

$$T_2(x) = x + e_0 \tag{6}$$

For others, we have

$$T_3(x) = (e_2 + 2e_1)x + e_0 \tag{7}$$

$$T_5(x) = (4e_1 + 3e_2 + 2e_3 + e_4)x^3 + (4e_1 + e_2 + e_3 + 4e_4)x^2$$
$$+ (4e_1 + 2e_2 + 3e_3 + e_4)x + e_0 \tag{8}$$

$$T_7(x) = (2e_0 + e_1 + 6e_3 + 5e_4 + 4e_5 + 3e_6)x^5$$
$$+ (4e_0 + 3e_1 + 2e_3 + 2e_4 + 3e_6)^4$$
$$+ (e_0 + 2e_3 + 2e_5 + 2e_6)x^3 + (2e_0 + e_1 + 5e_3 + 5e_4 + e_6)x^2$$
$$+ (4e_0 + 3e_1 + 6e_3 + 2e_4 + e_5 + 5e_6)x + e_0 \tag{9}$$

Problems

1. Find $T_7(x)$ for the permutation $\{0, 1, 2, 3, 4, 5, 6\} \rightarrow \{6, 5, 4, 3, 2, 1, 0\}$. Verify by substitution that the polynomial is correct.
2. Generalize problem 1 above and find $T_p(x)$ for the important reversing permutation $\{0, 1, 2, \ldots, p - 2, p - 1\} \rightarrow \{p - 1, p - 2, \ldots, 2, 1, 0\}$.

3.6. Reduced Residue Sets

We have seen that a complete residue set $\bmod (m)$ is the set $R = \{0, 1, 2, \ldots, m - 1\}$. If we now remove from R those members which are not relatively prime to m, we will have a small set Q which we term the Reduced Residue Set $\bmod (m)$. For example, if $m = 10$,

$$R = \{0, 1, 2, 3, 4, 5, 6, 7, 8, 9\}$$

$$Q = \{1, 3, 7, 9\}$$

The number of elements in Q is denoted by $\phi(m)$ and is

$$\phi(m) = \sum_{\substack{(a,m)=1 \\ 0<a\leq m}} 1 \tag{10}$$

3.7. The Euler–Fermat Theorem

Suppose we have a Reduced Residue Set Q for some number m. Let the elements of Q be as shown below:

$$Q = \{e_1, e_2, \ldots, e_{\phi(m)}\}$$

Now let us choose any a such that $(a, m) = 1$ and let us form

$$Q^* = \{ae_1, ae_2, \ldots, ae_{\phi(m)}\}$$

Recall that all elements in Q^* are understood to be reduced mod m. Is Q^* also a Reduced Residue Set? We can answer this question in only a few steps. First, because each element, e_i, of Q satisfies the condition $(e_i, m) = 1$ and since $(a, m) = 1$ it is clear that $(ae_i, m) = 1$. Second, no two elements in Q^* are identical for if they were then

$$ae_i \equiv ae_j \bmod (m), \qquad i \neq j \tag{11}$$

But we know from congruence theory that (11) implies that $e_i = e_j \bmod (m)$ which contradicts our hypothesis. In light of the two facts we have just uncovered, we can only conclude that Q^* is a Reduced Residue Set mod m and, in fact, merely a permutation of Q. Because Q^* is a permutation of Q, we can certainly write

$$e_1 e_2 \cdots e_{\phi(m)} \equiv (ae_1)(ae_2) \cdots (ae_{\phi(m)}) \bmod (m)$$

or

$$e_1 e_2 \cdots e_{\phi(m)} \equiv a^{\phi(m)} e_1 e_2 \cdots e_{\phi(m)} \bmod (m)$$

Because each $(e_i, m) = 1$, we have that

$$a^{\phi(m)} \equiv 1 \bmod (m) \tag{12}$$

The above is known as the Euler-Fermat Theorem. Note that if m is a prime, p, then

$$a^{p-1} \equiv 1 \bmod (p) \tag{13}$$

Note that if we write (12) as

$$a \cdot a^{\phi(m)-1} \equiv 1 \bmod (m) \tag{14}$$

we can consider $a^{\phi(m)-1}$ as the multiplicative inverse of $a \bmod (m)$.

Problem

Show that the elements of Q form a group (see p. 101) under multiplication mod (m).

3.8. Wilson's Theorem

The above problem has shown us that the elements of Q form a group. This means that each element in Q has an inverse also within Q. What elements are their own inverses, that is, for what elements does

$$x^2 \equiv 1 \bmod (m) \tag{15}$$

Rewriting (15) we see that we have asked for solutions to the equation

$$(x - 1)(x + 1) \equiv 0 \bmod (m) \tag{16}$$

Now let m be a prime p. Then $p|(x-1)(x+1)$ which implies that either $p|x-1$ or $p|x+1$. If $p \geq 3$ then either $p|x-1$ or $p|x+1$ but not both, and thus $x - 1 \equiv 0 \bmod (p)$ or $x + 1 \equiv 0 \bmod (p)$; therefore only 1 and $p - 1$ are their own inverses. We know that for a prime p, Q consists of $\{1, 2, 3, \ldots, p - 2, p - 1\}$ and thus the product of all entries in Q is $(p - 1)!$ Except for 1 and $p - 1$, which are their own inverses, all other elements of Q are multiplied by their inverses and thus

$$(p - 1)! \equiv (p - 1) \bmod (p) \tag{17}$$

and we have the interesting statement that "If p is a prime then it divides $1 + (p - 1)!$" (Note that this statement is true for $p = 2$; we mention this as we considered $p \geq 3$ in our argument.)

Now that we have shown a necessary consequence of p being prime, does it hold the other way around? If m divides $1 + (m - 1)!$, must m be prime? Suppose m is not prime and $m > 1$; then m is composite and $m = m_1 m_2$, where neither m_1 nor m_2 are 1 and it is clear that m_1 must divide $1 + (m - 1)!$ if m does. But because $(m - 1)!$ includes in its product all numbers smaller than m, m_1 will divide $(m - 1)!$ and in order that it divide $(m - 1)!$ and $(m - 1)! + 1$, m_1 must also divide 1 which is a contradiction. We thus have Wilson's Theorem which is succinctly stated as follows:

$$p \text{ is a prime if and only if it divides } 1 + (p - 1)!$$

Wilson's Theorem is one of the very few "if and only if" theorems governing prime numbers.

3.9. The Function ϕ

A number theoretic function is a function whose argument spans the natural numbers and whose range or mapping may be any number. The function $\phi(\ )$ is one of the most interesting and useful of the number theoretic functions. We investigate ϕ in some depth not only because of its innate importance but also to help the reader gain familiarity with number theoretic functions and operations involving them.

As a first step, let us see what we can devise from the simple definition of $\phi(m)$, that is, the number of natural numbers less than or equal to m that are relatively prime to m. Let m be expressed as the product of two integers: $m = f_1 f_2$. (If m is prime, one of the two factors is unity.) It is a direct consequence of ϕ's definition that the number of natural numbers less than or equal to m that have f_1 as their GCD when taken with m is $\phi(f_2)$. The proof of this is quite straightforward. The set of all the natural numbers that have the factor f_1 and are less than or equal to m is $\{f_1, 2f_1, 3f_1, \ldots, (f_2 - 1)f_1, f_2 f_1\}$. The number of multipliers of f_1 in this set that are relatively prime to f_2 is then $\phi(f_2)$. Let us call this result our elementary counting argument.

Now let $\{d_1, d_2, \ldots, d_k\}$ be the set of all the divisors of m. Let us also write an associated set of numbers $\{r_1, r_2, \ldots, r_k\}$ such that $\{r_1 = m/d_1, r_2 = m/d_2, \ldots, r_k = m/d_k\}$. Now consider the numbers $M = \{1, 2, 3, \ldots, m - 1, m\}$. We will partition these numbers into k classes by the following rule: x, from M, is placed in Class 1 if $(x, m) = r_1$, x is assigned to Class 2 if $(x, m) = r_2$, and so on. Our elementary counting argument tells us that Class 1 will have $\phi(d_1)$ members, Class 2 will have $\phi(d_2)$ members, and so on. Because each of the integers from M will be placed in one and only one class we have the remarkable result that

$$m = \sum_{d \mid m} \phi(d) \tag{18}$$

As an example, let us consider $m = 40$. The divisors of m are $1, 2, 4, 5, 8, 10, 20, 40$; $\phi(1) = 1$, $\phi(2) = 1$, $\phi(4) = 2$, $\phi(5) = 4$, $\phi(8) = 4$, $\phi(10) = 4$, $\phi(20) = 8$, $\phi(40) = 16$, and we have

$$\phi(1) + \phi(2) + \phi(4) + \phi(5) + \phi(8) + \phi(10) + \phi(20) + \phi(40) = 40$$

The identity (18) is a remarkable consequence and it can be shown that specifying a number theoretic function that possesses the property (18) is equivalent to specifying $\phi(\ )$, that is, $\phi(\ )$ is the only number theoretic function that fulfills (18).

Previously, we defined the f function, which, incidentally is also known as Euler's phi or "totient" function, as

$$\phi(m) = \sum_{\substack{(a,m)=1 \\ 0<a\leqslant m}} 1 \qquad (19)$$

Is there a more helpful definition—one that enables us to calculate $\phi(m)$ in a more efficient manner than (19) above? The answer is yes. In fact, we already know that if m is a prime p, then $\phi(p) = p - 1$. To gain some further insight consult Table 2, which gives $\phi(m)$ for $m = 1, 2, 3, \ldots, 200$. Let us consider just a few cases. If we pick $m_1 = 3$ and $m_2 = 10$, we see that $\phi(m_1) = 2$ and $\phi(m_2) = 4$. But look at $\phi(m_1 m_2) = 8$. For this pair, $m_1 m_2$, It is true that $\phi(m_1 m_2) = \phi(m_1)\phi(m_2)$. However, if we try $m_1 = 6$ and $m_2 = 10$ we find $\phi(m_1) = 2$ and $\phi(m_2) = 4$ but $\phi(m_1 m_2) = 16$ and for this case $\phi(m_1 m_2) \neq \phi(m_1)\phi(m_2)$. If you experiment with various m_1, m_2 pairs, you will come to the hypothesis that $\phi(m_1 m_2) = \phi(m_1)\phi(m_2)$ if and only if m_1 and m_2 are relatively prime, that is, $(m_1, m_2) = 1$. That this is so is easy to prove through a variety of approaches. We do it by induction. Let $(m_1, m_2) = 1$. We note that $\phi(m_1, m_2) = \phi(m_1)\phi(m_2)$ if $m_1 = m_2 = 1$. Let us assume that $\phi(m_1 m_2) = \phi(m_1)\phi(m_2)$ for all numbers up to $m_1 m_2 - 1$. Let $\{d_1, d_2, \ldots, d_k\}$ be the ordered, that is, $d_i < d_{i+1}$, set of divisors of m_1. (Note that $d_1 = 1$ and $d_k = m_1$.) Let the ordered set of divisors of m_2 be $\{e_1, e_2, \ldots, e_l\}$. (Again, $e_1 = 1$ and $e_l = m_2$.) Now consider the product

$$m_1 m_2 = \{\phi(d_1) + \phi(d_2) + \cdots + \phi(d_k)\}\{\phi(e_1) + \phi(e_2) + \cdots + \phi(e_l)\}$$

$$= \phi(d_1 e_1) + \phi(d_1 e_2) + \cdots + \phi(d_1 e_l)$$

$$\quad + \phi(d_2 e_1) + \cdots + \phi(d_k e_{l-1}) + \phi(d_k)\phi(e_l)$$

$$= \{\phi(d_1 e_1) + \phi(d_1 e_2) + \cdots + \phi(d_1 e_l)$$

$$\quad + \phi(d_2 e_1) + \cdots + \phi(d_k e_{l-1}) + \phi(d_k e_l)\}$$

$$\quad - \phi(d_k e_l) + \phi(d_k)\phi(e_l)$$

The term in brackets above is the sum of the phi function of all divisors of $m_1 m_2$ and therefore equal to $m_1 m_2$. We thus have $\phi(d_k e_l) = \phi(d_k)\phi(e_l)$ or $\phi(m_1 m_2) = \phi(m_1)\phi(m_2)$. A function, g, for which $g(xy) = g(x)g(y)$ is called multiplicative. If this is true only if $(x, y) = 1$, then g is called weakly multiplicative. We have shown that ϕ is a weakly multiplicative function.

If a function is weakly multiplicative then the key to calculating its value for any m is simply to know its value for the power of a prime. Let

Table 2. The Function $\phi(N)$ for $N = 1, 2, \ldots, 200$

N	ϕ	N	ϕ	N	ϕ	N	ϕ
1	1	51	32	101	100	151	150
2	1	52	24	102	32	152	72
3	2	53	52	103	102	153	96
4	2	54	18	104	48	154	60
5	4	55	40	105	48	155	120
6	2	56	24	106	52	156	48
7	6	57	36	107	106	157	156
8	4	58	28	108	36	158	78
9	6	59	58	109	108	159	104
10	4	60	16	110	40	160	64
11	10	61	60	111	72	161	132
12	4	62	30	112	48	162	54
13	12	63	36	113	112	163	162
14	6	64	32	114	36	164	80
15	8	65	48	115	88	165	80
16	8	66	20	116	56	166	82
17	16	67	66	117	72	167	166
18	6	68	32	118	58	168	48
19	18	69	44	119	96	169	156
20	8	70	24	120	32	170	64
21	12	71	70	121	110	171	108
22	10	72	24	122	60	172	84
23	22	73	72	123	80	173	172
24	8	74	36	124	60	174	56
25	20	75	40	125	100	175	120
26	12	76	36	126	36	176	80
27	18	77	60	127	126	177	116
28	12	78	24	128	64	178	88
29	28	79	78	129	84	179	178
30	8	80	32	130	48	180	48
31	30	81	54	131	130	181	180
32	16	82	40	132	40	182	72
33	20	83	82	133	108	183	120
34	16	84	24	134	66	184	88
35	24	85	64	135	72	185	144
36	12	86	42	136	64	186	60
37	36	87	56	137	136	187	160
38	18	88	40	138	44	188	92
39	24	89	88	139	138	189	108
40	16	90	24	140	48	190	72
41	40	91	72	141	92	191	190
42	12	92	44	142	70	192	64
43	42	93	60	143	120	193	192
44	20	94	46	144	48	194	96
45	24	95	72	145	112	195	96
46	22	96	32	146	72	196	84
47	46	97	96	147	84	197	196
48	16	98	42	148	72	198	60
49	42	99	60	149	148	199	198
50	20	100	40	150	40	200	80

us determine $\phi(p^\alpha)$ where p is a prime and $\alpha \geq 1$. Consider the set of natural numbers less than or equal to p^α: $\{1, 2, \ldots, p-1, p, p+1, \ldots, 2p, \ldots, p^\alpha - 1, p^\alpha\}$. Clearly, all the numbers in the set are prime to p^α except multiples of p. There are $p^\alpha/p = p^{\alpha-1}$ such multiples and therefore $\phi(p^\alpha) = p^\alpha - p^{\alpha-1} = p^{\alpha-1}(p-1)$. Thus, to compute $\phi(m)$ we perform the following steps:

1. Factor m into its (unique) product of powers of primes, that is, $m = p_1^{\alpha_1} p_2^{\alpha_2} \cdots p_r^{\alpha_r}$ where each $\alpha_i \geq 1$.
2. Use the fact that ϕ is a weakly multiplicative function and write $\phi(m) = \phi(p_1^{\alpha_1}) \phi(p_2^{\alpha_2}) \cdots \phi(p_r^{\alpha_r})$.
3. Sequentially evaluate all right-hand terms using the result $\phi(p_i^{\alpha_i}) = p_i^{\alpha_i - 1}(p_i - 1)$.

As an example, $\phi(540) = \phi(2^2 \cdot 3^3 \cdot 5) = \phi(2^2) \cdot \phi(3^3) \cdot \phi(5) = 2(2-1) \cdot 3^2(3-1) \cdot (5-1) = 144$.

Problem

Show that $\phi(m) = m(1 - 1/p_1)(1 - 1/p_2) \cdots (1 - 1/p_r)$ if $m = p_1^{\alpha_1} p_2^{\alpha_2} \cdots p_r^{\alpha_r}$.

3.9.1. Fermat's (Little) Theorem

Previously, we introduced the Euler–Fermat Theorem, namely, $a^{\phi(m)} \equiv 1 \bmod(m)$ if $(a, m) = 1$. There is a similar theorem that is occasionally useful. First, recall the multinomial expansion

$$(x_1 + x_2 + x_3 + \cdots + x_k)^n = \sum \binom{n}{r_1, r_2, \ldots, r_k} x_1^{r_1} x_2^{r_2} \cdots x_k^{r_k}$$

where $r_1 + r_2 + \cdots + r_k = n$ and

$$\binom{n}{r_1, r_2, \ldots, r_k} = \frac{n!}{r_1! r_2! \cdots r_k!}$$

Secondly, let n be a prime, p. Now consider

$$\binom{p}{r_1, r_2, \ldots, r_k} = \frac{p!}{r_1! r_2! \cdots r_k!} \tag{20}$$

Because p is a prime, (20) is divisible by p except when one of the r_i is p. Thus, we may write

$$(x_1 + x_2 + x_3 + \cdots + x_k)^p \equiv (x_1^p + x_2^p + x_3^p + \cdots + x_k^p) \bmod(p)$$

Now let $x_1 = x_2 = x_3 = \cdots = x_k = 1$ in the above expression. We obtain

$$k^p \equiv k \bmod (p) \qquad (21)$$

Note that (21) holds for *any* k as long as p is a prime; this is Fermat's Little Theorem.

3.10. Another Number Theoretic Function and the Concept of "Square-Free"

There are many important number theoretic functions but space limitations force us to consider only those of utmost relevance to our data engineering requirements. We have already developed Euler's phi function and now we take up another weakly multiplicative function, the mu (μ) function or Möbius function. It is defined as follows:

(i) $\mu(1) = 1$.
(ii) For $m = p_1^{\alpha_1} p_2^{\alpha_2} \cdots p_r^{\alpha_r}$, where each $\alpha_i \geqslant 1$,

$$\mu(m) = \begin{cases} 0, & \text{if any } \alpha_i > 1 \\ (-1)^r, & \text{otherwise} \end{cases}$$

The first part of definition component (ii) can be equivalently stated as follows: "$\mu(m)$ is zero if and only if m is divisible by a square other than unity." Thus, $\mu(m)$ is not zero if and only if m is "square-free."

Problems

1. What is $\mu(p^\alpha)$?
2. Prove that μ is a weakly multiplicative function.
3. Table 3 gives $\mu(m)$ for $m = 1, 2, 3, \ldots, 200$. Try the following experiment. Pick any number, m, square-free or not, and sum $\mu(d)$ over all the divisors, d, of m. The sum will be zero unless $m = 1$. Prove that this is so. [Hint: the proof depends on a simple combinatorial argument. You will need to consider only those divisors that have primes to the first power (why?) and the number of divisors formed by taking one of these primes at a time, two at a time, three at a time, and so on.]
4. If $m = p_1^{\alpha_1} p_2^{\alpha_2} \cdots p_k^{\alpha_k}$, where each $\alpha_i \geqslant 1$, show that $\sum_{d|m} |\mu(d)| = 2^k$. [Hint: Do problem 3 above first.] (Note that $d = 1$ and $d = m$ are included in the sum.)

Anent problem 3 above, it is a curious fact that if μ is *defined* as a number theoretic function having the property

$$\sum_{d|m} \mu(d) = \begin{cases} 1, & m = 1 \\ 0, & \text{otherwise} \end{cases}$$

then μ can be shown to be weakly multiplicative from this definition alone.

Table 3. The Function $\mu(N)$ for $N = 1, 2, \ldots, 200$

N	μ	N	μ	N	μ	N	μ
1	1	51	1	101	−1	151	−1
2	−1	52	0	102	−1	152	0
3	−1	53	−1	103	−1	153	0
4	0	54	0	104	0	154	−1
5	−1	55	1	105	−1	155	1
6	1	56	0	106	1	156	0
7	−1	57	1	107	−1	157	−1
8	0	58	1	108	0	158	1
9	0	59	−1	109	−1	159	1
10	1	60	0	110	−1	160	0
11	−1	61	−1	111	1	161	1
12	0	62	1	112	0	162	0
13	−1	63	0	113	−1	163	−1
14	1	64	0	114	−1	164	0
15	1	65	1	115	1	165	−1
16	0	66	−1	116	0	166	1
17	−1	67	−1	117	0	167	−1
18	0	68	0	118	1	168	0
19	−1	69	1	119	1	169	0
20	0	70	−1	120	0	170	−1
21	1	71	−1	121	0	171	0
22	1	72	0	122	1	172	0
23	−1	73	−1	123	1	173	−1
24	0	74	1	124	0	174	−1
25	0	75	0	125	0	175	0
26	1	76	0	126	0	176	0
27	0	77	1	127	−1	177	1
28	0	78	−1	128	0	178	1
29	−1	79	−1	129	1	179	−1
30	−1	80	0	130	−1	180	0
31	−1	81	0	131	−1	181	−1
32	0	82	1	132	0	182	−1
33	1	83	−1	133	1	183	1
34	1	84	0	134	1	184	0
35	1	85	1	135	0	185	1
36	0	86	1	136	0	186	−1
37	−1	87	1	137	−1	187	1
38	1	88	0	138	−1	188	0
39	1	89	−1	139	−1	189	0
40	0	90	0	140	0	190	−1
41	−1	91	1	141	1	191	−1
42	−1	92	0	142	1	192	0
43	−1	93	1	143	1	193	−1
44	0	94	1	144	0	194	1
45	0	95	1	145	1	195	−1
46	1	96	0	146	1	196	0
47	−1	97	−1	147	0	197	−1
48	0	98	0	148	0	198	0
49	0	99	0	149	−1	199	−1
50	0	100	0	150	0	200	0

3.11. The Möbius Inversion Formula

If $g(m) = \sum_{d|m} f(d)$, where f is a number theoretic function (f need not be weakly multiplicative), we can find f in terms of g. We can do this directly for any specific n by sequential evaluation as follows:

$$g(1) = \sum_{d|1} f(d) = f(1), \quad \text{therefore } f(1) = g(1)$$

$$g(2) = \sum_{d|2} f(d) = f(1) + f(2), \quad \text{therefore } f(2) = g(2) - g(1)$$

and so on. There is a general form for the inverse and it is known as the Möbius Inversion Formula:

$$f(m) = \sum_{d|m} g(d)\mu\left(\frac{m}{d}\right) \tag{22}$$

or, equivalently,

$$f(m) = \sum_{d|m} g\left(\frac{m}{d}\right)\mu(d) \tag{23}$$

The proof is by expansion. Consider form (23):

$$\sum_{d|m} g\left(\frac{m}{d}\right)\mu(d) = \sum_{d_1 d_2 = m} \mu(d_1) g(d_2) = \sum_{d_1 d = m} \mu(d_1) \sum_{d|d_2} f(d)$$

$$= \sum_{d_1 d|m} \mu(d_1) f(d) = \sum_{d|m} f(d) \sum_{d_1|m/d} \mu(d_1)$$

From problem 3 above we recognize that $\sum_{d_1|m/d} \mu(d_1)$ is zero except when $m/d = 1$, that is, when $d = m$; thus the identity is shown.

Problems

1. The Möbius Inversion Formula can be used to manipulate number theoretic forms and derive useful computational identities. Recall that we have shown that $m = \sum_{d|m} \phi(d)$. Solve for $\phi(m)$ using the Möbius Inversion

Formula and show that

$$\phi(m) = m \sum_{d|m} \frac{\mu(d)}{d}$$

2. An interesting "if and only if" result is the following. Let $g(m) = \sum_{d|m} f(d)$. (a) Show that g is weakly multiplicative if f is weakly multiplicative. (b) Show that f is weakly multiplicative if g is weakly multiplicative.

3.12. Primitive Roots

As we have seen, if $(a, m) = 1$ then the Euler–Fermat Theorem assures that $a^{\phi(m)} \equiv \mod(m)$. Let us now ask a related question. What is the smallest positive d such that $a^d \equiv 1 \mod(m)$? For motivation, consider that we choose $a = 3$ and $m = 7$. We note that $\phi(m) = 6$. Let us examine $\{a, a^2, \ldots, a^{\phi(m)}\}$ all reduced modulo 7. We have $\{3, 2, 6, 4, 5, 1\}$. For this case it is clear that $d = \phi(m)$. Now consider $a = 2, m = 7$. Forming $\{a, a^2, \ldots, a^{\phi(m)}\}$ for this case we find $\{2, 4, 1, 2, 4, 1\}$. For this case d is 3, not 6 as in the previous case. If d is the smallest positive integer such that $a^d \equiv 1 \mod(m)$, we say that "d is the exponent to which a belongs mod m." It is easy to see that $d \mid \phi(m)$. If $d = \phi(m)$, then a is said to be a primitive root of m.

Problem

If a is a primitive root of m, prove that $\{a, a^2, \ldots, a^{\phi(m)}\}$ is a reduced residue set mod m.

Not all numbers, m, will have primitive roots. Without proof we state that m will have primitive roots if and only if m is:

(a) 1
(b) 2
(c) 4
(d) p^α
(e) $2p^\alpha$

where p is any odd prime. If m is one of the five forms listed, then it will have $\phi(\phi(m))$ primitive roots. Table 4, derived from Beiler (1966), lists the primitive roots of all $m \leq 31$ that possess primitive roots.

Table 4. Primitive Roots

m	$\phi(m)$	$\phi(\phi(m))$	Primitive roots
1	1	1	1
2	1	1	1
3	2	1	2
4	2	1	3
5	4	2	2, 3
7	6	2	3, 5
9	6	2	2, 5
11	10	4	2, 6, 7, 8
13	12	4	2, 6, 7, 11
17	16	8	3, 5, 6, 7, 10, 11, 12, 14
19	18	6	2, 3, 10, 13, 14, 15
23	22	10	5, 7, 10, 11, 14, 15, 17, 19, 20, 21
25	20	8	2, 3 8, 12, 13, 17, 22, 23
27	18	6	2, 5, 11, 14, 20, 23
29	28	12	2, 3, 8, 10, 11, 14, 15, 18, 19, 21, 26, 27
31	30	8	3, 11, 12, 13, 17, 21, 22, 24

We already know that if $m = 2^\alpha$ and $\alpha > 2$, then m has no primitive roots. One very useful result, however, is that if $a \equiv 3$ or $5 \bmod (8)$, then the exponent to which a belongs $\bmod 2^\alpha$ is $d = 2^{\alpha-2}$. This is the basis behind many of the multiplicative, congruent, pseudorandom number generators in use today.

Problem

Check out the preceding statement by determining the exponent to which 5 belongs $\bmod 2^5$.

3.13. The Inverse Problem—Finding Discrete Logarithms

Exponentiation, computing the forward mapping, that is, finding α^n given α and n, is straightforward and quickly accomplished as we shall see later. The reverse operation, the "logarithm problem," that is, finding n given α and α^n, is apparently not a task that can in general be performed in a time comparable to exponentiation when performed in finite fields.

To find n given α and α^n we could, of course, try all possible exponents of α. The number of multiplications required on the average to find α^n will then be proportional to, or linear with, n. But we can do better. In fact, there has been substantial and profitable research on the question of how to do better. As an introduction only we present the simplest method that does better than the linear search. We call this method the "split-search" algorithm. It is described by Knuth (1973, p. 9, Problem 17) and it is applicable to any prime, p.

Let

$$q = \lceil \sqrt{p} \rceil n_1 + n_2 \qquad (24)$$

where if $[x]$ is the integer part of x, then

$$\lceil x \rceil = \begin{cases} x, & \text{if } x \in \{\text{integers}\} \\ [x] + 1, & \text{otherwise} \end{cases}$$

and $0 \le n_1, n_2 < \lceil \sqrt{p} \rceil$. It is clear that q will take on all values in the range $0 \le q \le p + 1$ which includes the range of interest $0 \le q \le p - 2$. Consider our logarithm problem to be that of recovering q given α and the equation $\alpha^q \equiv b \bmod (p)$. Substituting for q, we obtain

$$\alpha^{mn_1 + n_2} \equiv b \bmod (p) \qquad (25)$$

where $m = \lceil \sqrt{p} \rceil$. We rewrite this equation as

$$\alpha^{mn_1} \equiv b\alpha^{-n_2} \bmod (p) \qquad (26)$$

where $\alpha^{-1} \equiv \alpha^{p-2} \bmod (p)$ by Fermat's Theorem.

We now create two tables. The first table consists of $\alpha^{mn_1} \bmod (p)$ for $0 \le n_1 < m$. The second table consists of $b(\alpha^{-1})^{n_2}$ for $0 \le n_2 < m$. Because $q = mn_1 + n_2$ will span the entire range of possible exponents, there will be an entry in the first table that is the same as an entry in the second table. Finding these matching elements allows us to compute directly the unknown exponent q.

3.13.1. An Example of the Split-Search Algorithm

We choose $p = 127$. A primitive root of 127 is $\alpha = 3$. Therefore, for every b, $1 \le b \le 126$, there exists a q, $0 \le q \le 125$, such that $3^q \equiv b \bmod (127)$. Table 5 presents the residues of $3^q \bmod (127)$.

Let us determine q such that $3^q \equiv 100 \bmod (127)$. The three steps are:

1. We calculate $m = \lceil \sqrt{127} \rceil = \lceil 11.3 \rceil = 12$.
2. We calculate $\alpha^{-1} = 3^{-1} \equiv 85 \bmod (127)$.
3. We prepare Tables 6 and 7.

Upon examination of Tables 6 and 7 we find that the entry 47 is common to both of them. Thus, $n_1 = 5$ and $n_2 = 6$. We now compute $q = 12 \cdot 5 + 6 = 66$ which may be verified by looking up α^{66} in Table 5.

Table 5. Residues of 3^q Mod (127)

q	3^q	q	3^q	q	3^q	q	3^q	q	3^q	q	3^q
0	1	21	108	42	107	63	126	84	19	105	20
1	3	22	70	43	67	64	124	85	57	106	60
2	9	23	83	44	74	65	118	86	44	107	53
3	27	24	122	45	95	66	100	87	5	108	32
4	81	25	112	46	31	67	46	88	15	109	96
5	116	26	82	47	93	68	11	89	45	110	34
6	94	27	119	48	25	69	33	90	8	111	102
7	28	28	103	49	75	70	99	91	24	112	52
8	84	29	55	50	98	71	43	92	72	113	29
9	125	30	38	51	40	72	2	93	89	114	87
10	121	31	114	52	120	73	6	94	13	115	7
11	109	32	88	53	106	74	18	95	39	116	21
12	73	33	10	54	64	75	54	96	117	117	63
13	92	34	30	55	65	76	35	97	97	118	62
14	22	35	90	56	68	77	105	98	37	119	59
15	66	36	16	57	77	78	61	99	111	120	50
16	71	37	48	58	104	79	56	100	79	121	23
17	86	38	17	59	58	80	41	101	110	122	69
18	4	39	51	60	47	81	123	102	76	123	80
19	12	40	26	61	14	82	115	103	101	124	113
20	36	41	78	62	42	83	91	104	49	125	85

3.14. The Chinese Remainder Theorem

Consider that we have the end-around or "necklace" shift register shown in Figure 3.1. The shift register will exhibit two states: either the leftmost stage will contain a zero and the rightmost stage a one or vice

Table 6			Table 7		
n_1	$3^{12 n_1} \bmod (127)$		$100 \cdot 85^{n_2} \bmod (127)$	n_2	
0	1		100	0	
1	73		118	1	
2	122		124	2	
3	16		126	3	
4	25		42	4	
5	47		14	5	
6	2		47	6	
7	19		58	7	
8	117		104	8	
9	32		77	9	
10	50		68	10	
11	94		65	11	

CLOCK   Figure 3.1. Two-stage end-around shift register.

Table 8. Possible Values of T
Given End State

End state	T
01	Even; $T \equiv 0 \bmod 2$
10	Odd; $T \equiv 1 \bmod 2$

versa. Let us assume that the shift register is started in the state shown in Figure 3.1. Let us now clock the shift register an unknown number of times, T. If we observe the final state of the shift register, what can we conclude about T? Table 8 enumerates the possible conclusions.

Now let us make our experiment slightly more involved. Figure 3.2 depicts our new sequential machine and its initial state. The two end-around shift register sequential machine of Figure 3.2 may exhibit six states. What these states tell us about T is tabulated in Table 9.

Some observations are in order:

1. All of the six possible states in Table 9 can occur.
2. The number of stages (2) in the top register is relatively prime to the number of stages (3) in the bottom register.

Consider for a moment what would have resulted if the bottom register had four stages instead of three. Instead of six different *possible* states, there would have been only four. This is a direct consequence of two and four not being relatively prime.

In Table 9 we lexicographically listed all possible states of the two registers and the equivalent values of T modulo 6. How can we determine T if we do not prepare an entire table such as Table 9? Consider that instead of two registers of two and three stages, we had many long registers. It is clear that a table giving all possible states would quickly grow beyond

CLOCK   Figure 3.2. Two end-around shift registers of different lengths.

**Table 9. Possible Values of T
Given End State**

End state	T
$\begin{cases} 01 \\ 001 \end{cases}$	$T \equiv 0 \bmod 6$
$\begin{cases} 10 \\ 100 \end{cases}$	$T \equiv 1 \bmod 6$
$\begin{cases} 01 \\ 010 \end{cases}$	$T \equiv 2 \bmod 6$
$\begin{cases} 10 \\ 001 \end{cases}$	$T \equiv 3 \bmod 6$
$\begin{cases} 01 \\ 100 \end{cases}$	$T \equiv 4 \bmod 6$
$\begin{cases} 10 \\ 010 \end{cases}$	$T \equiv 5 \bmod 6$

**Table 10a. Values of T Given
End State of Top Register**

End state of top register	T
01	$T \equiv 0 \ (2)$
10	$T \equiv 1 \ (2)$

any reasonable bound. With this as motivation, we now consider the calculation of T from the component register tables only. What this means is that we prepare individual tables for the registers. For the machine of Figure 3.2, these tables are given as Tables 10a and 10b.

**Table 10b. Values of T Given
End State of Bottom Register**

End state of bottom register	T
001	$T \equiv 0 \ (3)$
100	$T \equiv 1 \ (3)$
010	$T \equiv 2 \ (3)$

Now let us suppose that the top register ended in state 10 and the bottom register ended in 010. We know from our exhaustive listing in Table 9 that $T \equiv 5 \bmod (6)$. Let us see if we can arrive at a similar result using only Tables 10a and 10b. From Table 10a we know that $T \equiv 1 \bmod (2)$ and therefore

$$T = 2x + 1 \tag{27}$$

From Table 10b we also know that $T \equiv 2 \bmod (3)$ and thus, using (27), we have

$$2x + 1 \equiv 2 \bmod (3) \tag{28}$$

Equation (28) is a simple congruence. We easily find that the first positive x that satisfies (28) is $x = 2$. Equation (27) then tells us that $T = 5$ is the smallest positive solution. We have thus recovered T without exhaustively examining all possible states.

The preceding has been a prelude to one of the most important, pervasive, and useful theorems of elementary number theory—the Chinese Remainder Theorem, so called because of ancient Chinese writings discussing similar problems involving simultaneous congruences. The Chinese Remainder Theorem is simply stated:

> If we have a set of linear congruences, the moduli of which are all relatively prime, then there exists a unique solution modulo the product of all the moduli.

For purposes of elaboration, consider that we have the following n congruences:

$$
\begin{aligned}
x &\equiv a_1 \bmod (m_1) \\
x &\equiv a_2 \bmod (m_2) \\
&\;\;\vdots \\
x &\equiv a_n \bmod (m_n)
\end{aligned}
\tag{29}
$$

If $(m_i, m_j) = 1$ for all $i \neq j$, then a solution for x exists modulo $\prod_{i=1}^{n} m_i$. If you encounter a case for which $(m_i, m_j) \neq 1$ for all $i \neq j$, then a solution may not exist. A solution will exist if and only if $(m_i, m_j) | (a_i - a_j)$ for all $i \neq j$.

We have seen how to find a solution to the set of equations in (29) by iteratively solving the congruences as shown on Equations (27) and (28).

There is a more direct, computer-oriented method. To do this we will need some definitions:

1. $M = \prod_{i=1}^{n} m_i$.
2. $M_j = M / m_j$. Note that $(m_j, M_j) = 1$

For every j, we then (easily) solve the congruence

$$\alpha_j \frac{M}{m_j} \equiv 1 \bmod (m_j) \tag{30}$$

A solution to the set of congruences is the following:

$$x \equiv \left(\alpha_1 a_1 \frac{M}{m_1} + \alpha_2 a_2 \frac{M}{m_2} + \cdots + \alpha_n a_n \frac{M}{m_n} \right) \bmod (M) \tag{31}$$

As an example, let us look at our previous problem involving the simultaneous congruences $T \equiv 1$ (2) and $T \equiv 2$ (3). For this case, $a_1 = 1$, $a_2 = 2$, $m_1 = 2$, $m_2 = 3$, $M = 6$, $M_1 = 3$, and $M_2 = 2$. We must now find α_1 and α_2 where

$$\alpha_1 \cdot 3 \equiv 1 \ (2) \tag{32a}$$

and

$$\alpha_2 \cdot 2 \equiv 1 \ (3) \tag{32b}$$

Equations (32a) and (32b) are quickly solved to yield $\alpha_1 = 1$ and $\alpha_2 = 2$. Inserting these values into (31) we find that

$$x \equiv 5 \ (6) \tag{33}$$

Another way of expressing the solution is to realize that what (30) is calling for is the multiplicative inverses of the $\{M/m_j\}$. By Fermat's Theorem we know that the inverse of M/m_j is

$$\left(\frac{M}{m_j} \right)^{\phi(m_j)-1} \tag{34}$$

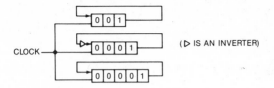

(▷ IS AN INVERTER)

CLOCK

Figure 3.3. Sequential machine.

Using (34), we can also write our solution as

$$x \equiv \sum_{j=1}^{n} a_j M_j^{\phi(m_j)-1} \bmod (M) \tag{35}$$

Problems

1. Show that (31) satisfies all the congruences in (29).
2. Consider that we have the following three end-around shift register sequential machine with initial states as shown in Figure 3.3. If the final shift register states are

$$010$$
$$1111$$
$$00010$$

what are x and y for the expression $T \equiv x \bmod (y)$?

3.15. Finding Primes: II

We recall that the Euler–Fermat Theorem states that $a^{\phi(m)} \equiv 1 \bmod (m)$. We also remember that if m is a prime, p, then $a^{p-1} \equiv 1 \bmod (p)$. What about the converse? That is, if $(a, m) = 1$ and m is such that $a^{m-1} \equiv 1 \bmod (m)$, is m necessarily a prime? If the converse were true, we would have a very simple method to determine whether or not m was prime. Unfortunately, the converse is not true in general. An example is easily produced by considering $m = 341$ and $a = 2$. Clearly $(a, m) = 1$. Our number m is composite; $m = 341 = 11 \cdot 31$. If we calculate 2^{340} and reduce modulo 341, we find that $2^{340} \equiv 1 \bmod (341)$. An odd composite number, m, for which $a^{m-1} \equiv 1 \bmod (m)$ is said to be a pseudoprime to the base a.

In calculating residues of $a^{m-1} \bmod m$, we presuppose that we can perform two important steps. First, we must establish that a and m are relatively prime. We can do this quickly by application of the Euclidean Algorithm. Second, we must be able to perform the exponentiation in an efficient manner. A suboptimal, but nonetheless efficient and easily implemented, exponentiation algorithm is the Binary Exponentiation Algorithm. The algorithm is shown in Figure 3.4. In the following table we use the Binary Exponentiation Algorithm to show that $2^{340} \equiv 1 \bmod (341)$.

	N	Y	Z	Multiplications (Cumulative)
Initialization	340	1	2	0
$N \leftarrow \left[\dfrac{N}{2}\right]$	170	1	2	0
$N \neq 0$; Square Z	170	1	4	1
$N \leftarrow \left[\dfrac{N}{2}\right]$	85	1	4	1
$N \neq 0$; Square Z	85	1	16	2
$Y \leftarrow Y * Z$	85	16	16	3
$N \leftarrow \left[\dfrac{N}{2}\right]$	42	16	16	3
$N \neq 0$; Square Z	42	16	256	4
$N \leftarrow \left[\dfrac{N}{2}\right]$	21	16	256	4
$N \neq 0$; Square Z	21	16	64**	5
$Y \leftarrow Y * Z$	21	1**	64	6
$N \leftarrow \left[\dfrac{N}{2}\right]$	10	1	64	6
$N \neq 0$; Square Z	10	1	4**	7
$N \leftarrow \left[\dfrac{N}{2}\right]$	5	1	4	7
$N \neq 0$; Square Z	5	1	16	8
$Y \leftarrow Y * Z$	5	16	16	9
$N \leftarrow \left[\dfrac{N}{2}\right]$	2	16	16	9
$N \neq 0$; Square Z	2	16	256	10
$N \leftarrow \left[\dfrac{N}{2}\right]$	1	16	256	10
$N \neq 0$; Square Z	1	16	64**	11
$Y \leftarrow Y * Z$	1	1**	64	12
$N \leftarrow \left[\dfrac{N}{2}\right]$	0	1	64	12
$N = 0$; Finished	0	$\boxed{1}$	64	12

** Reduction modulo 341

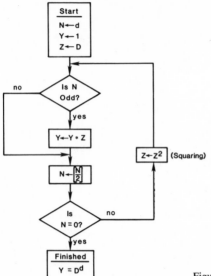

Figure 3.4. The Binary Exponentiation Algorithm.

Problems

1. Study the Binary Exponentiation Algorithm and explain, in your own words, why it works.

2. Show that the Binary Exponentiation Algorithm requires $\lfloor \log_2 m \rfloor + \sigma(m) - 1$ multiplications, where $\lfloor m \rfloor$ is the greatest integer less than or equal to m and $\sigma(m)$ is the number of ones in the normal binary representation of m. In Figure 3.4, $m = d$.

3. We stated that the Binary Exponentiation Algorithm is suboptimal in general. Note that we can calculate α^{15} with only five multiplications as follows:

Start: $\beta \leftarrow \alpha$
$\quad\quad \beta \leftarrow \beta^2$ (1 multiplication)
$\quad\quad \beta \leftarrow \beta \cdot \alpha$ (1 multiplication)
$\quad\quad \gamma \leftarrow \beta$ (saves α^3)
$\quad\quad \beta \leftarrow \beta^2$ (1 multiplication)
$\quad\quad \beta \leftarrow \beta^2$ (1 multiplication)
$\quad\quad \beta \leftarrow \beta \cdot \gamma$ (1 multiplication)
Finished: $\beta = \alpha^{15}$

Now go through the Binary Exponentiation Algorithm for α^{15} and show that an additional multiplication is required.

We have seen that the odd composite number 341 is a pseudoprime to the base 2. A logical question would be, "Is 341 a pseudoprime to other bases relatively prime to it, and, if not, could we test for primeness of 341, or m, in general, by checking to see whether or not $a^{m-1} \equiv 1 \bmod (m)$ for each a such that $(a, m) = 1$?" Unfortunately, we again meet with difficulty. There are odd composite numbers, called Carmichael numbers, which exhibit $a^{m-1} \equiv 1 \bmod (m)$ for every a such that $(a, m) = 1$. Not much is known about Carmichael numbers; it has not even been determined whether or not there are an infinite number of them, although many number theorists believe there are. The smallest Carmichael number is $561 = 3 \cdot 11 \cdot 17$; there are 43 Carmichael numbers less than one million and the first 24 are given in Table 11.

There is another approach, however, and it derives from a consideration of the calculation of $a^{m-1} \bmod (m)$. Because m is odd, $m - 1$ can be written $m - 1 = 2^s t$, where $s > 0$ and t is odd. We now introduce the term "strong pseudoprime." We say that m is a strong pseudoprime to the base a if either

$$a^t \equiv 1 \bmod (m) \tag{36a}$$

or

$$a^{2^r t} \equiv m - 1 \bmod (m) \quad \text{for some } r \text{ such that } 0 \leqslant r < s \tag{36b}$$

Either (36a) or (36b) must hold if m is prime.

What are the advantages to introducing strong pseudoprimes? First, it has been established that no odd composite number is a strong pseudoprime to all of the bases to which it is relatively prime. In other words, there are no "Carmichael-like" strong pseudoprimes. Second, the arithmetical calculations required by (36a) and (36b) for exponentiation and testing are essentially commensurate in effort with those required to form and test a^{m-1}. It has been shown that the smallest odd composite number that is a strong pseudoprime to *both* 2 and 3 is 1373653. Thus, we can easily test for

Table 11. The First 24 Carmichael Numbers and Their Factors

561	$3 \cdot 11 \cdot 17$	15841	$7 \cdot 31 \cdot 73$	101101	$7 \cdot 11 \cdot 13 \cdot 101$
1105	$5 \cdot 13 \cdot 17$	29341	$13 \cdot 37 \cdot 61$	115921	$13 \cdot 37 \cdot 241$
1729	$7 \cdot 13 \cdot 19$	41041	$7 \cdot 11 \cdot 13 \cdot 41$	126217	$7 \cdot 13 \cdot 19 \cdot 73$
2465	$5 \cdot 17 \cdot 29$	46657	$13 \cdot 37 \cdot 97$	162401	$17 \cdot 41 \cdot 233$
2821	$7 \cdot 13 \cdot 31$	52633	$7 \cdot 73 \cdot 103$	172081	$7 \cdot 13 \cdot 31 \cdot 61$
6601	$7 \cdot 23 \cdot 41$	62745	$3 \cdot 5 \cdot 47 \cdot 89$	188461	$7 \cdot 13 \cdot 19 \cdot 109$
8911	$7 \cdot 19 \cdot 67$	63973	$7 \cdot 13 \cdot 19 \cdot 37$	252601	$41 \cdot 61 \cdot 101$
10585	$5 \cdot 29 \cdot 73$	75361	$11 \cdot 13 \cdot 17 \cdot 31$	278545	$5 \cdot 17 \cdot 29 \cdot 113$

primes up to this number by two simple tests. As an example, let us consider $m = 4033 = 37 \cdot 109$. We happen to know that 4033 is a strong pseudoprime to base 2 but not, of course, to base 3. Let us test $m = 4033$ as specified by (36a) and (36b). First, we establish that $(2, 4033) = (3, 4033) = 1$. Second, we express $m - 1$ as $m - 1 = 2^6 \cdot 63$. Third, we proceed to calculate 2^{63} and 3^{63} using the Binary Exponentiation Algorithm as shown in the table below.

	N	Y_1	Z_1	Y_2	Z_2
Initialization	63	1	2	1	3
$Y \leftarrow Y * Z$	63	2	2	3	3
$N \leftarrow \left[\dfrac{N}{2}\right]$	31	2	2	3	3
$N \neq 0$; Square Z	31	2	4	3	9
$Y \leftarrow Y * Z$	31	8	4	27	9
$N \leftarrow \left[\dfrac{N}{2}\right]$	15	8	4	27	9
$N \neq 0$; Square Z	15	8	16	27	81
$Y \leftarrow Y * Z$	15	128	16	2187	81
$N \leftarrow \left[\dfrac{N}{2}\right]$	7	128	16	2187	81
$N \mp 0$; Square Z	7	128	256	2187	2528**
$Y \leftarrow Y * Z$	7	504**	256	3526**	2528
$N \leftarrow \left[\dfrac{N}{2}\right]$	3	504	256	3526	2528
$N \neq 0$; Square Z	3	504	1008**	3526	2512**
$Y \leftarrow Y * Z$	3	3907**	1008	844**	2512
$N \leftarrow \left[\dfrac{N}{2}\right]$	1	3907	1008	844	2512
$N \neq 0$; Square Z	1	3907	3781**	844	2532**
$Y \leftarrow Y * Z$	1	3521**	3781	3551**	2532
$N \leftarrow \left[\dfrac{N}{2}\right]$	0	3521	3781	3551	2532
$N = 0$; Finished	0	$\boxed{3521}$	3781	$\boxed{3551}$	2532

** Reduction Modulo 4033

We find that

$$2^{63} \equiv 3521 \bmod (4033) \tag{37}$$

$$3^{63} \equiv 3551 \bmod (4033) \tag{38}$$

Neither (37) nor (38) satisfy (36a). Let us now check to see whether there is an r, $0 \leqslant r < 6$, such that $2^{2^r \cdot 63} \equiv 4032 \bmod (4033)$:

$$2^{63} \equiv 3521 \bmod (4033)$$

$$(2^{63})^2 \equiv 4032 \bmod (4033)$$

When $r = 1$, we satisfy condition (36b) and we conclude that 4033 is either a prime or a strong pseudoprime to base 2. Now let us check to see whether there is an r, $0 \leqslant r < 6$, such that $3^{2^r \cdot 63} \equiv 4032 \bmod (4033)$:

$$3^{63} \equiv 3551 \bmod (4033)$$

$$(3^{63})^2 \equiv 2443 \bmod (4033)$$

$$(3^{63})^4 \equiv 3442 \bmod (4033)$$

$$(3^{63})^8 \equiv 2443 \bmod (4033)$$

$$(3^{63})^{16} \equiv 3442 \bmod (4033)$$

$$(3^{63})^{32} \equiv 2443 \bmod (4033)$$

Clearly, we have failed to meet condition (36b) and thus we have determined that 4033 can not be prime.

We can build on what we have just learned and present a test which is a "quick check" for primality for m, $m < 25 \cdot 10^9$. We use the fact that all the pseudoprimes for the bases $a = 2$, 3, and 5 have been determined. Surprisingly, only 13 odd composite numbers less than 25 billion are strong pseudoprimes to 2, 3, and 5; thus, the algorithm shown in Figure 3.5, due to Pomerance et al. (1980), can be employed to quickly check m, odd, and $m < 25 \cdot 10^9$ for primeness.

We now have a quick way to find primes below $25 \cdot 10^9$. All we need do is generate a random number less than 25 billion and then test by the preceding algorithm.

3.15.1. A Probabilistic Extension (Rabin, 1980)

In the previous section we mentioned that no odd composite number is a strong pseudoprime to all of the bases to which it is relatively prime. We can be a bit more quantitative. It has been shown that an odd composite number can be a strong pseudoprime to no more than one-quarter of the

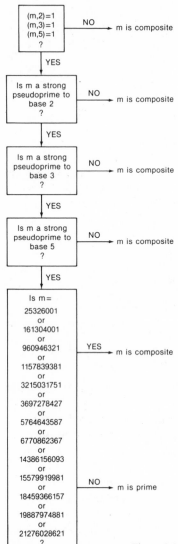

Figure 3.5. Algorithm for checking for primality of $m < 25 \cdot 10^9$.

relatively prime bases. We thus have the motivation for a probabilistic primality test. The algorithm is as follows:

1. Choose b different bases, $\{a_1, a_2, \ldots, a_b\}$.
2. Check *all* the bases *first* to ensure that $(a_i, m) = 1$ for each i.
3. Check to see that m behaves as a strong pseudoprime to each base.

If m passes step 3, then we can say that m is prime with probability at least $1 - 2^{-2b}$. This probability can be rendered insignificant with modest b.

3.15.2. Building Large Primes

In the chapter on privacy, we will encounter a need to construct large primes, p, of the form

$$p = 2p' + 1 \tag{39}$$

where p' is also prime. A method derived by Williams and Schmid (1979) makes construction of such primes a straightforward task.

The first step is to randomly select two primes of about the same size. Let these randomly selected primes be denoted by p_1 and p_2.

The second step is to find x such that

$$xp_2 \equiv -1 \bmod (p_1) \tag{40}$$

where we restrict x to the range $0 < x < p_1$.

The third step is to create the set of number pairs

$$\{(s_i, t_i)\} \tag{41}$$

where

$$s_i = 2p_1 p_2 i + 2p_2 x + 1 \tag{42a}$$

$$t_i = 2s_i + 1 \tag{42b}$$

Although there is no magic number for the size of the set (41) it is suggested that i run through the domain $i = 1, 2, 3, \ldots, L$, where

$$L = 1.5(\ln 4p_1 p_2)^2 \tag{43}$$

The fourth step is to sieve the set (43) in the following manner. Sequentially test each pair (s_i, t_i) for lack of primeness by using the 168 primes less than 1000. If either member is not prime, discard the pair. Most pairs will be discarded by this test.

The fifth step is to consider the surviving pairs sequentially and find the s_i such that the pair of congruences

$$3^{s_i - 1} \equiv 1 \bmod (s_i) \tag{44a}$$

and

$$3^{s_i} \equiv 1 \bmod (t_i) \tag{44b}$$

hold. When this happens we know that s_i and $t_i = 2s_i + 1$ are both prime.

We should, at this juncture, say a word about the second step. The solution of the equation

$$ax \equiv -1 \bmod (m) \tag{45}$$

can be approached in a number of ways. First, we can convert it to the equation

$$ay \equiv 1 \bmod (m) \tag{46}$$

by setting $y = m - x$. Equation (46) has an immediate solution. Recall that Fermat's "Little Theorem" guarantees that $a^{\phi(m)} \equiv 1 \bmod (m)$ for $(a, m) = 1$ and so

$$y \equiv a^{\phi(m)-1} \tag{47}$$

must be a solution. A computationally faster solution to (46) can often be obtained through the Euclidean Algorithm. A closer look at the Euclidean Algorithm will show us why this is so. Consider that we wish to determine the GCD of a and m. The Euclidean Algorithm would proceed as follows:

$$m = aq_1 + r_1$$
$$a = r_1 q_2 + r_2$$
$$r_1 = r_2 q_3 + r_3$$
$$\vdots$$
$$r_{i-1} = r_i q_{i+1} + 1$$

(The GCD will be unity for the above since we are interested in the case for which a is a prime.) What we can do with the Euclidean Algorithm is to thread our way from start to finish and eliminate the remainder terms $r_1, r_2, r_3, \ldots$ and obtain the solution as a function of $q_1, q_2, q_3, \ldots$. For example, let us suppose that the algorithm terminated in two steps:

$$m = aq_1 + r_1 \tag{48a}$$
$$a = r_1 q_2 + 1 \tag{48b}$$

Solving (48a) for r_1 we have $r_1 = m - aq_1$. Inserting r_1 into (48b) we obtain

$$a = (m - aq_1)q_2 + 1 \tag{49}$$

Regrouping (49) to isolate a and we find that

$$a(q_1 q_2 + 1) = mq_2 + 1 \tag{50}$$

Changing (50) to a congruence mod (m) we immediately have

$$a(q_1 q_2 + 1) \equiv 1 \bmod (m) \tag{51}$$

and thus the solution to (46) is

$$y \equiv (q_1 q_2 + 1) \bmod (m) \tag{52}$$

This "forward threading" process is easily extended; it is simple to program and run quickly on a computer. [When the process terminates in one step, that is, $m = aq_1 + 1$, use (47).]

As an example, let us use the above construction method to build a larger prime. We arbitrarily choose $p_1 = 3$ and $p_2 = 7$ for our first step. We must solve for x from $7x \equiv -1 \bmod (3)$. We quickly find that $x = 2$. The third step requires us to form the set of pairs $\{(s_i, t_i)\}$ according to

$$s_i = 42i + 29 \tag{53a}$$

and

$$t_i = 2s_i + 1 \tag{53b}$$

Instead of creating all the pairs up to L of (43), let us, for this example, generate and test the pairs one at a time over the first two odd primes versus all the primes less than 1000. The first pair is (71, 143). The second member of this pair is divisible by 11 and we therefore discard the pair. The second pair is (113, 227) and passes the sieve and so we go on to the fifth step and test to see if

$$3^{112} \stackrel{?}{\equiv} 1 \bmod (113) \tag{54a}$$

and

$$3^{113} \stackrel{?}{\equiv} 1 \bmod (227) \tag{54b}$$

We find that (54a) and (54b) hold and thus we have determined that 113 and 227 are two primes that, by construction, exhibit the relation (42b), that is, $227 = 2 \cdot 113 + 1$.

References

Beiler, A. (1966), *Recreations in the Theory of Numbers*, Dover Publications, New York.

Brent, R. (1976), Analysis of the Binary Euclidean Algorithm, in (Symposium on) *Algorithms and Complexity*, J. Traub, ed., pp. 321–355, Academic Press, New York.

Knuth, D. (1973), *The Art of Computer Programming, Vol. 1*, 2nd Ed., Addison-Wesley, Reading, MA.

Pomerance, C., J. Selfridge, and S. Wagstaff, Jr. (1980), The Pseudoprimes to $25 \cdot 10^9$, *Mathematics of Computation*, Vol. 35, pp. 1003–1026.

Rabin, M. (1980), Probabilistic Algorithm for Testing Primality, *Journal of Number Theory*, Vol. 12, pp. 128–138.

Williams, H. and B. Schmid (1979), Some Remarks Concerning the M.I.T. Public-Key Crypto Systems, *BIT*, Vol. 19, pp. 525–538.

Basic Concepts in Matrix Theory

4.1. Introduction

In this chapter we consider some of the basic concepts associated with matrix theory and its applications to digital communications, signal processing, and many others. To cast some of the basic ideas, let us consider first an analog signal, $x_a(t)$, and its sampled values

$$x_a(nT) = x_a(t)|_{t=nT} \tag{1}$$

where T is the sampling interval. For obvious reasons, we assume that $x_a(t)$ is time limited and therefore, without losing any generality, we can assume that $x_a(nT) = 0$ for $n < 0$ and $n > N - 1$, for some N. The sampling interval T is a constant and can be omitted. For simplicity, let the data signal be represented by

$$
\begin{aligned}
x(n) &= x_a(nT), & 0 \le n \le N - 1 \\
&= 0, & n < 0, n > N - 1
\end{aligned} \tag{2}
$$

thereby allowing us to handle digitized analog signals or discrete data in the same vein. In the digital arena, we may be interested in transmitting data, filtering data to eliminate noise, compressing data for efficient transmission and storage, extracting information from data, encrypting data for protection of privacy, and so on.

Some of the topics mentioned come under the category of digital signal processing concepts or coding. The basic concept here, given a set of data $x(n)$, $0 \le n \le N - 1$, is to transform or convert these data into another set of data, say $y(n)$, $0 \le n \le N - 1$. This process can best be illustrated in terms of matrix analysis.

The data in (1), by nature, are assumed to have infinite precision. In practice, we can only have finite precision, that is, in reality, we use quantized values,

$$x_q(n) = Q[x(n)]$$

which can be represented by a finite number of bits. Representation of the quantized data, processing the data, and coding the data are some of the primary goals of this book. Some aspects of processing the quantized data can again be illustrated by matrices.

Our presentation of matrix analysis is by examples. Theoretical aspects are avoided wherever possible. Before we introduce matrices, we should briefly study the nature of the data $x(n)$ and $x_q(n)$. We assume that $x(n)$ are real or complex numbers. Before we manipulate these we need to know the number system in which addition and multiplication are defined and are subjected to the usual laws of algebra. For the quantized data, we may have to use other number systems. For example, the binary data have only two numbers 0 or 1 (some consider -1 and 1) and therefore we need to investigate this representation. A type of system that generalizes our systems is a field (Perlis, 1952).

4.2. Concept of a Field and a Group

Each of the number systems we study here is a nonempty collection F of numbers that does not consist of the number zero alone. It is called a number field if and only if the sum, difference, product, and quotient of any two numbers of F are again numbers of F, with the exception of division by zero.

It is necessary to discuss the operations in a field explicitly. First, with each pair of elements u and v of F, a unique quantity called the sum of u and v is represented by $u + v$. Sums have the following properties for all u, v, and w in F:

1a. $(u + v) + w = u + (v + w)$.
1b. $u + v = v + u$.
1c. F has a unique element 0 such that $u + 0 = u$.
1d. For each u in F there is a unique v in F such that $u + v = 0$. v can be denoted by $-u$.

The last property allows us to deal with the difference of two elements u and v in F by

$$u - v = u + (-v)$$

Second, with each pair of elements u and v in F, a unique quantity called the product of u and v is represented by uv. Products have the following properties for all u, v, and w in F.

2a. $(uv)w = u(vw)$.
2b. $uv = vu$.
2c. F has a unique element 1, that is different from 0, such that $1u = u1 = u$.
2d. For each $u \neq 0$ that is in F, there is a unique v in F such that $uv = vu = 1$. v is denoted by u^{-1}.

This last property allows us to deal with the division of two elements u and v in F by

$$\frac{u}{v} = u(v^{-1})$$

The two operations, that is, the addition and multiplication, are connected by the distributive laws for all u, v, and w:

3a. $u(v + w) = uv + uw$.
3b. $(v + w)u = vu + wu$.

Any collection F of numbers on which additions and multiplications are defined such that they lie in F and obey the laws 1a-1d, 2a-2d, and 3a-3b is called a field. The four elementary operations, addition, subtraction, multiplication, and division are available in any field.

Next let us consider some examples. The set of all real numbers under ordinary addition and multiplication constitute a field, usually referred to as the real field. The set of all complex numbers under ordinary addition and multiplication constitute a field, usually referred to as the complex field. Some sets of numbers that are not fields under ordinary addition and multiplication are, for example, the set of all integers and the set of all positive real numbers. In the first case, the quotient of two integers of the set is not in the set for all cases. In the second case, the difference of two numbers of the set is not in the set for all cases.

The sets mentioned so far are most familiar to us. Next, let us discuss some sets that have only a finite number of elements. The set $z_2 = \{0, 1\}$ constitutes a field under the $+$ operation, usually referred to as mod 2 addition, defined by

$$
\begin{aligned}
0 + 0 &= 0 \\
0 + 1 &= 1 \\
1 + 0 &= 1 \\
1 + 1 &= 0
\end{aligned}
\tag{3}
$$

and ordinary multiplication. The ordinary multiplication for this set is defined by

$$0 \cdot 0 = 0$$
$$0 \cdot 1 = 0$$
$$1 \cdot 0 = 0 \qquad (4)$$
$$1 \cdot 1 = 1$$

In this case we have a finite field (or Galois Field). The notation GF(q) is generally used to denote the finite field of q elements. For our example here, $q = 2$, and therefore GF(2) is used. Some refer to this as a binary field also. The two arithmetic operations defined in (3) and (4) are usually referred to as modulo 2, or simply mod 2, arithmetic. Next let us consider a more complicated case with more than two elements, say a set of q elements defined by

$$z_q = \{0, 1, 2, \ldots, q - 1\} \qquad (5)$$

In such a case, we need to define some arithmetic operations, since the product of two numbers for all cases may not be in the set.

In the last chapter we have discussed the concept of congruence (see Section 3.4). Let us review this briefly here.

Two integers n and m are referred to as congruent modulo q if

$$n = m + kq \qquad (6)$$

where k is some integer and m is the modulus. Notationwise we have used

$$n \equiv m(\mathrm{mod}\ q) \quad \text{or} \quad n \equiv m(q) \qquad (7)$$

when there is no confusion. From this it follows that all integers can be reduced to one of the integers in the set z_q in (5). In other words, we say that all integers are congruent mod q to some integer in the set z_q.

Next let us consider some examples using mod 5 arithmetic.

Addition: $3 + 4 = 7 \equiv 2 \pmod 5$
Negation: $-3 \equiv -3 + 5 \equiv 2 \pmod 5$
Subtraction: $8 - 3 = 8 + (-3) = 8 + 2 \equiv 0 \pmod 5$
Multiplication: $8 \times 3 = 24 \equiv 4 \pmod 5$
Multiplicative Inverse: $3^{-1} \equiv 2 \pmod 5$

The multiplicative inverse of an integer n in z_q exists if and only if n and

q are relatively prime, that is, n and q have no common divisors. When the inverse exists, n^{-1} is found from

$$n \times n^{-1} \equiv 1 \,(\text{mod } q)$$

For division, $\frac{4}{3} = 4 \times (2) \equiv 3 \,(\text{mod } 5)$. Clearly, the ratio n/m exists if and only if m^{-1} exists.

Note that any integer in the set $\{1, 2, 3, 4\}$ and 5 are relatively prime and therefore the inverse of any integer in this set exists. It can be shown that $z_5 = \{0, 1, 2, 3, 4\}$ satisfies the conditions for a field under the operation of mod 5 addition and multiplication.

These results can be extended for other cases. For example, for mod 2 operations we have

$$5 \equiv 1 \,(\text{mod } 2)$$

Finite field mathematics is an important part of our work here.

Next let us consider another important and useful concept. Consider a set G of mathematical elements, each ordered pair (a, b) of which can be combined by an operation denoted here by 0 to give a unique result $a0b$ is called a group if and only if the following properties hold:

1. G is closed with respect to the operation 0. That is, if a and b are members of G, then $a0b$ is also a member of G.
2. The associative law holds. That is, for all a, b, c of G, $a0(b0c) = (a0b)0c$.
3. G contains an identity element. That is, there is an element e in G such that for all a of G, $a0e = e0a = a$.
4. For each member a of G, G contains a member a^{-1}, called the inverse of a, such that $a0a^{-1} = a^{-1}0a = 1$.

If the commutative law holds, that is, $a0b = b0a$, then G is called a commutative or Abelian group.

There are many examples that can be given for groups. The set of all complex numbers with the operation 0 being ordinary addition is an example. The binary field z_2 satifies the above properties [see (3)] when the operation is modulo 2 addition.

The next section deals with the basic definitions of matrices.

Problems

1. Show that the set of all rational numbers, that is, the set of all quotients of the form a/b, $b \neq 0$, where a and b are integers, constitute a field (usually referred to as a rational field).
2. Find the multiplicative inverses of the integers 1, 2, 3, 4 over GF(5).

4.3. Basic Definitions

An $m \times n$ matrix

$$A = (a_{ij}) \tag{8}$$

over a field is a rectangular array arranged in m rows and n columns. The entry a_{ij} is located in the ith row and the jth column. It is on the diagonal if $i = j$, off the diagonal if $i \neq j$, below the diagonal or subdiagonal if $i > j$, and above the diagonal or superdiagonal if $i < j$. The sum of the diagonal entries a_{ii} is the trace of A. That is,

$$\text{tr}\,(A) = \sum_i a_{ii} \tag{9}$$

The matrix A is called a square matrix if $m = n$; otherwise, it is called a rectangular matrix. If $m = 1$, then A is called a row vector

$$A = [a_{11} \quad a_{12} \quad \cdots \quad a_{1n}] \tag{10}$$

If $n = 1$ and m is arbitrary, then

$$A = \begin{bmatrix} a_{11} \\ a_{21} \\ \vdots \\ a_{m1} \end{bmatrix} \tag{11a}$$

is called a column vector. For simplicity, we also write (11a) in the form

$$A = \text{col}\,(a_{11}, a_{21}, \ldots, a_{m1}) \tag{11b}$$

There are some interesting classes of square matrices that are of interest. These are discussed next.

Let A be an $m \times m$ matrix. Then A is called diagonal if $a_{ij} = 0$ for $i \neq j$. A 3×3 example is

$$A = \begin{bmatrix} a_{11} & 0 & 0 \\ 0 & a_{22} & 0 \\ 0 & 0 & a_{33} \end{bmatrix} \tag{12}$$

A shorthand notation for this is

$$A = \text{dia}\,(a_{11}, a_{22}, a_{33}) \tag{13}$$

An identity matrix I is a diagonal matrix with all the diagonal entries equal to 1. A is an upper triangular matrix if $a_{ij} = 0$ for $i > j$.

$$A = \begin{bmatrix} a_{11} & a_{12} & a_{13} \\ 0 & a_{22} & a_{23} \\ 0 & 0 & a_{33} \end{bmatrix} \tag{14}$$

and is strictly upper triangular matrix if in addition $a_{ii} = 0$. A is a lower triangular matrix if $a_{ij} = 0$ for $i < j$.

$$A = \begin{bmatrix} a_{11} & 0 & 0 \\ a_{21} & a_{22} & 0 \\ a_{31} & a_{32} & a_{33} \end{bmatrix} \tag{15}$$

and is strictly lower triangular if in addition $a_{ii} = 0$. Finally, an $m \times n$ matrix A with all zero entries is called a null matrix.

The transpose of an $m \times n$ matrix $A = (a_{ij})$ is the $n \times m$ matrix where the ij entry is a_{ji}. That is,

$$A^T = \begin{bmatrix} a_{11} & a_{21} & \cdots & a_{m1} \\ a_{12} & a_{22} & \cdots & a_{m2} \\ \vdots & \vdots & & \vdots \\ a_{1n} & a_{2n} & \cdots & a_{mn} \end{bmatrix} \tag{16}$$

For the following discussion, we use a bar above the matrix, $\bar{A}$, to identify that all the entries of A, a_{ij}, are replaced by their conjugates, $\bar{a}_{ij}$. The matrix A^* denotes that

$$A^* = (\bar{A})^T = (\bar{A^T}) \tag{17}$$

That is, the superscript $*$ denotes that A^* corresponds to the transpose conjugate (or conjugate transpose) of the matrix A. It is clear that the matrix A is real if all its entries are real. That is, A is real if $A = \bar{A}$. A is symmetric if $A = A^T$. A is skew symmetric if $A = -A^T$. A is Hermitian if $A = A^*$ and skew Hermitian if $A = -A^*$. Examples of some of the above matrices are given below.

$$A = \begin{bmatrix} 2 & 1 & 3 \\ 1 & 4 & 2 \\ 3 & 2 & 3 \end{bmatrix} \qquad \text{symmetric matrix} \tag{18a}$$

$$A = \begin{bmatrix} 0 & -1 & -2 \\ 1 & 0 & 3 \\ 2 & -3 & 0 \end{bmatrix} \qquad \text{skew symmetric matrix} \tag{18b}$$

$$A = \begin{bmatrix} 1 & 1+j1 & 2-j2 \\ 1-j1 & 2 & 3 \\ 2+j2 & 3 & 0 \end{bmatrix} \quad \text{Hermitian matrix} \qquad (18c)$$

$$A = \begin{bmatrix} 0 & 1-j2 & 0 \\ -1-j2 & 0 & 2 \\ 0 & -2 & j2 \end{bmatrix} \quad \text{skew Hermitian matrix} \qquad (18d)$$

Problems

1. Show that the diagonal entries of a skew symmetric matrix are always equal to zero.
2. Compute A^* for each of the matrices given in (18).

4.4. Matrix Operations

In the following we consider scalars and matrices that have entries from a real or complex number field. At the end of the section, we consider the finite field case.

The matrices A and B are equal if the corresponding entries are equal. That is, $A = B$, if $a_{ij} = b_{ij}$. It is obviously implicit in this definition that the matrices A and B must be of the same size. The product of a matrix A by a scalar multiple c is

$$cA = (ca_{ij}) \qquad (19)$$

That is, every entry of A is multiplied by the scalar c. The addition of two matrices A and B is

$$A + B = (a_{ij} + b_{ij}) \qquad (20)$$

That is, the corresponding entries of A and B are added. Again, it is implicit that both A and B are of the same dimensions. Also, it is clear that

$$A + B = B + A \qquad (21)$$

since, for scalars, $a_{ij} + b_{ij} = b_{ij} + a_{ij}$.

4.4.1. Matrix Multiplication

Let A be an $m \times n$ matrix and B be a $p \times q$ matrix. Then the matrix product AB is defined only for $n = p$, that is, when the number of columns in A is equal to the number of rows in B. The dimensions of the matrix product is $m \times q$. The ikth entry in AB is given by

$$(AB)_{ik} = \sum_{j=1}^{n} a_{ij}b_{jk}, \qquad i = 1, 2, \ldots, m, \, k = 1, 2, \ldots, q \qquad (22)$$

In other words, to obtain the ikth element in the matrix product AB, multiply the elements of the ith row of A by the corresponding elements of the kth column in B and add the terms.

Example

Let

$$A = \begin{bmatrix} 1 & 2 & 3 \\ 0 & 1 & 2 \end{bmatrix}, \qquad B = \begin{bmatrix} 2 & 0 \\ 1 & 1 \\ 1 & 1 \end{bmatrix}$$

$$AB = \begin{bmatrix} 2+2+3 & 0+2+3 \\ 0+1+2 & 0+1+2 \end{bmatrix} = \begin{bmatrix} 7 & 5 \\ 3 & 3 \end{bmatrix} \qquad (23)$$

Similarly,

$$BA = \begin{bmatrix} 2+0 & 4+0 & 6+0 \\ 1+0 & 2+1 & 3+2 \\ 1+0 & 2+1 & 3+2 \end{bmatrix} = \begin{bmatrix} 2 & 4 & 6 \\ 1 & 3 & 5 \\ 1 & 3 & 5 \end{bmatrix} \qquad (24)$$

Note that AB is a 2×2 matrix and BA is a 3×3 matrix. In general, AB is not necessarily equal to BA. Furthermore, even if AB is defined, BA may not be defined. If

$$AB = BA \qquad (25)$$

then A is said to commute with B and vice versa. If both products are defined, then it can be shown that

$$\text{tr}\,(AB) = \text{tr}\,(BA) = \sum_{i=1}^{m} \sum_{j=1}^{n} a_{ij}b_{ji} \qquad (26)$$

The transpose of a product AB is given by

$$(AB)^T = B^T A^T \tag{27}$$

4.4.2. Determinants

The determinants of matrices are defined only for square matrices. For a matrix A of order 2, the determinant is defined by

$$|A| = \begin{vmatrix} a_{11} & a_{12} \\ a_{21} & a_{22} \end{vmatrix} = a_{11}a_{22} - a_{21}a_{12} \tag{28}$$

For matrices of higher dimension, define the determinants in terms of cofactors.

Let A be an $n \times n$ matrix and let M_{ij} denote the $(n-1) \times (n-1)$ submatrix of A obtained by deleting the ith row and the jth column. Then the scalar

$$c_{ij} = (-1)^{i+j}|M_{ij}| \tag{29}$$

is called the cofactor of a_{ij} in A. The $n \times n$ matrix

$$C^T = (c_{ji}) \tag{30}$$

is called the adjoint of A and is usually denoted by Adj (A). The determinant is related to the adjoint by the equations

$$|A| = \sum_{j=1}^{n} a_{ij}c_{ij} = \sum_{i=1}^{n} c_{ij}a_{ij} \tag{31}$$

for each value of $i = 1, 2, \ldots, n$ (or $j = 1, 2, \ldots, n$ in the second expression).

Example

Find the determinant of the following matrix:

$$A = \begin{bmatrix} 1 & 1 & 2 \\ 0 & 1 & 2 \\ 1 & 1 & 0 \end{bmatrix} \tag{32}$$

The cofactors are

$$c_{11} = (-1)^{1+1} \begin{vmatrix} 1 & 2 \\ 1 & 0 \end{vmatrix}, \qquad c_{12} = (-1)^{1+2} \begin{vmatrix} 0 & 2 \\ 1 & 0 \end{vmatrix}$$

$$c_{13} = (-1)^{1+3} \begin{vmatrix} 0 & 1 \\ 1 & 1 \end{vmatrix}$$

$$c_{21} = (-1)^{2+1} \begin{vmatrix} 1 & 2 \\ 1 & 0 \end{vmatrix}, \qquad c_{22} = (-1)^{2+2} \begin{vmatrix} 1 & 2 \\ 1 & 0 \end{vmatrix}$$

$$c_{23} = (-1)^{2+3} \begin{vmatrix} 1 & 1 \\ 1 & 1 \end{vmatrix}$$

$$c_{31} = (-1)^{3+1} \begin{vmatrix} 1 & 2 \\ 1 & 2 \end{vmatrix}, \qquad c_{32} = (-1)^{3+2} \begin{vmatrix} 1 & 2 \\ 0 & 2 \end{vmatrix}$$

$$c_{33} = (-1)^{3+3} \begin{vmatrix} 1 & 1 \\ 0 & 1 \end{vmatrix}$$

From these we have

$$\text{Adj} (A) = \begin{bmatrix} c_{11} & c_{21} & c_{31} \\ c_{12} & c_{22} & c_{32} \\ c_{13} & c_{23} & c_{33} \end{bmatrix} = \begin{bmatrix} -2 & 2 & 0 \\ 2 & -2 & -2 \\ -1 & 0 & 1 \end{bmatrix} \tag{32a}$$

$$|A| = a_{11}c_{11} + a_{12}c_{12} + a_{13}c_{13}$$

$$= 1 \times -2 + 1 \times 2 + 2 \times -1 = -2 \tag{32b}$$

From (31), one interesting aspect results. That is, for every square matrix A

$$A \, \text{Adj} (A) = \text{Adj} (A)A = |A|I \tag{33}$$

where I is an identity matrix.

For some special classes of matrices, determinants can be written by inspection. For example, the determinants of diagonal, upper triangular, and lower triangular matrices can be obtained by multiplying their respective diagonal entries. Two side results that come in handy are

$$|A^T| = |A| \tag{34}$$

and

$$|AB| = |A||B| \tag{35}$$

where A and B are assumed to be square matrices. Finding determinants by the above method is impractical. A practical method is to use the following elementary properties of determinants.

1. If B is obtained from A by interchanging two rows, then $|B| = -|A|$.
2. Let the jth row of B be given by the jth row of A plus a constant c times the ith row of A; furthermore, let the rows of B other than row j be the same as the rows in A. Then $|B| = |A|$.
3. If B is obtained from A by multiplying a row by a constant c, then $|B| = c|A|$.

These operations provide a simple way to evaluate determinants.

Example

Find the determinant of the following matrix using the above properties:

$$A = \begin{bmatrix} 1 & 1 & 2 \\ 0 & 1 & 2 \\ 1 & 1 & 0 \end{bmatrix} \tag{36}$$

Multiplying the first row by (-1) and adding it to the last row, we have

$$|A| = \begin{vmatrix} 1 & 1 & 2 \\ 0 & 1 & 2 \\ 0 & 0 & -2 \end{vmatrix} \tag{37}$$

Noting that the determinant of a triangular matrix is equal to the product of the diagonal entries, we have $|A| = -2$.

4.4.3. Finite Field Matrix Operations

From the above discussion, we see that the matrix operations discussed earlier can be broken down into elementwise operations. Therefore, we can use the modulo q arithmetic discussed in Section 4.2 to perform the matrix operations over $\text{GF}(q)$ for matrices having elements from the finite field $\text{GF}(q)$. Let us illustrate these aspects using some of the above examples.

Examples

If the field is $\text{GF}(5)$, the matrix product in (23) reduces to

$$AB = \begin{bmatrix} 7 & 5 \\ 3 & 3 \end{bmatrix} \equiv \begin{bmatrix} 2 & 0 \\ 3 & 3 \end{bmatrix} \pmod{5}$$

It is clear that the above result is obtained from first computing the result in the real field and then reducing the result. The same result could have been obtained by considering each entry in (23) and reducing at each stage.

If the field is GF(3), the determinant of A in (32) can be computed using (32b). That is,

$$|A| = -2 \equiv 1 \, (\text{mod} \, 3)$$

Similarly, we can use the elementary properties of determinants to compute the determinants. Since $-1 = 2 \, (\text{mod} \, 3)$, we can multiply row 1 by 2 and add to row 3 in (36) resulting in

$$|A| = \begin{vmatrix} 1 & 1 & 2 \\ 0 & 1 & 2 \\ 0 & 0 & 1 \end{vmatrix} = 1 \tag{38}$$

which we could have obtained from (37), since $-2 \equiv 1 \, (\text{mod} \, 3)$. There should not be any confusion if we omit the term $(\text{mod} \, q)$ in our work, if the operations are implicitly defined in GF(q). However, if there appears to be any confusion, it is best to identify explicitly the field (GF(q)) or the arithmetic $(\text{mod} \, q)$.

Problems

1. Given

$$A = \begin{bmatrix} 1 & 2 & 1 \\ 1 & 3 & 2 \end{bmatrix} \quad \text{and} \quad B = \begin{bmatrix} 2 & 0 & 1 \\ -1 & 3 & 1 \end{bmatrix}$$

find $A + B$.

2. For the given matrices in problem 1, find $A^T B$. Can you explain why AB is not defined?

3. Compute the determinants of the following matrices using real field operations. Do this by using the cofactors and by using the elementary properties of determinants:

$$|A| = \begin{vmatrix} 1 & 2 & 2 \\ 2 & 1 & 0 \\ 1 & 0 & 0 \end{vmatrix} \quad \text{and} \quad |B| = \begin{vmatrix} 1 & 2 & 1 \\ 0 & 1 & 1 \\ 2 & 0 & 4 \end{vmatrix}$$

4. Do problem 3 using the operations over GF(5).

5. Show that $|A| = 0$, when A is an odd order skew symmetric matrix.

Hint: Use $|A| = |A^T|$ and the elementary properties of determinants discussed in this section.

4.5. Partitioned Matrices

For large matrices it is convenient to use methods wherein we can do matrix operations using sub-blocks of matrices (or simply submatrices). Let us consider two partitioned matrices written in this form

$$P = \begin{bmatrix} P_{11} & P_{12} \\ P_{21} & P_{22} \end{bmatrix} \quad \text{and} \quad Q = \begin{bmatrix} Q_{11} & Q_{12} \\ Q_{21} & Q_{22} \end{bmatrix} \tag{39}$$

of arbitrary size. The sum $P + Q$ is defined provided P and Q have the same dimensions. If P_{ij} and Q_{ij} have the same dimensions, then we can write

$$P + Q = \begin{bmatrix} P_{11} + Q_{11} & P_{12} + Q_{12} \\ P_{21} + Q_{21} & P_{22} + Q_{22} \end{bmatrix} \tag{40}$$

Similarly, we can do the partitioned matrix multiplication PQ. First, the product (PQ) is defined only if the number of columns in P is equal to the number of rows in Q. Second, if the products $P_{ij}Q_{jk}$, for all i, j, and k, $1 \leq i, j, k \leq 2$, are defined, then

$$PQ = \begin{bmatrix} P_{11}Q_{11} + P_{12}Q_{21} & P_{11}Q_{12} + P_{12}Q_{22} \\ P_{21}Q_{11} + P_{22}Q_{21} & P_{21}Q_{12} + P_{22}Q_{22} \end{bmatrix} \tag{41}$$

4.5.1. Direct Sum of Matrices

If a matrix M has the form

$$M = \begin{bmatrix} M_1 & 0 & \cdots & 0 \\ 0 & M_2 & \cdots & 0 \\ \vdots & \vdots & & \vdots \\ 0 & 0 & \cdots & M_N \end{bmatrix} \tag{42}$$

where M_i are some matrices, then M is said to be a direct sum of N matrices. These have important applications in many areas. One area of interest is in the connectivity of communication networks (Seshu and Reed, 1961).

4.5.2. Connectivity in Terms of Matrices

For our present purposes, we define an $N \times N$ connectivity matrix, $M = (m_{ij})$, corresponding to an N-node network by the following rules. If there is a direct path from node i to node j, then $m_{ij} = 1$. On the other hand, if there is no direct path, then $m_{ij} = 0$. A simple example illustrates this.

Example

Given a three-node network (Figure 4.1), find the connectivity matrix
M. Now

$$
M = \begin{array}{c} \\ 1 \\ 2 \\ 3 \end{array}
\begin{array}{ccc} 1 & 2 & 3 \end{array}
\left[\begin{array}{ccc}
1 & 0 & 1 \\
1 & 1 & 0 \\
1 & 1 & 1
\end{array} \right] \tag{43}
$$

where the diagonal elements are 1. If the connectivity matrix can be
expressed as a direct sum of matrices by permuting rows and/or columns
if necessary, then it is clear that we can say that there are subnetworks
which are not connected together. The basic problem we consider here,
given a matrix M, is to determine whether the matrix M possesses a permuted
form that may be partitioned as a direct sum of matrices. If there exists
such a decomposition, we need to find that. Let us illustrate the problem
in terms of arbitrary matrices.

Example

We are given

$$
M = \begin{array}{c} \\ 1 \\ 2 \\ 3 \\ 4 \end{array}
\begin{array}{ccccc} 1 & 2 & 3 & 4 & 5 \end{array}
\left[\begin{array}{ccccc}
1 & 0 & 2 & 1 & 0 \\
0 & 1 & 0 & 0 & 1 \\
1 & 0 & 1 & 2 & 0 \\
0 & 0 & 2 & 1 & 0
\end{array} \right] \tag{44}
$$

which can be written

$$
M_p = \begin{array}{c} \\ 1 \\ 3 \\ 4 \\ 2 \end{array}
\begin{array}{ccccc} 1 & 3 & 4 & 2 & 5 \end{array}
\left[\begin{array}{ccccc}
1 & 2 & 1 & 0 & 0 \\
1 & 1 & 2 & 0 & 0 \\
0 & 2 & 1 & 0 & 0 \\
0 & 0 & 0 & 1 & 1
\end{array} \right] \tag{45}
$$

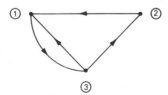

Figure 4.1. A three-node network.

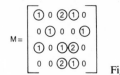

Figure 4.2. The encircling of all nonzero elements.

where M_p in (45) is obtained from (44) by permuting the second row to the fourth row and the second column to the fourth column, etc. This may look easy for a small matrix, but for large matrices, this gets complicated. There is a simple approach suggested by Demetriou and Wing (1964), which is discussed below.

The method is based on topology and the results can be obtained almost by inspection. The procedure is as follows:

1. Given the matrix, encircle all nonzero elements.
2. Considering the encircled elements as nodes of a topological pattern, connect them to one another to form a tree: (a) by drawing a link horizontally between two nodes and (b) by drawing a link vertically between nodes. The separate parts of the resulting pattern give the partition of the matrix.

Let us use this procedure for the above example. Figure 4.2 shows the results of applying the first step to the example matrix M. Figure 4.3 depicts the result of applying the second step to M. The first tree (solid lines) gives

$$M_1 = \begin{bmatrix} 1 & 2 & 1 \\ 1 & 1 & 2 \\ 0 & 2 & 1 \end{bmatrix}$$

and the second tree (dashed lines) gives

$$M_2 = \begin{bmatrix} 1 & 1 \end{bmatrix}$$

Thus,

$$M_p = \begin{array}{c} \\ 1 \\ 3 \\ 4 \\ 2 \end{array} \begin{array}{cccccc} 1 & 3 & 4 & 2 & 5 \\ \begin{bmatrix} 1 & 2 & 1 & 0 & 0 \\ 1 & 1 & 2 & 0 & 0 \\ 0 & 2 & 1 & 0 & 0 \\ 0 & 0 & 0 & 1 & 1 \end{bmatrix} \end{array} = \begin{bmatrix} M_1 & 0 \\ 0 & M_2 \end{bmatrix}$$

giving the decomposition corresponding to two trees. In general, if M_p

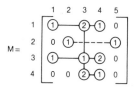

Figure 4.3. The partitioning of M.

has the general form given in (42), then N trees can be drawn on the matrix M. Also, if only one tree can be drawn, then

$$M_p = M_1 = M$$

Problems

1. Given the partitioned matrix A

$$A = \begin{bmatrix} 1 & 2 & 1 \\ 2 & 1 & 0 \\ 1 & 0 & 2 \end{bmatrix} = \begin{bmatrix} A_{11} & A_{12} \\ A_{21} & A_{22} \end{bmatrix}$$

where A_{11} is a 2×2 matrix, A_{12} is a 2×1 matrix, A_{21} is a 1×2 matrix, and A_{22} is a scalar, and a matrix B that is not partitioned,

$$B = \begin{bmatrix} 1 & 2 \\ 0 & 1 \\ 1 & 0 \end{bmatrix}$$

partition the matrix B such that A can be multiplied by B using block matrices. Find

$$C = AB = \begin{bmatrix} C_{11} \\ C_{21} \end{bmatrix}$$

and identify C_{11} and C_{21}.

2. Determine whether the matrix

$$M = \begin{bmatrix} 1 & 1 & 0 & 0 & 0 \\ 0 & 0 & 1 & 0 & 1 \\ 0 & 1 & 0 & 1 & 0 \\ 0 & 0 & 0 & 1 & 0 \end{bmatrix}$$

has a permuted form M_p that can be written as a direct sum of matrices.

4.6. Inverses of Matrices

The inverse of a matrix A, A^{-1}, is defined implicitly by the relation

$$AA^{-1} = A^{-1}A = I \qquad (46)$$

the identity matrix. From (33), it is clear that if $|A| \neq 0$, then for real or complex field applications

$$A^{-1} = \frac{1}{|A|} \text{Adj}(A) \qquad (47)$$

For GF(q) applications, we can use mod q arithmetic to first find $|A|$ in the appropriate field and then, if it is not equal to zero, (47) can be used to find the inverse.

The necessary and sufficient condition for the existence of the inverse of A is that $|A| \neq 0$ for real and complex fields and $(|A| \bmod (q)) \neq 0$ over GF(q). If $|A| \neq 0$, then A is nonsingular; otherwise, A is singular.

Example

Find the inverse of the matrix A in (32) over the real field.
From (32)–(33) using (47), we have

$$A^{-1} = \frac{-1}{2} \begin{bmatrix} -2 & 2 & 0 \\ 2 & -2 & -2 \\ -1 & 0 & 1 \end{bmatrix} \qquad (48)$$

Finding inverses using the adjoint matrix is impractical especially when the size of the matrix is large. In the following we consider a simpler method.

4.6.1. Inverses Using Elementary Operations

The basic idea is to first define a rectangular matrix

$$B = [A \quad I] \qquad (49)$$

where A is the matrix whose inverse over a given field is required. If the inverse of A, A^{-1}, exists, then

$$A^{-1}B = [I \quad A^{-1}] \qquad (50)$$

This gives a clue to finding the inverse of A by operating on the rows of A. Equivalently, this can be interpreted as premultiplying the matrix B by a matrix which is a product of elementary matrices (Frame, 1964). First, let us discuss the elementary operations.

4.6.2. Elementary Row Operations

1. Interchange of two rows.
2. Multiplication of a row by a scalar that has an inverse in the appropriate field.
3. Replacement of the ith row by the sum of the ith row and C times the jth row, where $j \neq i$ and C is any scalar in the appropriate field.

These operations can be used systematically on the matrix B in (49) to find the inverse of A. The following steps can be used in the appropriate field.

1. Permute the rows if necessary to get a nonzero entry in the (11) entry and later in an (ii) entry.
2. Multiply the first row by the inverse of the first entry and later the same operation to the ith row.
3. Multiply the first row by the negative of the $(j1)$ entry and add the resulting row to the jth row until all the entries in the first column are zero except for the (11) entry and later an (ii) entry. Start with the second row and go to step 1 above.

Let us illustrate these ideas.

Example

Use the elementary operations on the matrix B in (49), where A is given in (32), to find the inverse of A over the real field. Initially:

$$B = \begin{bmatrix} 1 & 1 & 2 & 1 & 0 & 0 \\ 0 & 1 & 2 & 0 & 1 & 0 \\ 1 & 1 & 0 & 0 & 0 & 1 \end{bmatrix} = [A \quad I] = B_0 \tag{51}$$

Multiply row 1 by -1 and add to row 3:

$$B_1 = \begin{bmatrix} 1 & 1 & 2 & 1 & 0 & 0 \\ 0 & 1 & 2 & 0 & 1 & 0 \\ 0 & 0 & -2 & -1 & 0 & 1 \end{bmatrix} \tag{52}$$

Multiply row 2 by -1 and add to row 1:

$$B_2 = \begin{bmatrix} 1 & 0 & 0 & 1 & -1 & 0 \\ 0 & 1 & 2 & 0 & 1 & 0 \\ 0 & 0 & -2 & -1 & 0 & 1 \end{bmatrix} \tag{53}$$

Divide the entries in row 3 by -2:

$$B_3 = \begin{bmatrix} 1 & 0 & 0 & 1 & -1 & 0 \\ 0 & 1 & 2 & 0 & 1 & 0 \\ 0 & 0 & 1 & \frac{1}{2} & 0 & -\frac{1}{2} \end{bmatrix} \tag{54}$$

Multiply the last row by -2 and add to the second row:

$$B_4 = \begin{bmatrix} 1 & 0 & 0 & 1 & -1 & 0 \\ 0 & 1 & 0 & -1 & 1 & 1 \\ 0 & 0 & 1 & \frac{1}{2} & 0 & -\frac{1}{2} \end{bmatrix} = [I \quad A^{-1}] \tag{55}$$

Therefore,

$$A^{-1} = \begin{bmatrix} 1 & -1 & 0 \\ -1 & 1 & 1 \\ \frac{1}{2} & 0 & -\frac{1}{2} \end{bmatrix} \tag{56}$$

It is interesting to point out that B_i in (51)–(54) can be obtained from B_{i-1} by premultiplying by an elementary matrix which can be generated according to an elementary operation. To explicitly represent these matrices, let I be the identity matrix and let ε_{ij} be the episilon matrix which has all its entries equal to zero except the i,jth entry, which is equal to 1.

4.6.3. Elementary Matrices

1. Interchange rows i and j.
The corresponding elementary matrix is given by

$$E_{ij} = I - \varepsilon_{ii} - \varepsilon_{jj} + \varepsilon_{ij} + \varepsilon_{ji} \tag{57}$$

Example

Let $i = 1$ and $j = 2$, then the corresponding 3×3 matrix is given by

$$E_{12} = \begin{bmatrix} 1 & 0 & 0 \\ 0 & 1 & 0 \\ 0 & 0 & 1 \end{bmatrix} - \begin{bmatrix} 1 & 0 & 0 \\ 0 & 0 & 0 \\ 0 & 0 & 0 \end{bmatrix} - \begin{bmatrix} 0 & 0 & 0 \\ 0 & 1 & 0 \\ 0 & 0 & 0 \end{bmatrix} + \begin{bmatrix} 0 & 1 & 0 \\ 0 & 0 & 0 \\ 0 & 0 & 0 \end{bmatrix}$$

$$+ \begin{bmatrix} 0 & 0 & 0 \\ 1 & 0 & 0 \\ 0 & 0 & 0 \end{bmatrix} = \begin{bmatrix} 0 & 1 & 0 \\ 1 & 0 & 0 \\ 0 & 0 & 1 \end{bmatrix}$$

2. Multiply row i by $C \neq 0$.
   The corresponding elementary matrix is given by

$$E_i(C) = I + (C - 1)\varepsilon_{ii} \tag{58}$$

3. Add C times row j to row i.

$$E_{ij}(C) = I + C\varepsilon_{ij} \tag{59}$$

For the example considered in (51), we have

$$B_1 = E_{31}(-1)B_0$$
$$B_2 = E_{12}(-1)B_1$$
$$B_3 = E_3(-\tfrac{1}{2})B_2$$
$$B_4 = E_{23}(-2)B_3$$

where

$$E_{31}(-1) = \begin{bmatrix} 1 & 0 & 0 \\ 0 & 1 & 0 \\ -1 & 0 & 1 \end{bmatrix}, \quad E_{12}(-1) = \begin{bmatrix} 1 & -1 & 0 \\ 0 & 1 & 0 \\ 0 & 0 & 1 \end{bmatrix}$$

$$E_3(-\tfrac{1}{2}) = \begin{bmatrix} 1 & 0 & 0 \\ 0 & 1 & 0 \\ 0 & 0 & -\tfrac{1}{2} \end{bmatrix}, \quad E_{23}(-2) = \begin{bmatrix} 1 & 0 & 0 \\ 0 & 1 & -2 \\ 0 & 0 & 1 \end{bmatrix}$$

and

$$B_4 = E_{23}(-2)E_3(-\tfrac{1}{2})E_{12}(-1)E_{31}(-1)B_0 = A^{-1}[A \quad I]$$

From this it follows that

$$A^{-1} = E_{23}(-2)E_3(-\tfrac{1}{2})E_{12}(-1)E_{31}(-1)$$

which illustrates the idea that every nonsingular matrix is expressible as a product of elementary matrices.

Next let us consider an example over GF(5).

Example

Find the inverse of the matrix A below, where the entries in A are assumed to be over GF(5):

$$A = \begin{bmatrix} 1 & 1 \\ 1 & 0 \end{bmatrix}$$

Using first the cofactors, we have

$$|A| = 1.0 - 1.1 = -1 \equiv 4 \,(\mathrm{mod}\,5)$$
$$\mathrm{Adj}\,(A) = \begin{bmatrix} 0 & -1 \\ -1 & 1 \end{bmatrix} \equiv \begin{bmatrix} 0 & 4 \\ 4 & 1 \end{bmatrix} \quad (\mathrm{mod}\,5)$$
$$(|A|)^{-1} = (4)^{-1} = 4 \,(\mathrm{mod}\,5)$$

Therefore,

$$A^{-1} = 4 \begin{bmatrix} 0 & 4 \\ 4 & 1 \end{bmatrix} = \begin{bmatrix} 0 & 1 \\ 1 & 4 \end{bmatrix}$$

Using (49), we have

$$B = \begin{bmatrix} 1 & 1 & 1 & 0 \\ 1 & 0 & 0 & 1 \end{bmatrix} = B_0$$

Since $(-1) = 4 \,(\mathrm{mod}\,5)$, multiply the first row by 4 and add to the second row:

$$B_1 = \begin{bmatrix} 1 & 1 & 1 & 0 \\ 0 & 4 & 4 & 1 \end{bmatrix}$$

Multiply row 2 by 1 and add to the first row. Then multiply row 2 by 4. Using mod (5) arithmetic, we have

$$B_2 = \begin{bmatrix} 1 & 0 & 0 & 1 \\ 0 & 1 & 1 & 4 \end{bmatrix} = [I \quad A^{-1}]$$

verifying the earlier result.

The methods presented so far are efficient for small and medium sized matrices. The following method is efficient for large matrices.

4.6.4. Inversion Using Partitioned Matrices

For large matrices, it is convenient to use methods wherein the inverse of a large matrix can be computed in terms of two smaller matrices. We use partitioned matrices discussed in the last section.

Let X be an $n \times n$ matrix and be partitioned as

$$X = \begin{array}{c} \\ r \\ (n-r) \end{array} \begin{array}{c} r \quad (n-r) \\ \begin{bmatrix} A & B \\ C & D \end{bmatrix} \end{array} \qquad (60)$$

where A is an $r \times r$ matrix, B is an $r \times (n-r)$ matrix, C is a $(n-r) \times r$ matrix, and D is an $(n-r) \times (n-r)$ matrix. Note the notation above and to the left of the matrix in (60).

Example

The matrix X below is partitioned as

$$X = \begin{bmatrix} 1 & 1 & 1 \\ 1 & 0 & 2 \\ \hline 1 & -1 & 3 \end{bmatrix} = \begin{bmatrix} A & B \\ C & D \end{bmatrix} \qquad (61)$$

where

$$A = \begin{bmatrix} 1 & 1 \\ 1 & 0 \end{bmatrix}, \qquad B = \begin{bmatrix} 1 \\ 2 \end{bmatrix}, \qquad C = [1 \quad -1], \qquad D = [3] \qquad (62)$$

4.6.5. A Useful Factorization of a Matrix

When the submatrix A in (60) is nonsingular, X can be expressed as

$$\begin{bmatrix} A & B \\ C & D \end{bmatrix} = \begin{bmatrix} I & 0 \\ CA^{-1} & I \end{bmatrix} \begin{bmatrix} A & 0 \\ 0 & D - CA^{-1}B \end{bmatrix} \begin{bmatrix} I & A^{-1}B \\ 0 & I \end{bmatrix} \qquad (63)$$

or

$$X = X_1 X_2 X_3 \qquad (64)$$

where

$$X_1 = \begin{bmatrix} I & 0 \\ CA^{-1} & I \end{bmatrix}, \qquad X_2 = \begin{bmatrix} A & 0 \\ 0 & D - CA^{-1}B \end{bmatrix}$$

$$X_3 = \begin{bmatrix} I & A^{-1}B \\ 0 & I \end{bmatrix}$$

Next let us verify (63). First,

$$X_1 X_2 = \begin{bmatrix} A & 0 \\ C & D - CA^{-1}B \end{bmatrix}$$

Second,

$$X_1 X_2 X_3 = \begin{bmatrix} A & AA^{-1}B \\ C & CA^{-1}B + D - CA^{-1}B \end{bmatrix} = X$$

thus verifying the result.

Interestingly, the inverses of X_1 and X_3 always exist and can be written by inspection. These are given by

$$X_1^{-1} = \begin{bmatrix} I & 0 \\ -CA^{-1} & I \end{bmatrix} \tag{65}$$

$$X_3^{-1} = \begin{bmatrix} I & -A^{-1}B \\ 0 & I \end{bmatrix} \tag{66}$$

Note that the inverses of X_1 and X_3 are obtained by simply changing the sign of the off-diagonal entries. If the matrix $(D - CA^{-1}B)$ is nonsingular, then

$$X_2^{-1} = \begin{bmatrix} A^{-1} & 0 \\ 0 & (D - CA^{-1}B)^{-1} \end{bmatrix} \tag{67}$$

It can be shown that if X and A are nonsingular, then $(D - CA^{-1}B)$ is nonsingular. It is clear that the inverse of X_2 is obtained by computing the inverses of two submatrices of smaller dimensions.

Next the inverse of X is given by

$$X^{-1} = X_3^{-1} X_2^{-1} X_1^{-1} \tag{68}$$

which can be verified by showing $X^{-1}X = I$.

Partitioned matrices can also be used in finding the determinants. First,

$$|X| = |X_1||X_2||X_3| \qquad (69)$$

Second,

$$|X_1| = 1 \quad \text{and} \quad |X_3| = 1 \qquad (70)$$

Third,

$$|X_2| = |A||D - CA^{-1}B| \qquad (71)$$

provided $|A| \neq 0$. In computing (69) and (71), we have used the properties of determinants of triangular matrices and determinants of the direct sum of matrices.

Problems

1. Find the inverse of the following matrix by using the cofactors and by using elementary operations:

$$X = \begin{bmatrix} 4 & 2 & 1 \\ 2 & 4 & 2 \\ 1 & 2 & 4 \end{bmatrix}$$

2. Find the inverse of X by first partitioning X as

$$X = \left[\begin{array}{cc|c} 4 & 2 & 1 \\ 2 & 4 & 2 \\ \hline 1 & 2 & 4 \end{array}\right]$$

and then using (68).
3. Find the inverse of the following matrix, assuming that the entries are over GF(2):

$$X = \begin{bmatrix} 1 & 1 \\ 1 & 0 \end{bmatrix}$$

References

Demetriou, P. and O. Wing (1964), Partitioning a Matrix by Inspection, *IEEE Transactions on Circuit Theory*, Vol. CT-11, p. 162.

Frame, J. S. (1964), Matrix Functions and Applications, *IEEE Spectrum*, Vol. 1, Part I, March, pp. 208–220; Part II, April, pp. 102–108; Part III, May, pp. 100–109; Part IV, June, pp. 123–131; Part V, July, pp. 103–109.

Perlis, S. (1982), *Theory of Matrices*, Addison-Wesley, Reading, MA.

Seshu, S. and M. B. Reed (1961), *Linear Graphs and Electrical Networks*, Addison-Wesley, Reading, MA.

Matrix Equations and Transformations

5.1. Introduction

In this chapter we consider solutions of a linear set of equations using some of the concepts discussed in Chapter 4. In addition, we introduce the concepts of a vector space, rank of a matrix, and so on. We end the chapter with a discussion of various transformations that are popular in the digital signal processing area.

5.2. Linear Vector Spaces

A set of M numbers $(x_{i1}, x_{i2}, \ldots, x_{im})$ over a field defines a vector, say $\mathbf{x}_i = \text{col}(x_{i1}, x_{i2}, \ldots, x_{im})$. For example, $(1, 0, 1)$ could be considered a vector over the real field or over the finite field $GF(2)$. It is clear that a column vector of dimension m is an $m \times 1$ matrix and a row vector of dimension m is a $1 \times m$ matrix. Therefore, the matrix operations discussed in Chapter 4 are applicable for vectors. The two operations that are useful here are the addition of two vectors $\mathbf{x}_i$ and $\mathbf{x}_j$,

$$\mathbf{x}_i^T + \mathbf{x}_j^T = [x_{i1} + x_{j1} \quad x_{i2} + x_{j2} \quad \cdots \quad x_{im} + x_{jm}] \tag{1}$$

and the multiplication of a vector $\mathbf{x}_i$ by a scalar c,

$$c\mathbf{x}_i^T = [cx_{i1} \quad cx_{i2} \quad \cdots \quad cx_{im}] \tag{2}$$

It is clear that the appropriate additions and multiplications must be done over the appropriate field. Now we are ready to define a vector space over a field F.

The set of all vectors, say x_i, is said to form a linear vector space, or simply vector space, V_m of dimension m if and only if for all the vectors x_i belonging to V_m, cx_i and $x_i + x_j$ also belong to V_m, where c is an arbitrary scalar from a field and the additions and multiplications must be done over the appropriate field.

Next we need to ask the important question: How many vectors are there in a vector space over a field F?

Example

Consider the binary field, GF(2). We can define four vectors

$$x_1 = \begin{bmatrix} 0 \\ 0 \end{bmatrix}, \qquad x_2 = \begin{bmatrix} 0 \\ 1 \end{bmatrix}, \qquad x_3 = \begin{bmatrix} 1 \\ 0 \end{bmatrix}, \qquad x_4 = \begin{bmatrix} 1 \\ 1 \end{bmatrix} \qquad (3)$$

that give all possible combinations. Are these independent? For example, $x_4 = x_2 + x_3$ is a linear combination of vectors, and $x_1 = x_2 + x_2$ is a null vector. It appears from a cursory study that two vectors x_2 and x_3 are important. Other possibilities are x_2, x_4 and x_3, x_4.

Next let us consider the problem in more general terms. All operations are assumed to be over the appropriate field. The vector

$$y = c_1 x_1 + c_2 x_2 + \cdots + c_k x_k \qquad (4)$$

is a linear combination of the vectors $x_1, x_2, \ldots, x_k$. The set of vectors $x_1, x_2, \ldots, x_k$ is said to be linearly independent if

$$c_1 x_1 + c_2 x_2 + \cdots + c_k x_k = 0 \qquad (5)$$

only if $c_i = 0$, $i = 1, 2, \ldots, k$. Otherwise, the set of vectors is said to be dependent. The set of vectors obtained by all possible linear combinations of $x_1, x_2, \ldots, x_k$ forms a linear subspace V.

It is implicit that the maximum number of linearly independent vectors in V_m is m. Any such set of m linearly independent vectors, say $y_1, y_2, \ldots, y_m$, forms a basis for V_m. From this it follows that every vector x in V_m can be expressed as

$$x = a_1 y_1 + a_1 y_2 + \cdots + a_m y_m \qquad (6)$$

where $a_1, a_2, \ldots, a_m$ are called the coordinates of $\mathbf{x}$ with respect to the basis $\mathbf{y}_1, \mathbf{y}_2, \ldots, \mathbf{y}_m$. The simplest example of a basis over the real field R_m is given by the vectors

$$
\begin{aligned}
\mathbf{e}_1^T &= [1 \quad 0 \quad 0 \quad \cdots \quad 0] \\
\mathbf{e}_1^T &= [0 \quad 1 \quad 0 \quad \cdots \quad 0] \\
&\;\;\vdots \\
\mathbf{e}_m^T &= [0 \quad 0 \quad 0 \quad \cdots \quad 1]
\end{aligned}
\tag{7}
$$

and the vector $\mathbf{x}^T = [x_1 \quad x_2 \quad \cdots \quad x_m]$ can be obtained from the basis vectors in (7) by

$$
\mathbf{x} = x_1 \mathbf{e}_1 + x_2 \mathbf{e}_2 + \cdots + x_m \mathbf{e}_m
$$

The above concepts can be discussed in terms of matrices. First, (6) can be written in the form

$$
\mathbf{x} = [\mathbf{y}_1 \quad \mathbf{y}_2 \quad \cdots \quad \mathbf{y}_m]
\begin{bmatrix} a_1 \\ a_2 \\ \vdots \\ a_m \end{bmatrix}
\tag{8}
$$

or simply

$$
\mathbf{x} = T\mathbf{a}
\tag{9}
$$

where $\mathbf{x}$ and $\mathbf{a}$ are m-dimensional vectors and T is an $m \times m$ matrix consisting of column vectors $\mathbf{y}_1, \mathbf{y}_2, \ldots, \mathbf{y}_m$.

Next we consider the independence of vectors using matrix concepts. Equation (5) can be written in a matrix form as

$$
[\mathbf{x}_1 \mathbf{x}_2 \quad \cdots \quad \mathbf{x}_k]
\begin{bmatrix} c_1 \\ c_2 \\ \vdots \\ c_k \end{bmatrix} = 0
\tag{10}
$$

or

$$
A\mathbf{c} = 0
$$

where

$$A = [\mathbf{x}_1 \quad \mathbf{x}_2 \quad \cdots \quad \mathbf{x}_k]$$

is an $m \times k$ matrix. The independence of the vectors $\mathbf{x}_k$, now the columns in A, can be determined using matrix operations. We now discuss this in terms of rank of matrices.

5.2.1. Rank of Matrices

The column rank r_c and row rank r_r of an $m \times n$ matrix A are, respectively, the maximum number of columns and rows that are linearly independent. We state here without proof that the column rank is equal to the row rank, that is, $r_c = r_r$, and is simply referred to as the rank of A, r. One of the easiest ways to determine the rank of a matrix is by using elementary operations. This is discussed below.

Let B be any $m \times m$ nonsingular matrix and C be any $n \times n$ nonsingular matrix, then we state without proof that

$$\text{Rank of } [A] = \text{Rank of } [BAC] \tag{11}$$

In Section 4.6 we discussed the fact that any nonsingular matrix is expressible as a product of elementary matrices. Also, an elementary operation on rows corresponds to premultiplying the matrix by an elementary matrix. Obviously, an elementary operation on columns corresponds to postmultiplying the matrix by an elementary matrix. The rank of a matrix can be computed by using elementary operations on rows and columns. This is illustrated by the following example.

Example

Find the rank of the following matrix over the real field:

$$A = \begin{bmatrix} 1 & 1 & 2 & 3 & -1 \\ 1 & 1 & 3 & -1 & 4 \\ 2 & 2 & 5 & 2 & 3 \end{bmatrix}$$

$$A = A_0 \sim A_1 = \begin{bmatrix} 1 & 1 & 2 & 3 & -1 \\ 0 & 0 & 1 & -4 & 5 \\ 2 & 2 & 5 & 2 & 3 \end{bmatrix} \sim A_2 = \begin{bmatrix} 1 & 1 & 2 & 3 & -1 \\ 0 & 0 & 1 & -4 & 5 \\ 0 & 0 & 1 & -4 & 5 \end{bmatrix}$$

$$\sim A_3 = \begin{bmatrix} 1 & 1 & 2 & 3 & -1 \\ 0 & 0 & 1 & -4 & 5 \\ 0 & 0 & 0 & 0 & 0 \end{bmatrix} \sim A_4 = \begin{bmatrix} 1 & 1 & 0 & 11 & -11 \\ 0 & 0 & 1 & -4 & 5 \\ 0 & 0 & 0 & 0 & 0 \end{bmatrix}$$

where the sign $\sim$ indicates that the ranks are the same. Note that we have used row operations to obtain A_i from A_{i-1}. For example, A_1 is obtained from A_0 by replacing row 2 in A_0 by a row obtained by subtracting row 2 from row 1 in A_0, and so on. From A_4, it is clear that there are only two rows that are independent. Therefore, the row rank $r_r = 2$. What about the column rank? We can start with A_4 and use elementary operations on columns:

$$A_4 \sim A_5 = \begin{bmatrix} 1 & 0 & 1 & 11 & -11 \\ 0 & 1 & 0 & -4 & 5 \\ 0 & 0 & 0 & 0 & 0 \end{bmatrix}$$

where A_5 is obtained from A_4 by replacing the second column by the third column and vice versa. Continuing the process, we eventually obtain

$$A_5 \sim \begin{bmatrix} 1 & 0 & 0 & 0 & 0 \\ 0 & 1 & 0 & 0 & 0 \\ 0 & 0 & 0 & 0 & 0 \end{bmatrix}$$

indicating clearly that the column rank $r_c = 2$ also. Earlier it was pointed out that the rank of a matrix $r = r_c = r_r$ and is equal to 2 in our example.

Problems

Find the rank of the following matrix over the real field and GF(5):

$$\begin{bmatrix} 1 & 2 & 1 \\ 2 & 1 & 3 \\ 0 & 1 & 3 \end{bmatrix}$$

5.3. Gram–Schmidt Process

In many real field applications we are interested in generating orthogonal vectors from a set of independent vectors. Before we discuss this procedure, let us briefly review the principle of orthogonality of vectors over the real field. Two n-dimensional vectors, say $\mathbf{y}_i$ and $\mathbf{y}_j$, are said to be orthogonal if

$$\mathbf{y}_i^T \mathbf{y}_j = 0, \qquad i \neq j \tag{12}$$

The length of the vector $\mathbf{y}_i = \text{col}\,[\,y_{i1}, \ldots, y_{in}]$ is defined by

$$\|\mathbf{y}_i\| = (\mathbf{y}_i^T \mathbf{y}_i)^{1/2} = \left(\sum_{j=1}^{n} y_{ij}^2 \right)^{1/2} \tag{13}$$

Now let us define the problem.

Given a set of independent vectors,

$$\mathbf{x}_1, \mathbf{x}_2, \ldots, \mathbf{x}_m$$

find an orthogonal set of vectors,

$$\mathbf{y}_1, \mathbf{y}_2, \ldots, \mathbf{y}_m$$

such that for each i, $\mathbf{y}_i$ is a linear combination of

$$\mathbf{x}_1, \mathbf{x}_2, \ldots, \mathbf{x}_m$$

We begin the process by defining $\mathbf{y}_1 = \mathbf{x}_1$. Next we define a vector $\mathbf{y}_2$ which is orthogonal to $\mathbf{y}_1$. That is,

$$\mathbf{y}_2 = \mathbf{x}_2 + c_{21}\mathbf{y}_1 \tag{14}$$

where c_{21} is obtained from

$$\mathbf{y}_1^T \mathbf{y}_2 = 0 = \mathbf{y}_1^T \mathbf{x}_2 + c_{21}(\mathbf{y}_1^T \mathbf{y}_1) \tag{15}$$

or

$$-c_{21} = \frac{\mathbf{y}_1^T \mathbf{x}_2}{\mathbf{y}_1^T \mathbf{y}_1} \tag{16}$$

Similarly, we define $\mathbf{y}_3$ which is orthogonal to $\mathbf{y}_1$ and $\mathbf{y}_2$ by

$$\mathbf{y}_3 = \mathbf{x}_3 + c_{31}\mathbf{y}_1 + c_{32}\mathbf{y}_2 \tag{17}$$

where c_{31} and c_{32} can be computed from

$$\mathbf{y}_1^T \mathbf{y}_3 = 0 = \mathbf{y}_1^T \mathbf{x}_3 + c_{31}\mathbf{y}_1^T \mathbf{y}_1 + c_{32}\mathbf{y}_1^T \mathbf{y}_2$$
$$\mathbf{y}_2^T \mathbf{y}_3 = 0 = \mathbf{y}_2^T \mathbf{x}_3 + c_{31}\mathbf{y}_2^T \mathbf{y}_1 + c_{32}\mathbf{y}_2^T \mathbf{y}_2$$

From these it follows that

$$c_{31} = -\frac{\mathbf{y}_1^T \mathbf{x}_3}{\mathbf{y}_1^T \mathbf{y}_1} \tag{18}$$

$$c_{32} = -\frac{\mathbf{y}_2^T \mathbf{x}_3}{\mathbf{y}_2^T \mathbf{y}_2} \tag{19}$$

and

$$\mathbf{y}_3 = \mathbf{x}_3 - \frac{\mathbf{y}_1^T \mathbf{x}_3}{\mathbf{y}_1^T \mathbf{y}_1} \mathbf{y}_1 - \frac{\mathbf{y}_2^T \mathbf{x}_3}{\mathbf{y}_2^T \mathbf{y}_2} \mathbf{y}_2 \tag{20}$$

This can be generalized to

$$\mathbf{y}_i = \mathbf{x}_i - \sum_{j=1}^{i-1} \frac{\mathbf{y}_j^T \mathbf{x}_i}{\mathbf{y}_j^T \mathbf{y}_j} \mathbf{y}_j \tag{21}$$

To get the normalized vectors, replace $\mathbf{y}_i$ by $\mathbf{y}_i / \|\mathbf{y}_i\|$. These normalized vectors are called orthonormal vectors. The above process is usually referred to as the Gram–Schmidt orthogonalization process.

Example

Given the independent vectors

$$\mathbf{x}_1 = \begin{bmatrix} 1 \\ 1 \end{bmatrix} \quad \text{and} \quad \mathbf{x}_2 = \begin{bmatrix} 1 \\ 2 \end{bmatrix}$$

find a set of orthonormal vectors that can be expressed as a linear combination of these vectors.

Starting with $\mathbf{y}_1 = \mathbf{x}_1$, we have

$$\mathbf{y}_1 = \begin{bmatrix} 1 \\ 1 \end{bmatrix}$$

Using (14) and (16), we have

$$\mathbf{y}_2 = \begin{bmatrix} 1 \\ 2 \end{bmatrix} - \frac{3}{2}\begin{bmatrix} 1 \\ 1 \end{bmatrix} = \begin{bmatrix} -\frac{1}{2} \\ \frac{1}{2} \end{bmatrix}$$

Orthonormal vectors are given by

$$\frac{1}{\sqrt{2}}\begin{bmatrix} 1 \\ 1 \end{bmatrix} \quad \text{and} \quad \frac{1}{\sqrt{2}}\begin{bmatrix} -1 \\ 1 \end{bmatrix}$$

We can also define Gram–Schmidt orthogonalization techniques over a complex field. We need to use transpose conjugates (*) rather than transposes (T) in our algorithm. Use of this algorithm for finite fields is of little use and is not discussed here.

Problems

Given the vectors

$$\mathbf{x}_1 = \begin{bmatrix} 1 \\ 2 \end{bmatrix} \quad \text{and} \quad \mathbf{x}_2 = \begin{bmatrix} 2 \\ 1 \end{bmatrix}$$

find a set of orthonormal vectors $\mathbf{y}_1$ and $\mathbf{y}_2$ using the Gram–Schmidt process.

5.4. Solutions of Equations

In this section, we consider solutions of linear equations. The concept of rank of matrices, discussed in Section 5.2, allows us to investigate the possibility of multiple solutions. As before, the problem can be considered over any field and we assume that the arithmetic used in our analysis is in the appropriate field.

Example

Consider the following equations over the real field:

$$
\begin{aligned}
x_1 + x_2 + 2x_3 &= y_1 = 1 \\
x_2 + 2x_3 &= y_2 = 0 \\
x_1 + x_2 \quad\quad &= y_3 = -1
\end{aligned}
\tag{22}
$$

These equations can be written in a matrix form,

$$
\begin{bmatrix} 1 & 1 & 2 \\ 0 & 1 & 2 \\ 1 & 1 & 0 \end{bmatrix}
\begin{bmatrix} x_1 \\ x_2 \\ x_3 \end{bmatrix}
=
\begin{bmatrix} y_1 \\ y_2 \\ y_3 \end{bmatrix}
=
\begin{bmatrix} 1 \\ 0 \\ -1 \end{bmatrix}
\tag{23}
$$

or in a compact form,

$$A\mathbf{x} = \mathbf{y} \tag{24}$$

where A is the coefficient matrix on the left in (23), $\mathbf{x}$ is a column vector, and $\mathbf{y}$ is some known column vector. If A is nonsingular, we can premultiply the equation in (24) by A^{-1} and get

$$A^{-1}A\mathbf{x} = A^{-1}\mathbf{y} \tag{25}$$

or

$$Ix = x = A^{-1}y \tag{26}$$

Even though the above method looks conceptually simple, it is computationally intensive. A simpler method is to use elementary operations on both the coefficient matrix A and the known vector y simultaneously. This is discussed next.

5.4.1. Solutions Using Elementary Operations

Let us write (24) in the form

$$[A \quad y]\begin{bmatrix} x \\ -1 \end{bmatrix} = 0 \tag{27}$$

or

$$\begin{bmatrix} 1 & 1 & 2 & \vdots & 1 \\ 0 & 1 & 2 & \vdots & 0 \\ 1 & 1 & 0 & \vdots & -1 \end{bmatrix}\begin{bmatrix} x_1 \\ x_2 \\ x_3 \\ -1 \end{bmatrix} = 0 \tag{28}$$

Using the row elementary operations discussed in Chapter 4, we have successively

$$\begin{bmatrix} 1 & 1 & 2 & 1 \\ 0 & 1 & 2 & 0 \\ 0 & 0 & -2 & -2 \end{bmatrix}\begin{bmatrix} x_1 \\ x_2 \\ x_3 \\ -1 \end{bmatrix} = 0 \quad \begin{array}{c} \textit{Respective Operations} \\ \text{Multiply row 1 by } -1 \text{ and add} \\ \text{it to the third row in (28)} \end{array} \tag{29}$$

$$\begin{bmatrix} 1 & 0 & 0 & 1 \\ 0 & 1 & 2 & 0 \\ 0 & 0 & -2 & -2 \end{bmatrix}\begin{bmatrix} x_1 \\ x_2 \\ x_3 \\ -1 \end{bmatrix} = 0 \quad \begin{array}{c} \text{Multiply row 2 by } -1 \text{ and add} \\ \text{it to row 1 in (29)} \end{array} \tag{30}$$

$$\begin{bmatrix} 1 & 0 & 0 & 1 \\ 0 & 1 & 2 & 0 \\ 0 & 0 & 1 & +1 \end{bmatrix}\begin{bmatrix} x_1 \\ x_2 \\ x_3 \\ -1 \end{bmatrix} = 0 \quad \begin{array}{c} \text{Divide the entries in row 3 by} \\ -2 \text{ in (30)} \end{array} \tag{31}$$

$$\begin{bmatrix} 1 & 0 & 0 & 1 \\ 0 & 1 & 0 & -2 \\ 0 & 0 & 1 & +1 \end{bmatrix}\begin{bmatrix} x_1 \\ x_2 \\ x_3 \\ -1 \end{bmatrix} = 0 \quad \begin{array}{c} \text{Multiply the third row by } -2 \\ \text{and add to the second row} \\ \text{in (31)} \end{array} \tag{32}$$

From this we have

$$x_1 = +1$$
$$x_2 = -2 \tag{33}$$
$$x_3 = +1$$

It is clear that the method presented above is applicable for finite field applications also. Equation (26) can be written in the form of (32) and is

$$[I \quad A^{-1}\mathbf{y}]\begin{bmatrix} \mathbf{x} \\ -1 \end{bmatrix} = 0 \tag{34}$$

where we assumed that the inverse of A exists and the solution in (26) is unique. What can we say if the matrix A is singular? Before we answer this question, let us consider (24) again wherein A may be singular and may be even a rectangular matrix.

Example

Consider

$$\begin{bmatrix} 1 & 1 & 2 \\ 0 & 1 & 2 \end{bmatrix}\begin{bmatrix} x_1 \\ x_2 \\ x_3 \end{bmatrix} = \begin{bmatrix} 1 \\ 1 \end{bmatrix} \tag{35}$$

or

$$\begin{bmatrix} 1 & 1 & 2 & 1 \\ 0 & 1 & 2 & 1 \end{bmatrix}\begin{bmatrix} x_1 \\ x_2 \\ x_3 \\ -1 \end{bmatrix} = 0 \tag{36}$$

By using a set of elementary row operations (in this case one operation), we can write (36) as

$$\begin{bmatrix} 1 & 0 & 0 & 0 \\ 0 & 1 & 2 & 1 \end{bmatrix}\begin{bmatrix} x_1 \\ x_2 \\ x_3 \\ -1 \end{bmatrix} = 0 \tag{37}$$

which can be expressed in the form

$$\begin{bmatrix} x_1 \\ x_2 \end{bmatrix} = -\begin{bmatrix} 0 \\ 2 \end{bmatrix} x_3 + \begin{bmatrix} 0 \\ 1 \end{bmatrix} \tag{38}$$

indicating that there are infinite number of solutions (x_3 is arbitrary).

Next let us consider an example where there may not be any solutions.

Example

Consider the equations over the real field

$$\begin{bmatrix} 1 & 1 \\ 2 & 2 \end{bmatrix}\begin{bmatrix} x_1 \\ x_2 \end{bmatrix} = \begin{bmatrix} 1 \\ 3 \end{bmatrix} \tag{39}$$

which can be written in the form

$$\begin{bmatrix} 1 & 1 & 1 \\ 2 & 2 & 3 \end{bmatrix}\begin{bmatrix} x_1 \\ x_2 \\ -1 \end{bmatrix} = 0$$

These can be reduced to

$$\begin{bmatrix} 1 & 1 & 1 \\ 0 & 0 & 1 \end{bmatrix}\begin{bmatrix} x_1 \\ x_2 \\ -1 \end{bmatrix} = 0$$

indicating that $-1 = 0$ and, therefore, that no solution to the equations in (39) exists. The existence of the solutions in (38) and the nonexistence of a solution for (39) can be stated in terms of rank of A and the rank of the augmented matrix $[A \, y]$. We state this in the form of a theorem (Hohn, 1960) without proof.

5.4.2. The Fundamental Theorem

A system of m linear equations over a field F

$$A\mathbf{x} = \mathbf{y}$$

in n unknowns is consistent if and only if the matrix A and the augmented matrix $[A \, \mathbf{y}]$ have the same rank.

In the example above [see (39)]

$$A = \begin{bmatrix} 1 & 1 \\ 2 & 2 \end{bmatrix}$$

has a rank of 1 and the augmented matrix

$$[A \quad \mathbf{y}] = \begin{bmatrix} 1 & 1 & 1 \\ 2 & 2 & 3 \end{bmatrix}$$

has a rank of 2. This violates the fundamental theorem and therefore there is no solution.

Finally, we want to discuss the solution of a set of homogeneous equations. This is an important topic, as it has several applications. These are discussed in later chapters.

5.4.3. Solutions of Homogeneous Equations

Consider the set of equations

$$A\mathbf{x} = 0 \tag{40}$$

where A is an $n \times n$ matrix and $\mathbf{x}$ is an n-dimensional vector. It is clear that when A is nonsingular, then A^{-1} exists and $\mathbf{x} = 0$ which is usually referred to as the trivial solution. For a nontrivial solution to exist, the matrix A must be singular. In other words,

$$|A| = 0 \tag{41}$$

in order for a solution to exist. If the rank of the matrix A is $r < n$, then there are r independent rows in A and the remaining $(n - r)$ rows can be expressed as a linear combination of the r independent rows. This means that we can solve for r of the variables in $\mathbf{x}$ in terms of the remaining $(n - r)$ variables. Let us illustrate this by an example.

Example

We are given

$$\begin{bmatrix} 1 & 2 & 1 & 3 \\ 0 & 1 & 2 & 1 \\ 1 & 3 & 3 & 4 \\ 1 & 4 & 5 & 5 \end{bmatrix} \begin{bmatrix} x_1 \\ x_2 \\ x_3 \\ x_4 \end{bmatrix} = 0 \tag{42}$$

By using elementary operations on the rows, we can reduce (42) to

$$\begin{bmatrix} 1 & 0 & -3 & 1 \\ 0 & 1 & 2 & 1 \\ 0 & 0 & 0 & 0 \\ 0 & 0 & 0 & 0 \end{bmatrix} \begin{bmatrix} x_1 \\ x_2 \\ x_3 \\ x_4 \end{bmatrix} = 0 \tag{43}$$

where the last two equations are trivial. Therefore, (43) simply corresponds to

$$
\begin{bmatrix} 1 & 0 & -3 & 1 \\ 0 & 1 & 2 & 1 \end{bmatrix}
\begin{bmatrix} x_1 \\ x_2 \\ x_3 \\ x_4 \end{bmatrix} = 0
\tag{44}
$$

or

$$
\begin{bmatrix} x_1 \\ x_2 \end{bmatrix} = -\begin{bmatrix} -3 & 1 \\ 2 & 1 \end{bmatrix}\begin{bmatrix} x_3 \\ x_4 \end{bmatrix}
$$

where x_1 and x_2 are solved in terms of x_3 and x_4. It is clear that any set of nonzero values for x_3 and/or x_4 will give nontrivial solutions.

Next let us pick a solution, say $x_3 = 0$ and $x_4 = 1$. Then the corresponding solution vector **x** is given by, say

$$
\mathbf{x}_1 = \begin{bmatrix} -1 \\ -1 \\ 0 \\ 1 \end{bmatrix}
\tag{45a}
$$

Another solution might be when $x_3 = 1$ and $x_4 = 1$. Then the corresponding solution vector is given by

$$
\mathbf{x}_2 = \begin{bmatrix} 2 \\ -3 \\ 1 \\ 1 \end{bmatrix}
\tag{45b}
$$

Can we pick different values for x_3 and x_4 and get other solutions? Of course we can. But we do not need them. The reason is that any other solution can be expressed as a linear combination of the two vectors computed above. At this point let us make some general comments.

We can state that in the solution of the homogeneous equations in (40), the number of independent x_i that can be selected arbitrarily is given by $(n - r)$ where r is the rank of the $n \times n$ coefficient matrix A. The number $(n - r)$ is known as the nullity or the degeneracy of A when A is a square matrix. Obviously, the methods discussed in this section are applicable for equations in the finite field domain also. Let us illustrate this by a simple example.

Example

Consider the system of equations over GF(2) and solve for the unknowns:

$$\begin{bmatrix} 1 & 1 & 0 \\ 1 & 0 & 1 \\ 0 & 1 & 1 \end{bmatrix} \begin{bmatrix} x_1 \\ x_2 \\ x_3 \end{bmatrix} = \begin{bmatrix} 0 \\ 0 \\ 0 \end{bmatrix}$$

Using the elementary operations, we can compute the only nontrivial solution, which is

$$x_1 = 1, \qquad x_2 = 1, \qquad x_3 = 1$$

Solutions of overdetermined systems are discussed in Section 5.5.

Problems

1. Find the solutions of the following systems of equations over the real field:

(a) $\begin{bmatrix} 1 & 1 \\ 0 & 1 \end{bmatrix} \begin{bmatrix} x_1 \\ x_2 \end{bmatrix} = \begin{bmatrix} 1 \\ 2 \end{bmatrix}$

(b) $\begin{bmatrix} 4 & 2 & 1 \\ 2 & 4 & 2 \\ 1 & 2 & 4 \end{bmatrix} \begin{bmatrix} x_1 \\ x_2 \\ x_3 \end{bmatrix} = \begin{bmatrix} 1 \\ 1 \\ 1 \end{bmatrix}$

2. Use the fundamental theorem to show that there is no solution for the following system of equations:

$$\begin{bmatrix} 1 & 1 & 1 \\ 2 & 1 & 3 \\ 3 & 2 & 4 \end{bmatrix} \begin{bmatrix} x_1 \\ x_2 \\ x_3 \end{bmatrix} = \begin{bmatrix} 1 \\ 2 \\ 0 \end{bmatrix}$$

3. In the example in (42), assume a solution $x_3 = 1$ and $x_4 = -1$ and solve for $\mathbf{x}$, say $\mathbf{x}_3$. Show that the vector $\mathbf{x}_3$ is a linear combination of $\mathbf{x}_1$ and $\mathbf{x}_2$ given in (45a) and (45b) respectively.

5.5. Solutions of Overdetermined Systems

In this section we consider a system of equations where the number of equations is larger than the number of unknowns (Rao and Mitra, 1971).

Such equations have important applications in many areas, such as speech processing, seismic signal processing, and numerical analysis.

Consider the system of equations

$$\sum_{j=1}^{N} a_{ij}y_j = d_i, \qquad i = 1, 2, \ldots, Q \tag{46a}$$

which can be expressed in compact form as

$$Ay = d \tag{46b}$$

where $y = \text{col}\,(y_1, y_2, \ldots, y_N)$ is an N-dimensional vector, $d = \text{col}\,(d_1, d_2, \ldots, d_Q)$ is a Q-dimensional vector, and A is a $Q \times N$ matrix. We assume that $Q > N$ and the rank of A is N. The system in (46) is usually referred to as an overdetermined system because there are more equations than unknowns. Obviously, the methods we have discussed so far can not be used to solve these equations. Let us look at the problem in a slightly different way. The following discussion is applicable over the real field only.

For a vector y with a set of given values, we always get a unique vector, $z = \text{col}\,(z_1, z_2, \ldots, z_Q)$ or

$$z = Ay \tag{47}$$

Now find a vector y such that the vector z is as close as possible to the desired vector d. Closeness of vectors generally have different connotations. Let

$$e = d - z = \text{col}\,((d_1 - z_1), (d_2 - z_2), \ldots, (d_Q - z_Q)) \tag{48}$$

be the error vector and the square of the length of the error vector be represented by

$$E = \sum_{i=1}^{Q} (d_i - z_i)^2 = (d - z)^T(d - z) \tag{49}$$

Minimization of this error with respect to the unknown values y_i, $i = 1, 2, \ldots, N$, gives us a way to solve the problem. Clearly, the desired vector d is known. Substituting (47) in (49), we have

$$E = (d^T - y^T A^T)(d - Ay)$$
$$= d^T d - d^T Ay - y^T A^T d + y^T(A^T A)y \tag{50}$$

Since $\mathbf{d}^T A \mathbf{y}$ and its transpose $\mathbf{y}^T A \mathbf{d}$ are scalars, it follows that they are equal. Therefore,

$$E = \mathbf{d}^T \mathbf{d} - 2\mathbf{d}^T A \mathbf{y} + \mathbf{y}^T A^T A \mathbf{y} \tag{51}$$

In the minimization process, we need to take partials of E with respect to $y_j, j = 1, 2, \ldots, N$. Let

$$\text{Grad}_y E = \begin{bmatrix} \partial E / \partial y_1 \\ \vdots \\ \partial E / \partial y_N \end{bmatrix} \tag{52}$$

Therefore,

$$\text{Grad}_y E = \text{Grad}_y (\mathbf{d}^T \mathbf{d}) - 2 \text{Grad}_y (\mathbf{d}^T A \mathbf{y}) + \text{Grad}_y (\mathbf{y}^T (A^T A) \mathbf{y}) \tag{53}$$

where the first term is zero. Let us illustrate the gradients of the last two terms by a simple example.

Example

Let

$$\mathbf{d}^T A = [b_1 \quad b_2]$$

$$\mathbf{y} = \begin{bmatrix} y_1 \\ y_2 \end{bmatrix}$$

and

$$A^T A = \begin{bmatrix} c_{11} & c_{12} \\ c_{12} & c_{22} \end{bmatrix}$$

Find the partials in (53).
    First,

$$\mathbf{d}^T A \mathbf{y} = b_1 y_1 + b_2 y_2$$

and

$$\mathbf{y}^T (A^T A) \mathbf{y} = c_{11} y_1^2 + 2c_{12} y_1 y_2 + c_{22} y_2^2$$

Second,

$$\begin{bmatrix} \partial(\mathbf{d}^T A\mathbf{y})/\partial y_1 \\ \partial(\mathbf{d}^T A\mathbf{y})/\partial y_2 \end{bmatrix} = \begin{bmatrix} b_1 \\ b_2 \end{bmatrix} = A^T\mathbf{d}$$

That is,

$$\text{Grad}_\mathbf{y} (\mathbf{d}^T A\mathbf{y}) = A^T\mathbf{d} \qquad (54)$$

Third,

$$\begin{bmatrix} \partial(\mathbf{y}^T(A^TA)\mathbf{y})/\partial y_1 \\ \partial(\mathbf{y}^T(A^TA)\mathbf{y})/\partial y_2 \end{bmatrix} = \begin{bmatrix} 2c_{11}y_1 + 2c_{12}y_2 \\ 2c_{12}y_1 + 2c_{22}y_2 \end{bmatrix}$$

$$= 2\begin{bmatrix} c_{11} & c_{12} \\ c_{12} & c_{22} \end{bmatrix}\begin{bmatrix} y_1 \\ y_2 \end{bmatrix} \qquad (55)$$

Therefore, from this we see that

$$\text{Grad}_\mathbf{y} (y^T(A^TA)\mathbf{y}) = 2(A^TA)\mathbf{y} \qquad (56)$$

Using (54) and (55) in (53), we have

$$\text{Grad}_\mathbf{y} E = -2A^T\mathbf{d} + 2(A^TA)\mathbf{y}$$

Equating the gradient to zero and rewriting, we have

$$(A^TA)\mathbf{y} = (A^T\mathbf{d}) \qquad (57)$$

These equations are sometimes referred to as normal equations. Since A is of full rank, it follows that (A^TA) is a nonsingular matrix. In a later section we will see that (A^TA) is a positive definite matrix. Solving for $\mathbf{y}$, we have over the real field

$$\mathbf{y} = (A^TA)^{-1}A^T\mathbf{d} \qquad (58)$$

This solution is called the least squares solution. The matrix

$$A_{pI} = (A^TA)^{-1}A^T \qquad (59)$$

is of dimension $N \times Q$ and is sometimes called the pseudoinverse of the rectangular matrix A of dimension $Q \times N$. Note that for A_{pI} to exist (A^TA) must be nonsingular. Unfortunately, there is no equivalent to (58) that is of use over the finite field.

Let us illustrate the least squares solution by a simple example.

em of equations over the real field

$$\begin{bmatrix} 2 & 1 \\ 0 & 1 \\ 1 & 2 \end{bmatrix} \begin{bmatrix} y_1 \\ y_2 \end{bmatrix} = \begin{bmatrix} 1 \\ 1 \\ 1 \end{bmatrix} \tag{60}$$

$$A \qquad y \quad = \quad d$$

find the least squares solution.
   First,

$$A^T A = \begin{bmatrix} 2 & 0 & 1 \\ 1 & 1 & 2 \end{bmatrix} \begin{bmatrix} 2 & 1 \\ 0 & 1 \\ 1 & 2 \end{bmatrix} = \begin{bmatrix} 5 & 4 \\ 4 & 6 \end{bmatrix}$$

Second,

$$(A^T A)^{-1} = \tfrac{1}{14} \begin{bmatrix} 6 & -4 \\ -4 & 5 \end{bmatrix}$$

Third,

$$A^T d = \begin{bmatrix} 3 \\ 4 \end{bmatrix}$$

Fourth, the least squares solutions is given by

$$y = (A^T A)^{-1} A^T d = \tfrac{1}{14} \begin{bmatrix} 2 \\ 8 \end{bmatrix}$$

   The least squares approach can be applied to underdetermined systems of equations also. Consider the following equations:

$$\begin{bmatrix} 1 & 0 & 1 \\ 0 & 1 & 1 \end{bmatrix} \begin{bmatrix} y_1 \\ y_2 \\ y_3 \end{bmatrix} = \begin{bmatrix} 1 \\ 1 \end{bmatrix} \tag{61}$$

$$A \qquad y \quad = \quad d$$

From (61) we see that there are two equations in three unknowns, indicating that there are infinite number of solutions. Now let us put an additional restriction that the energy in the solution vector is to be minimum. That is,

$$E = \mathbf{y}^T \mathbf{y} = \sum_{i=1}^{3} y_i^2 \qquad (62)$$

is minimum. It is clear from (61) that

$$\begin{bmatrix} y_1 \\ y_2 \\ y_3 \end{bmatrix} = \begin{bmatrix} -1 \\ -1 \\ 1 \end{bmatrix} y_3 + \begin{bmatrix} 1 \\ 1 \\ 0 \end{bmatrix} \qquad (63)$$

and the minimization in (62) involves the solution of the overdetermined system of equations:

$$\begin{bmatrix} 1 \\ 1 \\ -1 \end{bmatrix} y_3 = \begin{bmatrix} 1 \\ 1 \\ 0 \end{bmatrix}$$

Using (58) we have

$$y_3 = \tfrac{2}{3}$$

Substituting this in (63), we have

$$y_1 = \tfrac{1}{3}, \qquad y_2 = \tfrac{1}{3}, \qquad y_3 = \tfrac{2}{3}$$

The same solution can be obtained by using the equation

$$\mathbf{y} = A^T (AA^T)^{-1} \mathbf{d} \qquad (64)$$

where A is given in (61). Note the similarity between (58) and (64).

So far we have been concerned with minimization of the squared error [see (49)]. Next we consider a more general problem—the minimization of the error:

$$E_p = \sum_{i=1}^{Q} |d_i - z_i|^p \qquad (65)$$

Of course when $p = 2$, (65) reduces to (49). The solutions obtained by minimizing (65) are sometimes referred to as L_p-based solutions. Unfortunately, there is no analytic solution for all values of p. For example, the function $f(x) = |x|$ is not differentiable at $x = 0$. Therefore, we can not use partials of E in (65) to compute an L_1 solution. However, there are iterative techniques for solving general L_p solutions. Convergence is guaranteed for $1 \leq p \leq 2$. We will assume this constraint in the following.

5.5.1. Iteratively Reweighted Least Squares (IRLS) Algorithm

The IRLS algorithm (Byrd and Payne, 1979) solves a set of equations of the form

$$A\mathbf{y} = \mathbf{d} \tag{66}$$

where A is a $Q \times N$ matrix, $\mathbf{y}$ is an N-dimensional vector, and $\mathbf{d}$ is a known Q-dimensional vector, for a particular p, by iteratively computing

$$\mathbf{y}(k + 1) = (A^T W(k) A)^{-1} A^T W(k) \mathbf{d} \tag{67}$$

where $W(k)$ is a diagonal matrix with its diagonal entries $W_{ii}(k)$ are given by

$$W_{ii}(k) = \begin{cases} |r_i(k)|^{p-2}, & |r_i(k)| > \varepsilon \\ \varepsilon^{p-2}, & |r_i(k)| \leq \varepsilon \end{cases} \tag{68}$$

where $r_i(k)$ corresponds to the ith entry in the residual vector

$$\mathbf{r}(k) = \mathbf{d} - A\mathbf{y}(k) \tag{69}$$

and ε is some small positive number. It is clear from (67) and (68) that the solution in (67) reduces to the least squares solution when $p = 2$. Let us illustrate the proposed method for $p = 1$ for the example in (60).

Starting with the initial vector $\mathbf{y}(0)$ as the least squares solution, the residual vector $\mathbf{r}(0)$, the weight matrix $W(0)$, and $\mathbf{y}(1)$ respectively, are given below for the system of equations in (60):

$$\mathbf{r}(0) = \begin{bmatrix} \frac{1}{7} \\ \frac{3}{7} \\ -\frac{2}{7} \end{bmatrix}, \qquad W(0) = \begin{bmatrix} 7 & 0 & 0 \\ 0 & \frac{7}{3} & 0 \\ 0 & 0 & \frac{7}{2} \end{bmatrix}$$

$$\mathbf{y}(1) = \begin{bmatrix} 0.222 \\ 0.5 \end{bmatrix}$$

Correspondingly,

$$\mathbf{r}(1) = \begin{bmatrix} 0.0777 \\ 0.5 \\ -0.222 \end{bmatrix}$$

Computing the error

$$E_1(k) = \sum_{t=1}^{Q} |r_i(k)|$$

we have

$$E_1(0) = \tfrac{1}{7}(1 + 3 + 2) = \tfrac{6}{7} = 0.857$$

$$E_1(1) = 0.077 + 0.5 + 0.222 = 0.799$$

Repeating the iteration scheme we eventually obtain an L_1 solution

$$\mathbf{y}^T = [0.333 \quad 0.333] \tag{70}$$

with

$$E_1 = 0.6667 \tag{71}$$

It is interesting to point out that the solution in (70) satisfies the first and the third equation in (60). The usual situation is that an L_1 solution of Q equations in $N(Q \geqslant N)$ unknowns satisfies N equations exactly (Claerbout and Muir, 1973). The L_1 solution is not always unique and the iterative algorithms, such as IRLS, do not always give a solution that satisfies N equations exactly. The L_1 solutions that satisfy N equations are useful, since if there are outliers, they are discarded in finding these L_1 solutions. These solutions can be found by using linear programming methods. The L_2 solution gives the same weight for all points. An interesting remark by Claerbout and Muir (1973) illustrates this point: "When a traveler reaches a fork in the road, the L_1 norm tells him to take either one way or the other, but the L_2 norm instructs him to head off into the bushes." Generally, the L_1 solutions are considered to be robust when compared to L_2 solutions (Yarlagadda et al., 1985).

The IRLS algorithm is computationally intensive, as we need to find the inverse of the matrix $(A^T W(k)A)$ of order N at each stage. When N is large, this becomes a limiting factor. A simpler algorithm is the residual steepest descent (RSD) algorithm (Byrd and Payne, 1979; Yarlagadda et al., 1985). This is discussed briefly below.

5.5.2. Residual Steepest Descent (RSD) Algorithm

The RSD algorithm solves (66) by iteratively computing

$$\mathbf{y}(k+1) = \mathbf{y}(k) - \Delta_k (A^T A)^{-1} A^T \mathbf{x}(k) \tag{72}$$

where

$$\mathbf{x}^T(k) = [x_1(k) \quad x_2(k) \quad \cdots \quad x_Q(k)] \tag{73}$$

with

$$x_i(k) = |(A\mathbf{y}(k) - \mathbf{d})_i|^{p-1} \operatorname{sgn}(A\mathbf{y}(k) - \mathbf{d})_i \tag{74}$$

where $\operatorname{sgn}(t) = +1(-1)$ if $t > 0 \, (t < 0)$. When $t = 0$, one can arbitrarily choose $\operatorname{sgn}(t)$ to be either 1 or -1. The scale factor Δ_k in (72) is to be determined by minimizing

$$E(k) = \| -\mathbf{d} + A\mathbf{y}(k) - \Delta_k A (A^T A)^{-1} A^T \mathbf{x}(k) \|_p \tag{75}$$

with respect to Δ_k in the L_p sense. Note that Δ_k is a scalar and $(\mathbf{d} - A\mathbf{y}(k))$ and $[A(A^T A)^{-1} A^T \mathbf{x}(k)]$ are vectors. Reasonably fast convergence can be obtained by using the IRLS algorithm to compute Δ_k in the minimization of $E(k)$ in (75).

Earlier it was pointed out that the RSD algorithm is the computationally less intensive. The main problem in the RSD algorithm is the computation of $(A^T A)^{-1}$. Note that this is also the same problem in L_2 solutions. We use these solutions in the deconvolution problem in a later chapter, where $(A^T A)$ takes a special form, that is, a Toeplitz form. Such special cases are discussed next.

5.5.3. Solutions of Equations that Involve Toeplitz Matrices

A symmetric Toeplitz matrix A_T of order N is given by

$$A_T = \begin{bmatrix} b_0 & b_1 & b_2 & \cdots & b_{N-1} \\ b_1 & b_0 & b_1 & \cdots & b_{N-2} \\ b_2 & b_1 & b_0 & \cdots & b_{N-3} \\ \vdots & \vdots & \vdots & & \vdots \\ b_{N-1} & b_{N-2} & b_{N-3} & \cdots & b_0 \end{bmatrix} \tag{76}$$

which has the general form that a typical entry in A_T, a_{ij} is given by

$$a_{ij} = b_{|i-j|} \tag{77}$$

where $b_k, k = 0, 1, \ldots, N - 1$ are some real numbers. It is clear that b_k defines the Toeplitz matrix in (76). These matrices have important applications in many areas, including speech analysis, time series, and statistical communication theory. In a later chapter we use these matrices in the deconvolution problem.

Next let us consider the persymmetry property of these matrices. A real matrix A is said to be persymmetric if

$$EA^TE = A \tag{78}$$

where E, the exchange matrix with 1's along the cross diagonal and 0's elsewhere, is given by

$$E = \begin{bmatrix} & & & & 1 \\ & & & \cdot{}^{\cdot{}^{\cdot}} & \\ & & 1 & & \\ 1 & & & & \end{bmatrix} \tag{79}$$

A symmetric Toeplitz matrix is persymmetric as $EA_TE = A_T$. Another interesting property of this matrix is that successive sizes of Toeplitz matrices can be expressed in a nice structure. For example, if $A_T(k)$ is a Toeplitz matrix of size k,

$$A_T(k) = \begin{bmatrix} b_0 & b_1 & \cdots & b_{k-1} \\ b_1 & b_0 & \cdots & b_{k-2} \\ \vdots & \vdots & & \vdots \\ b_{k-1} & b_{k-2} & \cdots & b_0 \end{bmatrix} \tag{80}$$

and

$$\mathbf{t}(k) = -\begin{bmatrix} b_1 \\ b_2 \\ \vdots \\ b_k \end{bmatrix} \tag{81}$$

then

$$A_T(k+1) = \begin{bmatrix} A_T(k) & -E\mathbf{t}(k) \\ -\mathbf{t}^T(k)E & b_0 \end{bmatrix} \tag{82}$$

Note that going from $A_T(k)$ to $A_T(k + 1)$, we need to know only one value and that is b_k. Let us illustrate some of these aspects.

Example

   Given

$$A_T(3) = \begin{bmatrix} 3 & 2 & 1 \\ 2 & 3 & 2 \\ 1 & 2 & 3 \end{bmatrix} \tag{83}$$

then

$$A_T(2) = \begin{bmatrix} 3 & 2 \\ 2 & 3 \end{bmatrix}, \qquad -E\mathbf{t}(2) = \begin{bmatrix} 1 \\ 2 \end{bmatrix}$$

and

$$\mathbf{t}(2) = \begin{bmatrix} -2 \\ -1 \end{bmatrix}$$

With this introduction, we are ready to talk about the solutions of equations involving Teoplitz matrices using two steps. In the first stage, we consider the solutions of equations

$$A_T(N)\mathbf{a}(N) = \mathbf{t}(N) \tag{84}$$

   The solutions of this set can be computed from a recursive manner using the following method (Durbin, 1960).

5.5.4. Durbin's Algorithm

   We assume that for some k the solution of

$$A_T(k)\mathbf{a}(k) = \mathbf{t}(k) \tag{85}$$

is known. The solution of

$$A_T(k + 1)\mathbf{a}(k + 1) = \mathbf{t}(k + 1) \tag{86}$$

is given by

$$\mathbf{a}(k + 1) = \begin{bmatrix} \mathbf{a}_1(k + 1) \\ a_2(k + 1) \end{bmatrix} = \begin{bmatrix} \mathbf{a}(k) \\ 0 \end{bmatrix} + c_{k+1} \begin{bmatrix} E\mathbf{a}(k) \\ 1 \end{bmatrix} \tag{87}$$

where

$$c_{k+1} = \frac{b_{k+1} - \mathbf{a}^T(k)E\mathbf{t}(k)}{-b_0 + \mathbf{a}^T(k)\mathbf{t}(k)} \tag{88}$$

Let us illustrate this method by an example.

Example

Find the solution of the following equations using the above method:

$$\begin{bmatrix} 3 & 2 & 1 \\ 2 & 3 & 2 \\ 1 & 2 & 3 \end{bmatrix} \begin{bmatrix} a_1 \\ a_2 \\ a_3 \end{bmatrix} = \begin{bmatrix} -2 \\ -1 \\ 0 \end{bmatrix} \tag{89}$$

We solve this problem in three stages. First,

$$A_T(1)a(1) = t(1) = -2 \tag{90a}$$

or

$$3a_1(1) = -2$$

giving

$$a_1(1) = -\tfrac{2}{3} \tag{90b}$$

Second, solve the equation

$$\begin{bmatrix} 3 & 2 \\ 2 & 3 \end{bmatrix} \begin{bmatrix} a_1(2) \\ a_2(2) \end{bmatrix} = \begin{bmatrix} -2 \\ -1 \end{bmatrix} = \mathbf{t}(2) \tag{91}$$

Using (88), we have

$$c_2 = \frac{1 - (-\tfrac{2}{3})(-2)}{-3 + (\tfrac{4}{3})} = \tfrac{1}{5}$$

Therefore,

$$a_2(2) = \tfrac{1}{5}$$

Now from the first equation in (87), we have

$$a_1(2) = a_1(1) + c_2 a_1(1) = -\tfrac{4}{5}$$

Third, solve

$$\begin{bmatrix} 3 & 2 & 1 \\ 2 & 3 & 2 \\ 1 & 2 & 3 \end{bmatrix} \begin{bmatrix} a_1(3) \\ a_2(3) \\ a_3(3) \end{bmatrix} = \begin{bmatrix} -2 \\ -1 \\ 0 \end{bmatrix}$$

using the above result. To compute c_3, we need

$$\mathbf{a}^T(2)\mathbf{t}(2) = \tfrac{7}{5}$$
$$\mathbf{a}^T(2)E\mathbf{t}(2) = \tfrac{2}{5}$$

From these

$$c_3 = \frac{0 - \tfrac{2}{5}}{-3 + \tfrac{7}{5}} = \tfrac{1}{4}$$

Using (87), we have

$$\mathbf{a}_1(3) = \mathbf{a}(2) + c_3 E\mathbf{a}(2) = \begin{bmatrix} -\tfrac{3}{4} \\ 0 \end{bmatrix}$$

$$a_2(3) = c_3 = \tfrac{1}{4}$$

The solution of (89) is given by

$$\begin{bmatrix} a_1 \\ a_2 \\ a_3 \end{bmatrix} = \begin{bmatrix} \mathbf{a}_1(3) \\ a_2(3) \end{bmatrix} = \begin{bmatrix} -\tfrac{3}{4} \\ 0 \\ \tfrac{1}{4} \end{bmatrix}$$

This approach of solving the equations in (84) is usually referred to as Durbin's Algorithm. Next we consider a more general system of equations involving Toeplitz matrices. Consider

$$A_T \hat{\mathbf{a}} = \mathbf{y} \tag{92}$$

where $\mathbf{y} = \text{col}(y_1, y_2, \ldots, y_N)$, an arbitrary vector, and A_T is a Toeplitz matrix of order N. We discuss a method that is usually referred to as Levinson's Algorithm (Robinson, 1967).

5.5.5. Levinson's Algorithm

This algorithm uses Durbin's Algorithm discussed above and can be stated in the following manner. We assume that we know the solution for

$$A_T(k)\hat{\mathbf{a}}(k) = \mathbf{y}(k) \tag{93}$$

where

$$\mathbf{y}(k) = \begin{bmatrix} y_1 \\ y_2 \\ \vdots \\ y_k \end{bmatrix} \tag{94}$$

We will solve for the solution of

$$A_T(k+1)\hat{\mathbf{a}}(k+1) = \mathbf{y}(k+1) \tag{95}$$

where

$$\hat{\mathbf{a}}(k+1) = \begin{bmatrix} \hat{\mathbf{a}}_1(k+1) \\ \hat{a}_2(k+1) \end{bmatrix}$$

using the computed solutions $\hat{\mathbf{a}}_1(k)$ and $\mathbf{a}(k)$, where $\mathbf{a}(k)$ is computed in Durbin's Algorithm. The solution is given by

$$\begin{aligned} \hat{\mathbf{a}}_1(k+1) &= \hat{\mathbf{a}}(k) + w_{k+1}E\mathbf{a}(k) \\ \hat{a}_2(k+1) &= w_{k+1} \end{aligned} \tag{96}$$

with

$$w_{k+1} = \frac{y_{k+1} + \mathbf{t}^T(k)E\hat{\mathbf{a}}(k)}{b_0 - \mathbf{a}^T(k)\mathbf{t}(k)} \tag{97}$$

Let us illustrate this by a simple example.

Example

Solve the following equations using Levinson's Algorithm:

$$\begin{bmatrix} 3 & 2 \\ 2 & 3 \end{bmatrix}\begin{bmatrix} \hat{a}_1 \\ \hat{a}_2 \end{bmatrix} = \begin{bmatrix} 1 \\ 0 \end{bmatrix} \tag{98}$$

First, using (93) with $k = 1$, we have

$$3\hat{a}_1(1) = y(1) = 1$$

and

$$\hat{a}_1(1) = \tfrac{1}{3}$$

Second, using (97), we have

$$w_2 = \frac{0 + (-2)(1)\frac{1}{3}}{3 - (-\frac{2}{3})(-2)} = -\frac{2}{5} \tag{99}$$

where we have used the results for $\mathbf{a}(1)$ and $\mathbf{t}(1)$ from (90). From the first equation in (96), we have

$$\hat{a}_1(2) = \frac{1}{3} + (-\frac{2}{5})(-\frac{2}{3}) = \frac{3}{5} \tag{100}$$

Therefore, from (96), (99), and (100), we have the solution

$$\begin{bmatrix} \hat{a}_1 \\ \hat{a}_2 \end{bmatrix} = \begin{bmatrix} \hat{a}_1(2) \\ \hat{a}_2(2) \end{bmatrix} = \begin{bmatrix} \frac{3}{5} \\ -\frac{2}{5} \end{bmatrix}$$

which satisfies the given equations in (98).

It is clear that the examples presented above are to illustrate the algorithms. The number of computations, that is, multiplications and additions, required to solve equations involving symmetric Toeplitz matrices is of the order of N^2, $O(N^2)$, compared to $O(N^3)$ by conventional methods. Obviously, for large Toeplitz matrices, the approaches presented here can provide a significant saving in computational time. This is especially important in many applications where on-line processing is desired.

Problem

Solve the following overdetermined system of equations using least squares and Levinson's Algorithm:

$$\begin{bmatrix} 1 & 0 & 0 \\ 2 & 1 & 0 \\ 1 & 2 & 1 \\ 0 & 1 & 2 \\ 0 & 0 & 1 \end{bmatrix} \begin{bmatrix} a_1 \\ a_2 \\ a_3 \end{bmatrix} = \begin{bmatrix} 1 \\ 1 \\ 1 \\ 1 \\ 1 \end{bmatrix}$$

In a later section we discuss some linear transformations that are of interest to communication engineers. These transformations involve some special classes of matrices. One such class is normal matrices and this is discussed next.

5.6. Normal Matrices

A complex square matrix A is called normal if

$$AA^* = A^*A \qquad (101)$$

where A^* is the conjugate transpose of A. Some examples of these matrices are Hermitian ($B = B^*$), skew Hermitian ($B = -B^*$), real symmetric ($B = B^T$), real skew ($B = -B^T$), unitary ($BB^* = B^*B = I$), orthogonal ($B^TB = BB^T = I$), diagonal, and all matrices unitarily similar to normal matrices. We have not defined the last class of matrices that involve similarity. This will be discussed in a later chapter.

In Chapter 4 we gave examples for Hermitian, skew Hermitian, symmetric, and skew symmetric matrices. Now let us give some examples for the other cases.

For real field applications, an example of an orthogonal matrix is

$$B = \frac{1}{\sqrt{2}}\begin{bmatrix} 1 & 1 \\ 1 & -1 \end{bmatrix}$$

For complex field applications, an example of a unitary matrix is

$$B = \frac{1}{\sqrt{3}}\begin{bmatrix} 1+j & 1 \\ -1 & 1-j \end{bmatrix}$$

We can also define orthogonal matrices in a finite field domain. An example of a 4×4 matrix over GF(2) is given by

$$B = \begin{bmatrix} 1 & 1 & 1 & 0 \\ 0 & 1 & 1 & 1 \\ 1 & 0 & 1 & 1 \\ 0 & 1 & 1 & 1 \end{bmatrix}$$

and

$$BB^T = \begin{bmatrix} 3 & 2 & 2 & 2 \\ 2 & 3 & 2 & 2 \\ 2 & 2 & 3 & 2 \\ 2 & 2 & 2 & 3 \end{bmatrix} \equiv \begin{bmatrix} 1 & 0 & 0 & 0 \\ 0 & 1 & 0 & 0 \\ 0 & 0 & 1 & 0 \\ 0 & 0 & 0 & 1 \end{bmatrix} \pmod 2$$

One has to be very careful in coming up with orthogonal matrices over a finite field domain. There are not many of these for each size of the matrix. For example, there is only one 2×2 matrix over GF(2) that is orthogonal— the identity matrix. Obviously, this is not true over the real field, as one can give infinite number of 2×2 orthogonal matrices. For example,

$$B = \begin{bmatrix} a & \sqrt{1-a^2} \\ \sqrt{1-a^2} & -a \end{bmatrix} \tag{102}$$

is an orthogonal matrix in the real field for any value of $a, 0 \le a \le 1$. We do not discuss finite field orthogonal matrices any further here. However, we do discuss some special orthogonal and unitary matrices in real and complex fields later.

5.6.1. Use of Symmetric Matrices in Testing the Property of Positive Definite and Semidefinite Matrices

In many applications, we are interested in the existence of inverses of matrices for solutions for real field applications. For example, when we discussed the solutions of overdetermined systems we were interested in inverses of matrices of the form $A^T A$. These symmetric matrices have some special properties. The concept of definiteness evolved from quadratic forms. Some of these concepts are discussed next.

First, any quadratic form in several variables, say $x_1, x_2, \ldots, x_n$, over the field of real numbers,

$$f(x_1, x_2, \ldots, x_n) = \sum_{i=1}^{n} \sum_{j=1}^{n} b_{ij} x_j x_i \tag{103}$$

can be expressed in the matrix notation

$$f(x_1, x_2, \ldots, x_n) = \mathbf{x}^T B \mathbf{x} \tag{104}$$

where

$$\mathbf{x}^T = \begin{bmatrix} x_1 & x_2 & \cdots & x_n \end{bmatrix}$$

and

$$B = \begin{bmatrix} b_{11} & b_{12} & \cdots & b_{1n} \\ b_{21} & b_{22} & \cdots & b_{2n} \\ \vdots & \vdots & & \vdots \\ b_{n1} & b_{n2} & \cdots & b_{nn} \end{bmatrix} \tag{105}$$

There are other expressions that may be more convenient than the one in (104). One such expression uses the property that any real matrix B can be expressed as

$$B = S + K \tag{106}$$

where S is a symmetric matrix $(S = S^T)$ and K is a skew symmetric matrix $(K = -K^T)$. Using these ideas, we can write

$$S = \tfrac{1}{2}(B + B^T) \tag{107}$$

and

$$K = \tfrac{1}{2}(B - B^T) \tag{108}$$

From these, it follows that from (104)

$$\mathbf{x}^T B \mathbf{x} = \tfrac{1}{2}\mathbf{x}^T(B + B^T)\mathbf{x} + \tfrac{1}{2}\mathbf{x}^T(B - B^T)\mathbf{x} \tag{109}$$

Noting that $\mathbf{x}^T K \mathbf{x} = \tfrac{1}{2}\mathbf{x}^T(B - B^T)\mathbf{x}$ is a scalar, we have, $\mathbf{x}^T K^T \mathbf{x} = -\mathbf{x}^T K \mathbf{x}$. From this, it follows that $\mathbf{x}^T K \mathbf{x} = 0$, and

$$\mathbf{x}^T B \mathbf{x} = \mathbf{x}^T S \mathbf{x} \tag{110}$$

where S is a symmetric matrix given in (107).

Second, we consider the principal topic of this part of the section. A matrix B is called a positive definite (semidefinite) matrix if

$$\mathbf{x}^T B \mathbf{x} > 0 \ (\geq 0) \tag{111}$$

for all vectors $\mathbf{x} \neq 0$. This definition is obviously impractical to test for the definiteness of a matrix. An efficient method for the evaluation is the use of Sylvester's criterion, which is based on (110): that is, the definiteness test depends on S.

Sylvester's criterion (Hohn, 1960) states that the symmetric matrix S is positive definite if the principal submatrices satisfy

$$s_{11} > 0$$

$$\begin{vmatrix} s_{11} & s_{12} \\ s_{12} & s_{22} \end{vmatrix} > 0, \quad \begin{vmatrix} s_{11} & s_{12} & s_{13} \\ s_{12} & s_{22} & s_{23} \\ s_{13} & s_{23} & s_{33} \end{vmatrix} > 0, \dots, \tag{112}$$

$$|S| > 0$$

For positive semidefinite, we need to replace the strict inequality (>0) by inequality (≥ 0).

There are some special cases wherein the positive definiteness (or semidefiniteness) can be stated from inspection. One such example is if the symmetric matrix S is generated from

$$S = C_{n \times q} C_{q \times n}^T \tag{113}$$

where C and C^T are some real matrices with $n \leq q$. If the rank of C is $n(<n)$, then S is a positive definite (positive semidefinite) matrix. Obviously, if $q < n$, then S is a positive semidefinite matrix. One useful result is that if B is a positive definite matrix, then B is a nonsingular matrix. However, the converse is not true.

Problems

1. Check the positive definiteness (positive semidefiniteness) of the following matrices:

$$B = \begin{bmatrix} 1 & -2 \\ 1 & -3 \end{bmatrix} \quad \text{and} \quad B = \begin{bmatrix} 5 & 2 & 3 \\ 2 & 5 & 2 \\ 3 & 2 & 5 \end{bmatrix}$$

2. Given

$$C = \begin{bmatrix} 1 & 2 & 1 \\ 0 & 1 & 2 \end{bmatrix}$$

use Sylvester's criterion to validate that

$$S = CC^T$$

is positive definite.

In the next section we use some classes of normal matrices and discuss their use in transform theory.

5.7. Discrete Transforms

In this section we discuss some of the basic discrete transforms that are of interest to digital communication engineers. We cast these concepts in terms of linear transformations in the following manner. Given a set of data, say, $x(0)$, $x(1)$, $x(2), \ldots, x(N-1)$, transform this data set to $y(0)$, $y(1), \ldots, y(N-1)$ via the transformation

$$
\begin{bmatrix} y(0) \\ y(1) \\ \vdots \\ y(N-1) \end{bmatrix} = \begin{bmatrix} t_{11} & t_{12} & \cdots & t_{1N} \\ t_{21} & t_{22} & \cdots & t_{2N} \\ \vdots & & & \vdots \\ t_{N1} & t_{N2} & \cdots & t_{NN} \end{bmatrix} \begin{bmatrix} x(0) \\ x(1) \\ \vdots \\ x(N-1) \end{bmatrix} \tag{114a}
$$

or

$$
\mathbf{y} = T\mathbf{x} \tag{114b}
$$

where

$$
T = [\mathbf{T}_1 \quad \mathbf{T}_2 \quad \cdots \quad \mathbf{T}_N] \tag{114c}
$$

with

$$
T_i^T = [t_{1i} \quad t_{2i} \quad \cdots \quad t_{Ni}] \tag{114d}
$$

We require that the vectors $\mathbf{T}_i$ are linearly independent, that is, the matrix T is nonsingular, so that the data vector $\mathbf{x}$ can be recovered from the transformed vector $\mathbf{y}$ via the operation

$$
\mathbf{x} = T^{-1}\mathbf{y} \tag{115}
$$

where T^{-1} is the inverse of the matrix T. As we have seen earlier, the inverses of large matrices are difficult to compute and if the transformation T is orthogonal (or unitary) then the inverse of T is simply the transpose (conjugate transpose) of T. This makes it easy to find (115). Later on we further restrict T so that (114) can best be implemented on a computer and also that the transform coefficients represent pertinent information about the data in terms of the basis vectors $\mathbf{T}_i$.

In the following we consider a few of the transforms that are of major importance at the present time.

5.7.1. Discrete Fourier Transform (DFT)

One of the most important transforms is the Fourier transform of a signal $x(t)$, where t is a continuous variable, such as time or distance. The continuous transform is defined by

$$
X(\Omega) = \int_{-\infty}^{\infty} x(t)\, e^{-j\Omega t}\, dt \tag{116}
$$

where Ω is a continuous frequency variable measured in radians per second. The inverse transform is

$$x(t) = \frac{1}{2\pi} \int_{-\infty}^{\infty} X(\Omega)\, e^{j\Omega t}\, d\Omega \tag{117}$$

Discretization of the expressions in (116) and (117) results in discrete Fourier and discrete inverse Fourier transforms. By use of the sampling theorem (Papoulis, 1962), $x(t)$ can be recovered from the samples $x(nT)$ taken at equal intervals of T provided that

$$X(\Omega) = 0 \quad \text{for } |\Omega| \geq \frac{\pi}{T} \tag{118}$$

That is, the spectrum $x(\Omega)$ is band limited to the radian frequency $\Omega = \pi/T$. The integral in (116) can be evaluated by using the following procedure:

1. Divide the time axis into equal intervals of T.
2. Assume the value of the integrand $x(t)\, e^{-j\Omega t} \cong x(nT)\, e^{-j\Omega nT}$ over the interval $nT \leq t < (n+1)T$.
3. Approximate the integral of the function in the interval $nT \leq t < (n+1)T$ by $Tx(nT)\, e^{-j\Omega nT}$.
4. Sum over all the intervals.

Now, using the above steps, we can express $X(\Omega)$ as

$$X(\Omega) \approx T \sum_{n=-\infty}^{\infty} x(nT)\, e^{-j\Omega nT} \tag{119}$$

which can be shown to be periodic with period $2\pi/T$. That is,

$$X\left(\Omega + \frac{2\pi}{T}\right) = X(\Omega) \tag{120}$$

Most signals are assumed to be nonzero only for a finite interval. Assuming that $x(t) = 0$ for $t < 0$ and $t > NT$, where N is some integer, we can express $X(\Omega)$ as

$$X(\Omega) = T \sum_{n=0}^{N-1} x(nT)\, e^{-j\Omega nT} \tag{121}$$

Since $X(\Omega)$ is periodic with period $2\pi/T$, we can sample $X(\Omega)$ at intervals of $2\pi/NT$, that is,

$$X\left(k\frac{2\pi}{NT}\right) = T \sum_{n=0}^{N-1} x(nT)\, e^{-j(2\pi/N)kn}, \qquad k = 0, 1, \ldots, N-1 \tag{122}$$

Since T and $2\pi/NT$ are constants, they are usually omitted and the discrete Fourier transform is expressed as

$$X(k) = \sum_{n=0}^{N-1} x(n)\, e^{-j(2\pi/N)kn}, \qquad k = 0, 1, \ldots, N-1 \qquad (123)$$

Using the analysis similar to the above, we can also write the inverse transform as

$$x(n) = \frac{1}{N} \sum_{k=0}^{N-1} X(k)\, e^{j(2\pi/N)kn}, \qquad n = 0, 1, \ldots, N-1 \qquad (124)$$

The expression in (123) is usually referred to as the forward discrete Fourier transform (or simply DFT) and (124) is referred to as the inverse discrete Fourier transform (or simply IDFT).

The two expressions in (123) and (124) can be conveniently expressed in a matrix form:

$$\mathbf{X} = A_{\text{DFT}}\mathbf{x} \qquad (125a)$$

$$\mathbf{x} = \frac{1}{N} A_{\text{DFT}}^{*}\mathbf{X} \qquad (125b)$$

where $\mathbf{x} = \text{col}\,(x(0), x(1), \ldots, x(N-1))$, $\mathbf{X} = \text{col}\,(X(0), X(1), \ldots, X(N-1))$, and the typical entries in A_{DFT} and A_{DFT}^{*}, respectively, are given by

$$A_{\text{DFT}}(k, n) = e^{-j(2\pi/N)kn}$$
$$A_{\text{DFT}}^{*}(k, n) = e^{j(2\pi/N)kn} \qquad k, n = 0, 1, \ldots, N-1$$

The matrix A_{DFT} has some interesting properties. First, it is symmetric ($A_{\text{DFT}} = A_{\text{DFT}}^{T}$). Second, $(1/\sqrt{N})A_{\text{DFT}}$ is a unitary matrix, that is,

$$\left(\frac{1}{\sqrt{N}} A_{\text{DFT}}\right)\left(\frac{1}{\sqrt{N}} A_{\text{DFT}}^{*}\right) = I$$

where $*$ represents complex conjugate transpose and I is the identity matrix.

Example

Illustrate (125) for $N = 4$.

Noting (123) and (124), we have

$$
\begin{bmatrix} X(0) \\ X(1) \\ X(2) \\ X(3) \end{bmatrix} = \begin{bmatrix} 1 & 1 & 1 & 1 \\ 1 & -j & -1 & j \\ 1 & -1 & 1 & -1 \\ 1 & j & -1 & -j \end{bmatrix} \begin{bmatrix} x(0) \\ x(1) \\ x(2) \\ x(3) \end{bmatrix} \tag{126a}
$$

$$
\begin{bmatrix} x(0) \\ x(1) \\ x(2) \\ x(3) \end{bmatrix} = \frac{1}{4} \begin{bmatrix} 1 & 1 & 1 & 1 \\ 1 & j & -1 & -j \\ 1 & -1 & 1 & -1 \\ 1 & -j & -1 & j \end{bmatrix} \begin{bmatrix} X(0) \\ X(1) \\ X(2) \\ X(3) \end{bmatrix} \tag{126b}
$$

It is clear from (125) that the DFT and the IDFT are linear transformations and they convert one set of data from one domain to another. The transformation in (125a) requires N^2 complex multiplications and additions, which is prohibitively large for N large. Various fast algorithms have been developed to reduce the number of computations. One algorithm, that is considered to be classical, is given by Cooley and Tukey (1965) and is applicable where N is a power of 2, that is, $N = 2^L$, with L being an integer. This algorithm is sometimes referred to as a fast Fourier transform (FFT) algorithm. An outline of this algorithm is given below.

The idea is to express an N-point discrete Fourier transform (DFT) by two $N/2$-point DFTs. That is, if

$$
A(k) = \sum_{i=0}^{N/2-1} x(2i)\, e^{-j[2\pi/(N/2)]ik}, \qquad k = 0, 1, 2, \ldots, \frac{N}{2} - 1 \tag{127a}
$$

$$
B(k) = \sum_{i=0}^{N/2-1} x(2i+1)\, e^{-j[2\pi/(N/2)]ik}, \qquad k = 0, 1, 2, \ldots, \frac{N}{2} - 1 \tag{127b}
$$

then $X(k)$ can be expressed by

$$
X(k) = A(k) + B(k)\, e^{-j(2\pi/N)k}, \qquad k = 0, 1, 2, \ldots, N - 1 \tag{127c}
$$

where we can use the fact that $A(k)$ and $B(k)$ are periodic with period $N/2$. From (127a) and (127b), it is clear that $A(k)$ and $B(k)$ are DFTs of even and odd sample points, respectively.

Example

Illustrate the above algorithm for $N = 4$. Using (127a) and (127b), we have

$$\begin{bmatrix} A(0) \\ A(1) \\ B(0) \\ B(1) \end{bmatrix} = \begin{bmatrix} 1 & 1 & 0 & 0 \\ 1 & -1 & 0 & 0 \\ 0 & 0 & 1 & 1 \\ 0 & 0 & 1 & -1 \end{bmatrix} \begin{bmatrix} x(0) \\ x(2) \\ x(1) \\ x(3) \end{bmatrix} \tag{128a}$$

Using (127c), we have

$$\begin{bmatrix} X(0) \\ X(1) \\ X(2) \\ X(3) \end{bmatrix} = \begin{bmatrix} 1 & 0 & 1 & 0 \\ 0 & 1 & 0 & -j \\ 1 & 0 & -1 & 0 \\ 0 & 1 & 0 & j \end{bmatrix} \begin{bmatrix} A(0) \\ A(1) \\ B(0) \\ B(1) \end{bmatrix} \tag{128b}$$

where we have used the fact that $A(k)$ and $B(k)$ are periodic with period 2. From (128), we see that the algorithm simply corresponds to matrix multiplication of vectors.

Note that the data vector $x(n)$, $n = 0, 1, 2, 3$, is not in the natural order in (128a). It is in the bit reversed order (Oppenheim and Schafer, 1975). For clarity, let $x_0(0) = x(0)$, $x_0(1) = x(2)$, $x_0(2) = x(1)$, and $x_0(3) = x(3)$. The argument n' in $x_0(n')$, $n' = 0, 1, 2, 3$, is obtained from n in $x(n)$ by bit reversing. This is explained below for $N = 2^L$.

5.7.2. Bit Reversion

1. Identify the data location, say n'.
2. Write it in binary form,

$$n' = (n'_{L-1}, n'_{L-2}, \ldots, n'_0)$$

corresponding to L bits needed to represent any value n', $0 \le n' \le (N - 1)$.
3. Write n' in the bit reversed order:

$$(n'_0, n'_1, \ldots, n'_{L-1})$$

4. Find the decimal equivalent of the bit reversed value in Step 3 and equate it to n.

Example

Obtain the data vector in (128a) in the bit reversed order.

Following the above steps, we have

n'	n	Decimal Equivalent of n	Data in Bit Reversed Order
$x_0(0)$	(00)	(00)	0	$x(0)$
$x_0(1)$	(01)	(10)	2	$x(2)$
$x_0(2)$	(10)	(01)	1	$x(1)$
$x_0(3)$	(11)	(11)	3	$x(3)$

The reduction in the number of computations in the FFT algorithm can be seen from the following argument. Noting that an N-point DFT uses N^2 complex operations, we have that reducing the N-point DFT to two $N/2$-point DFTs obviously reduces the number of operations approximately by half. Successive application of this idea reduces the number of operations to $N \log_2 N$ as compared to N^2 by direct computation of DFT. The method discussed here is for the case of $N = 2^L$. There are various other algorithms for this case and for the case of arbitrary N (Oppenheim and Schafer, 1975). It is interesting to point out that the FFT algorithm for DFT can be used for IDFT also, since the only difference in the DFT and the IDFT is that the entries in the DFT and the IDFT matrices are complex conjugates within the scale factor of N.

The DFT has become a basis for many areas, including signal processing and communication theory. The uses of this transform are numerous aside from the fact that the DFT coefficients give the spectral values of the signal. Other uses include its application in data compression, convolution, deconvolution, and so on. In the data compression application, for example, instead of transmitting the raw data, we could transform the raw data using the DFT first and then transmit only the coefficients that are larger than a preset threshold.

The DFT is based on sinusoidal functions and the basis vectors; that is, the column vectors in the DFT matrix are obtained from

$$e^{-j(2\pi/N)kn} = \cos([2\pi/N]kn) - j\sin([2\pi/N]kn)$$

Therefore, if the data can be represented in terms of a few sinusoids, we need to keep only the corresponding transform coefficients, and a consequence of this of course is the reduction of data for data transmission or storage.

The DFT is a complex transform, that is, the field is the complex field and obviously not the answer for every signal. There are other transforms that may be of interest. A transform that has become very popular in the data compression area is the discrete cosine transform (Ahmed and Rao, 1975), which is based on Chebyshev polynomials. This is discussed next.

5.7.3. Discrete Cosine Transform (DCT)

Given a set of data $x(0), \ldots, x(N-1)$, the DCT of this data set is given by

$$y(0) = \frac{1}{\sqrt{N}} \sum_{n=0}^{N-1} x(n)$$

$$y(k) = \sqrt{\frac{2}{N}} \sum_{n=0}^{N-1} x(n) \cos\left(\frac{2n+1}{2N} k\pi\right), \qquad k = 1, 2, \ldots, N-1$$

$$(129)$$

In matrix form, (129) can be written

$$\mathbf{y} = A_{\text{DCT}} \mathbf{x} \tag{130}$$

where the matrix A_{DCT} can be shown to be an orthogonal matrix. Therefore, the inverse discrete cosine transform (IDCT) can be expressed as

$$\mathbf{x} = A_{\text{DCT}}^T \mathbf{y} \tag{131}$$

where

$$x(n) = \frac{1}{\sqrt{N}} y(0) + \sqrt{\frac{2}{N}} \sum_{k=1}^{N-1} y(k) \cos\left(\frac{2n+1}{2N} k\pi\right),$$

$$n = 0, 1, \ldots, N-1 \quad (132)$$

Fast algorithms have been developed to compute the DCT and the IDCT using fast Fourier transform algorithms. These are based on expressing (129) in the form

$$y(0) = \frac{1}{\sqrt{N}} \sum_{n=0}^{N-1} x(n)$$

$$(133)$$

$$y(k) = \sqrt{\frac{2}{N}} \operatorname{Re}\left[e^{-jk\pi/2N} \sum_{n=0}^{2N-1} x(n)\, e^{-(j2\pi/2N)kn}\right], \qquad k = 1, 2, \ldots, N-1$$

where $x(n) = 0$ for $n = N, N+1, \ldots, 2N-1$, and $\operatorname{Re}[\ ]$ denotes the real part of the term enclosed. From (133), it is clear that the DCT coefficients can be computed using a $2N$-point FFT. In a similar manner, the IDCTs in (132) can be expressed as

$$x(n) = \frac{1}{\sqrt{N}} y'(0) + \sqrt{\frac{2}{N}} \operatorname{Re}\left[\sum_{k=1}^{2N-1} y'(k)\, e^{(j2\pi/2N)nk}\right],$$

$$n = 0, 1, \ldots, N-1 \quad (134)$$

$$= \begin{cases} y(k)\, e^{jk\pi/2N}, & k = 0, 1, \ldots, N - 1 \\ 0, & k = N, N + 1, \ldots, 2N - 1 \end{cases}$$

clear that the IDCTs can be computed using a $2N$-point IFFT.

Example

Illustrate the DCT given in (129) for $N = 4$.
Using (129), we have

$$\begin{bmatrix} y(0) \\ y(1) \\ y(2) \\ y(3) \end{bmatrix} = \begin{bmatrix} 0.5 & 0.5 & 0.5 & 0.5 \\ 0.6533 & 0.2705 & -0.2705 & -0.6533 \\ 0.5 & -0.5 & -0.5 & 0.5 \\ 0.2705 & -0.6533 & 0.6533 & -0.2705 \end{bmatrix} \begin{bmatrix} x(0) \\ x(1) \\ x(2) \\ x(3) \end{bmatrix}$$

Most of the discrete transforms require multiplications, which are computationally expensive compared to additions. This is especially true for fixed-point computers. The following gives a comparison of how expensive the additions, multiplications, and transfers are for fixed- and floating-point machines.

	Multiplication	Addition	Transfer
Fixed point	10	1	0.5
Floating point	2	1	0.5

In this table we assumed that 1 unit of expense corresponds to an addition and is compared with other operations. Note that this table is used for general comparison rather than particular machine comparison.

A transform that avoids multiplications is the Walsh–Hadamard transform, which is discussed next for N a power of 2.

5.7.4. Walsh–Hadamard Transform

The "standard" class of Hadamard matrices of order $N = 2^L$ can be generated by the recursive relation

$$H_k = \begin{bmatrix} H_{k-1} & H_{k-1} \\ H_{k-1} & -H_{k-1} \end{bmatrix}, \qquad k = 1, 2, 3, \ldots \tag{135}$$

with

$$H_0 = 1$$

Example

For $k = 2$ we have

$$H_0 = 1 \qquad H_1 = \begin{bmatrix} 1 & 1 \\ 1 & -1 \end{bmatrix} \quad \text{and} \quad H_2 = \begin{bmatrix} 1 & 1 & 1 & 1 \\ 1 & -1 & 1 & -1 \\ 1 & 1 & -1 & -1 \\ 1 & -1 & -1 & 1 \end{bmatrix}$$

Interestingly, the recursive description in (135) is sometimes expressed in terms of Kronecker products (Bellman, 1960).

5.7.5. Kronecker Products

Let $A = (a_{ij})$ be an $m \times n$ matrix and $B = (b_{ij})$ be a kq matrix, then the Kronecker products are defined by

$$A \otimes B = \begin{bmatrix} a_{11}B & a_{12}B & \cdots & a_{1n}B \\ a_{21}B & a_{22}B & \cdots & a_{2m}B \\ \vdots & \vdots & & \vdots \\ a_{m1}B & a_{m2}B & \cdots & a_{mn}B \end{bmatrix} \tag{136}$$

$$B \otimes A = \begin{bmatrix} Ab_{11} & Ab_{12} & \cdots & Ab_{1q} \\ Ab_{21} & Ab_{22} & \cdots & Ab_{2q} \\ \vdots & \vdots & & \\ Ab_{k1} & Ab_{k2} & \cdots & Ab_{kq} \end{bmatrix} \tag{137}$$

where $\otimes$ denotes the Kronecker product operator. These give convenient representations. Now from (136) and (137), we can write

$$H_k = H_1 \otimes H_{k-1} \tag{138}$$

An interesting aspect of Hadamard matrices is that the entries of H_k are exclusively $+1$ and -1. Also, it can be shown that $(1/\sqrt{N})H_L$ is a symmetric orthogonal matrix. That is, $[(1/\sqrt{N})H_L]^2 = I$, an identity matrix.

The Walsh–Hadamard transform can now be defined and is

$$\mathbf{y} = H_L \mathbf{x} \tag{139}$$

where $\mathbf{x}$ and $\mathbf{y}$ are $N = 2^L$ dimensional vectors. The inverse Walsh–Hadamard transform is given by

$$\mathbf{x} = \frac{1}{N} H_L \mathbf{y} \tag{140}$$

Various fast algorithms are available to implement (139) which use the basic recursive structure

$$Q \ H_k = \begin{bmatrix} I & I \\ I & -I \end{bmatrix} \begin{bmatrix} H_{k-1} & 0 \\ 0 & H_{k-1} \end{bmatrix} \tag{141}$$

where I is an identity matrix of order $_2(k-1)$ and 0 is a null matrix of order $_2(k-1) \times _2(k-1)$. It is clear that successive use of the decomposition allows for the computation of the transform in $N \log_2 N$ additions and subtractions. Another interesting algorithm is based on the relation

$$H_L = A^L \tag{142}$$

where

$$A = \begin{bmatrix} 1 & 1 & 0 & 0 & \cdots & 0 & 0 \\ 0 & 0 & 1 & 1 & \cdots & 0 & 0 \\ \vdots & \vdots & \vdots & \vdots & & \vdots & \vdots \\ 0 & 0 & 0 & 0 & \cdots & 1 & 1 \\ 1 & -1 & 0 & 0 & \cdots & 0 & 0 \\ 0 & 0 & 1 & -1 & \cdots & 0 & 0 \\ \vdots & \vdots & \vdots & \vdots & & \vdots & \vdots \\ 0 & 0 & 0 & 0 & \cdots & 1 & -1 \end{bmatrix} \tag{143}$$

The reader is encouraged to study the algorithm presented in Figure 1.11 in Chapter 1 and the transform in (139) with H_L given in (142). Use of (142) allows repeated use of one structure and therefore this approach makes it easier for special-purpose hardware implementation. This approach is particularly useful when the random access memory is not available.

Example

The four-point Walsh-Hadamard transform is given by

$$\begin{bmatrix} y(0) \\ y(1) \\ y(2) \\ y(3) \end{bmatrix} = \begin{bmatrix} 1 & 1 & 1 & 1 \\ 1 & -1 & 1 & -1 \\ 1 & 1 & -1 & -1 \\ 1 & -1 & -1 & 1 \end{bmatrix} \begin{bmatrix} x(0) \\ x(1) \\ x(2) \\ x(3) \end{bmatrix} \tag{144}$$

Using (142) and (143), we can compute (144) in two steps. First, let

$$\begin{bmatrix} x_1(0) \\ x_1(1) \\ x_1(2) \\ x_1(3) \end{bmatrix} = \begin{bmatrix} 1 & 1 & 0 & 0 \\ 0 & 0 & 1 & 1 \\ 1 & -1 & 0 & 0 \\ 0 & 0 & 1 & -1 \end{bmatrix} \begin{bmatrix} x(0) \\ x(1) \\ x(2) \\ x(3) \end{bmatrix}$$

Second

$$\begin{bmatrix} y(0) \\ y(1) \\ y(2) \\ y(3) \end{bmatrix} = \begin{bmatrix} 1 & 1 & 0 & 0 \\ 0 & 0 & 1 & 1 \\ 1 & -1 & 0 & 0 \\ 0 & 0 & 1 & -1 \end{bmatrix} \begin{bmatrix} x_1(0) \\ x_1(1) \\ x_1(2) \\ x_1(3) \end{bmatrix}$$

5.7.6. Use of the Matrix A in (143) for DFT Implementation

The matrix structure in (135) has been used to implement $N = 2^L$ point DFTs also (Yarlagadda and Hershey, 1981). The structure is illustrated below for an $N = 2^2$ point DFT where $x(n)$ corresponds to the data, $X(k)$ corresponds to the transformed values, and $x_1(n)$ corresponds to the intermediate values.

$$\begin{bmatrix} x_1(0) \\ x_1(1) \\ x_1(2) \\ x_1(3) \end{bmatrix} = \begin{bmatrix} 1 & 0 & 0 & 0 \\ 0 & 1 & 0 & 0 \\ 0 & 0 & 1 & 0 \\ 0 & 0 & 0 & -j \end{bmatrix} \begin{bmatrix} 1 & 1 & 0 & 0 \\ 0 & 0 & 1 & 1 \\ 1 & -1 & 0 & 0 \\ 0 & 0 & 1 & -1 \end{bmatrix} \begin{bmatrix} x(0) \\ x(2) \\ x(1) \\ x(3) \end{bmatrix} \tag{145a}$$

$$\begin{bmatrix} X(0) \\ X(1) \\ X(2) \\ X(3) \end{bmatrix} = \begin{bmatrix} 1 & 1 & 0 & 0 \\ 0 & 0 & 1 & 1 \\ 1 & -1 & 0 & 0 \\ 0 & 0 & 1 & -1 \end{bmatrix} \begin{bmatrix} x_1(0) \\ x_1(1) \\ x_1(2) \\ x_1(3) \end{bmatrix} \tag{145b}$$

This can be seen from (128) by noting that

$$x_1(0) = A(0)$$
$$x_1(1) = B(0)$$
$$x_1(2) = A(1)$$
$$x_1(3) = (-j)B(1)$$

The example above can be generalized and can be implemented by the recursion (Yarlagadda and Hershey, 1981)

$$\mathbf{x}_i = D_i A \mathbf{x}_{i-1}, \qquad i = 1, 2, \ldots, L - 1$$

where $\mathbf{x}_0$ is the data vector written in the bit reversed order, $D_L = I$, an identity matrix, and D_i are diagonal matrices that depend on i, the iteration stage, and N, the number of data points. For more information, see Singleton (1969) and Ahmed and Rao (1975). The matrix A is common in each step and it has a special structure. The method allows for sequential data acquisition, perhaps using multistorage units, and even allows for multiprocessors to compute DFTs.

So far we have discussed transforms in only real and complex fields. Obviously, transforms can be defined in the finite field domain also. These include some error correction coding (Wiggert, 1978), number theoretic transforms, and many others. Obviously, the topic is wide and therefore we consider it only in terms of examples.

5.7.7. An Example of a Number Theoretic Transform (NTT) (Agarwal and Burrus, 1975)

The algorithms that use integer arithmetic obviously do not introduce any round-off error. Furthermore, if the proposed transform has a fast algorithm, improvement in accuracy and speed can be achieved. These transforms can be used in computing convolutions.

We define an N-point transform that has the general DFT structure

$$F(k) = \sum_{n=0}^{N-1} x(n) r^{nk}, \qquad k = 0, 1, \ldots, N - 1 \qquad (146)$$

where r is called a root of unity of order N if N is the least positive integer such that

$$r^N = 1 \ (\text{mod } M)$$

We consider M to be a prime number of the form $2^{2^t} + 1$ (Fermat number). It can be shown that the inverse transform is given by

$$x(n) = N^{-1} \sum_{k=0}^{N-1} F(k) r^{-nk}, \qquad n = 0, 1, \ldots, N - 1 \qquad (147)$$

It is clear that (146) and (147) have the same general form as the DFT and IDFT. In computing (146) and (147), we have to use mod M arithmetic. As an example, consider $N = 4$ with $t = 2$. Therefore,

$$M = 2^{2^2} + 1 = 17$$

and $r = 4$. Then (146) can be written

$$F = T\mathbf{x}$$

where

$$T = \begin{bmatrix} 1 & 1 & 1 & 1 \\ 1 & 4 & 4^2 & 4^3 \\ 1 & 4^2 & 4^4 & 4^6 \\ 1 & 4^3 & 4^6 & 4^9 \end{bmatrix} \equiv \begin{bmatrix} 1 & 1 & 1 & 1 \\ 1 & 4 & 16 & 13 \\ 1 & 16 & 1 & 16 \\ 1 & 13 & 16 & 4 \end{bmatrix} \pmod{17} \qquad (148)$$

The inverse transform is given by

$$\mathbf{x} = T^{-1}\mathbf{F} \qquad (149)$$

where

$$T^{-1} = 4^{-1}\begin{bmatrix} 1 & 1 & 1 & 1 \\ 1 & 4^{-1} & 4^{-2} & 4^{-3} \\ 1 & 4^{-2} & 4^{-4} & 4^{-6} \\ 1 & 4^{-3} & 4^{-6} & 4^{-9} \end{bmatrix} \equiv 13\begin{bmatrix} 1 & 1 & 1 & 1 \\ 1 & 13 & 16 & 4 \\ 1 & 16 & 1 & 16 \\ 1 & 4 & 16 & 13 \end{bmatrix} \pmod{17} \quad (150)$$

It can be verified that $TT^{-1} = I$ using again mod 17 arithmetic. We use this example in computing discrete convolutions in a later chapter.

In this section we discussed various transforms to illustrate that discrete transforms can be dealt with in the general context of transformations.

Problems

1. Write the DFT matrix for an eight-point DFT. Use the FFT algorithm illustrated in (127) to show that the number of operations is reduced by using this procedure when compared to the direct computation.
2. Write the IDCT matrix for a four-point IDCT. Show that the eight-point DFT can be used to compute the four-point IDCT.

References

Agarwal, R. C. and C. S. Burrus (1975), Number Theoretic Transforms to Implement Fast Digital Convolutions, *Proceedings of the IEEE*, Vol. 63, pp. 550–560.

Ahmed, N. and K. R. Rao (1975), *Orthogonal Transforms for Digital Signal Processing*, Springer-Verlag, New York.

Bellman, R. (1960), *Introduction to Matrix Analysis*, McGraw-Hill, New York.

Byrd, R. H. and D. A. Payne, Convergence of the Iteratively Reweighted Least Squares Algorithm for Robust Regression, The Johns Hopkins University, Baltimore, MD, Technical Report 313, 1979.

Claerbout, J. R. and F. Muir (1973), Robust Modelling with Erratic Data, *Geophysics*, Vol. 38, pp. 826–844.

Cooley, J. W. and R. W. Tukey (1965), An Algorithm for Machine Computation of Complex Fourier Series, *Mathematics of Computation*, Vol. 19, 297–301.

Durbin, J. (1960), The Fitting of Time Series Models, *Rev. Inst. Int. Statist.*, Vol. 28, 233–243.

Hohn, F. E. (1960), *Elementary Matrix Algebra*, Macmillan, New York.

Oppenheim, A. V. and R. W. Schafer (1975), *Digital Signal Processing*, Prentice-Hall, Englewood Cliffs, NJ.

Papoulis, A. (1962), *The Fourier Integral and Its Applications*, McGraw-Hill, New York.

Rao, C. R. and S. K. Mitra (1970), *Generalized Inverse of Matrices and Its Applications*, John Wiley, New York.

Robinson, E. A. (1967), *Multichannel Time Series Analysis with Digital Computer Programs*, Holden-Day, San Francisco.

Singleton, R. C. (1969), An Algorithm for Computing the Mixed Radix Fast Fourier Transform, *IEEE Transactions on Audio Electroacoustics*, Vol. AU-17, pp. 93–109.

Wiggert, D. (1978), *Error-Control Coding and Applications*, Artech House, Dedham, MA.

Yarlagadda, R. and J. Hershey (1981), Architecture of the Fast Walsh–Hadamard and Fast Fourier Transforms with Charge Transfer Devices, *International Journal of Electronics*, Vol. 51, pp. 669–681.

Yarlagadda, R., J. B. Bednar, and T. Watt (1985), Fast Algorithms for L_P Deconvolution, *IEEE Transactions on Acoustics, Speech, and Signal Processing*, Vol. 33, pp. 174–182.

Matrix Representations

6.1. Introduction

In some applications we are interested in functions of matrices such as a kth power of A. Even though it is conceivable that the power of a matrix can be found by repeated application, it may be desired to analytically represent the power of a matrix. It would be nice if the matrix could be expressed in a form that is amenable for such a procedure. In this chapter we discuss a matrix representation in terms of its eigenvalues and eigenvectors (Wilkinson, 1965). This approach is valid for real or complex field applications. In the later part of the chapter we discuss different approaches that allow for finite field matrix representations.

6.2. Eigenvalue Problem

Given a square matrix A of order n, determine scalars λ and the corresponding nonzero vectors that satisfy the equation

$$A\mathbf{x} = \lambda\mathbf{x} \tag{1}$$

which is called the eigenvalue problem.

The problem in (1) can be reduced to solving

$$(A - \lambda I)\mathbf{x} = 0 \tag{2}$$

where I is the identity matrix. From Chapter 5, we know that (2) has a nontrivial solution if and only if

$$|A - \lambda I| = 0 \tag{3}$$

The polynomial

$$f(\lambda) = |\lambda I - A| = 0 \tag{4}$$

is called the characteristic polynomial. The roots of this equation over the complex field are called the characteristic values or the eigenvalues of the matrix A. The corresponding solutions in (2) are called the eigenvectors. Let us illustrate the eigenvalues first by a simple example.

Example

Given the matrix

$$A = \begin{bmatrix} 0 & 1 \\ -3 & -4 \end{bmatrix} \tag{5}$$

find the eigenvalues.

The characteristic polynomial is given by

$$|\lambda I - A| = \begin{vmatrix} \lambda & -1 \\ 3 & \lambda + 4 \end{vmatrix} = \lambda^2 + 4\lambda + 3$$

and the roots are

$$\lambda_1 = -3 \quad \text{and} \quad \lambda_2 = -1 \tag{6}$$

Transferring these concepts into the finite field is not always possible. Let us consider another example.

Example

Given the matrix

$$A = \begin{bmatrix} 0 & 1 \\ 1 & 1 \end{bmatrix}$$

the characteristic polynomial over GF(2) is given by

$$|\lambda I + A| = \lambda^2 + \lambda + 1 \tag{7}$$

which obviously can not be factored over $GF(2)$. Therefore, the eigenvalue-eigenvector approach is not suitable over a finite field. However, the characteristic polynomial plays an important role for finite field matrices. We postpone this discussion to a later section, where we will concentrate on efficient methods for computing characteristic polynomials and their use for matrices over a given field. Next let us continue with the example in (5) to find the eigenvectors. From here on, in this section, we assume that the field of operation is over the complex field.

Example

Find a set of eigenvectors for the matrix in (5) over the complex field. The eigenvectors can be computed from

$$(A - \lambda_1 I)\mathbf{x}_1 = 0 \tag{8}$$

or

$$\begin{bmatrix} -3 & -1 \\ 3 & 1 \end{bmatrix} \begin{bmatrix} x_{11} \\ x_{21} \end{bmatrix} = 0 \tag{9}$$

We can use the procedure discussed in Chapter 5 for the solution of a homogeneous system of equations. Using elementary operations, we can reduce (9) to

$$\begin{bmatrix} 1 & \frac{1}{3} \\ 0 & 0 \end{bmatrix} \begin{bmatrix} x_{11} \\ x_{21} \end{bmatrix} = 0 \tag{10}$$

indicating that there is only one independent solution and that it can be obtained from

$$x_{11} = -\tfrac{1}{3} x_{21} \tag{11}$$

by selecting any arbitrary nonzero value for x_{21}. Taking $x_{21} = -3$, we have

$$\mathbf{x}_1 = \begin{bmatrix} 1 \\ -3 \end{bmatrix} \tag{12}$$

Similarly, we can compute the vector $\mathbf{x}_2$ from

$$(A - \lambda_2 I)\mathbf{x}_2 = 0 \tag{13}$$

or

$$\begin{bmatrix} -1 & -1 \\ +3 & 3 \end{bmatrix} \begin{bmatrix} x_{12} \\ x_{22} \end{bmatrix} = 0 \tag{14}$$

Again, we have only one independent solution which is

$$x_{12} = -x_{22} \tag{15}$$

Assuming $x_{22} = 1$, we have the second eigenvector

$$\mathbf{x}_2 = \begin{bmatrix} -1 \\ 1 \end{bmatrix} \tag{16}$$

What do we do with these eigenvectors? Recall that our goal is to find a simplified representation of the matrix A using the eigenvectors. Let us proceed with this using the above example. The two equations given in (8) and (13) can be written in the form

$$A\mathbf{x}_1 = \mathbf{x}_1\lambda_1$$
$$A\mathbf{x}_2 = \mathbf{x}_2\lambda_2 \tag{17}$$

Now we can combine (17) and write

$$A[\mathbf{x}_1\mathbf{x}_2] = [\mathbf{x}_1\lambda_1 \quad \mathbf{x}_2\lambda_2] = [\mathbf{x}_1\mathbf{x}_2]\begin{bmatrix} \lambda_1 & 0 \\ 0 & \lambda_2 \end{bmatrix} \tag{18}$$

Let us identify

$$X = [\mathbf{x}_1 \quad \mathbf{x}_2] \tag{19}$$

and

$$\Lambda = \begin{bmatrix} -3 & 0 \\ 0 & -1 \end{bmatrix} \tag{20}$$

Equation (18) can be written in the compact form

$$AX = X\Lambda \tag{21}$$

Noting that the vectors $\mathbf{x}_1$ and $\mathbf{x}_2$ are independent, we can say that X is a nonsingular matrix. Therefore, we can write

$$A = X\Lambda X^{-1} \tag{22}$$

For our example

$$X = \begin{bmatrix} 1 & -1 \\ -3 & 1 \end{bmatrix}, \quad X^{-1} = -\frac{1}{2}\begin{bmatrix} 1 & 1 \\ 3 & 1 \end{bmatrix}, \quad \Lambda = \begin{bmatrix} -3 & 0 \\ 0 & -1 \end{bmatrix} \tag{23}$$

The decomposition in (22) is very useful and is based on the concept of similarity.

Similarity

Two matrices A and B are said to be similar if

$$B = S^{-1}AS \tag{24}$$

where S is any nonsingular matrix. Note that from (22) and (24), we can say that A and Λ are similar and Λ is sometimes referred to as the spectral matrix of A. An interesting result is that A and B have the same characteristic polynomial. This can be seen from

$$|\lambda I - B| = |\lambda I - S^{-1}AS| = |S^{-1}(\lambda I - A)S|$$
$$= |S^{-1}||\lambda I - A||S|$$
$$= |\lambda I - A| \tag{25}$$

Therefore, A and B have the same eigenvalues, determinant, and rank.

Problem

Given the matrix A in (5), find the eigenvectors and eigenvalues for A^T and show that A and A^T are similar.

6.3. Diagonal Representation of Normal Matrices

In Chapter 5, we defined a normal matrix A by the property

$$AA^* = A^*A$$

This is a very important class of matrices. In this section, through an example, we illustrate that normal matrices are diagonable over the complex field.

Example

We are given the symmetric matrix

$$A = \begin{bmatrix} 1 & 2 & 1 \\ 2 & 4 & 2 \\ 1 & 2 & 1 \end{bmatrix} \tag{26}$$

Compute its eigenvalues and eigenvectors. The characteristic polynomial is given by

$$f(\lambda) = |\lambda I - A| = \begin{vmatrix} \lambda - 1 & -2 & -1 \\ -2 & \lambda - 4 & -2 \\ -1 & -2 & \lambda - 1 \end{vmatrix}$$

$$= (\lambda - 1) \begin{vmatrix} \lambda - 4 & -2 \\ -2 & \lambda - 1 \end{vmatrix} + 2 \begin{vmatrix} -2 & -1 \\ -2 & \lambda - 1 \end{vmatrix}$$

$$- 1 \begin{vmatrix} -2 & -1 \\ \lambda - 4 & -2 \end{vmatrix}$$

$$= \lambda^2(\lambda - 6)$$

indicating that the eigenvalues are

$$\lambda_1 = 0, \qquad \lambda_2 = 0, \qquad \lambda_3 = 6 \tag{27}$$

From this example, we can see that computing a characteristic polynomial is difficult, especially when the size of the matrix is large. In a later section, we will consider an elegant method for computing the characteristic polynomial. Next, let us consider the eigenvectors. First, for $\lambda_1 = 0$,

$$A\mathbf{x}_1 = 0\mathbf{x}_1 = 0 \tag{28}$$

Using the procedure for the solution of a homogeneous system of equations, we can reduce (28) to

$$\begin{bmatrix} 1 & 2 & 1 \\ 0 & 0 & 0 \\ 0 & 0 & 0 \end{bmatrix} \begin{bmatrix} x_{11} \\ x_{21} \\ x_{31} \end{bmatrix} = 0 \tag{29}$$

which indicates that $(A - \lambda_1 I)$ is of rank of 1 and therefore there are $(n - 1) = (3 - 1) = 2$ independent solutions that can be obtained from (29). First, from (29), we have

$$x_{11} = -(2x_{21} + x_{31}) \tag{30}$$

Using $x_{21} = -1$ and $x_{31} = 0$, we have a vector

$$\mathbf{x}_1 = \begin{bmatrix} 2 \\ -1 \\ 0 \end{bmatrix} \tag{31}$$

Using $x_{21} = 0$ and $x_{31} = -1$, we have another vector

$$\mathbf{x}_2 = \begin{bmatrix} 1 \\ 0 \\ -1 \end{bmatrix} \tag{32}$$

It can be shown that the vectors are independent. Does this example imply that if there are r equal eigenvalues, we get r independent solutions by this procedure? This is not always true for arbitrary matrices. However, for normal matrices, this is always true. In our example, A is symmetric and therefore it falls in that category. Before we discuss this class, let us complete the example. For the eigenvalue $\lambda_3 = 6$, we have

$$A\mathbf{x}_3 = 6\mathbf{x}_3 \tag{33}$$

or

$$(A - 6I)\mathbf{x}_3 = 0$$

with

$$\begin{bmatrix} -5 & 2 & 1 \\ 2 & -2 & 2 \\ 1 & 2 & -5 \end{bmatrix} \begin{bmatrix} x_{13} \\ x_{23} \\ x_{33} \end{bmatrix} = 0 \tag{34}$$

Again, using elementary operations, we can reduce (34) to

$$\begin{bmatrix} 0 & 0 & 0 \\ 0 & 1 & -2 \\ 1 & 0 & -1 \end{bmatrix} \begin{bmatrix} x_{13} \\ x_{23} \\ x_{33} \end{bmatrix} = 0 \tag{35}$$

indicating that the coefficient matrix, $(\lambda_3 I - A)$ has a rank of 2. Therefore, the homogeneous system of equations in (35) has only one independent solution and can be computed from

$$x_{13} = x_{33}$$
$$x_{23} = 2x_{33} \tag{36}$$

Arbitrarily selecting $x_{33} = 1$ (obviously, we cannot pick that to be zero, as if we do, we get the trivial solution), we have

$$\mathbf{x}_3 = \begin{bmatrix} 1 \\ 2 \\ 1 \end{bmatrix} \tag{37}$$

In the example above, we have obtained three vectors that are independent. For the first two eigenvectors we could have other possibilities. Suppose we decide to have the first two eigenvectors orthogonal to each other. This can be achieved by using Gram–Schmidt orthogonalization discussed in Chapter 5. That is, let

$$\mathbf{y}_1 = \mathbf{x}_1$$
$$\mathbf{y}_2 = a\mathbf{y}_1 + \mathbf{x}_2 \tag{38}$$

with

$$\mathbf{y}_1^T\mathbf{y}_2 = 0 = a\mathbf{y}_1^T\mathbf{y}_1 + \mathbf{y}_1^T\mathbf{x}_2$$

or

$$a = -\frac{\mathbf{y}_1^T\mathbf{x}_2}{\mathbf{y}_1^T\mathbf{y}_1} = -\frac{2}{5}$$

Therefore,

$$\mathbf{y}_2 = \begin{bmatrix} \frac{1}{5} \\ \frac{2}{5} \\ -1 \end{bmatrix} \tag{39}$$

The vector $\mathbf{y}_2$ is a valid eigenvector since

$$A\mathbf{y}_2 = A(a\mathbf{y}_1 + \mathbf{x}_2) = aA\mathbf{y}_1 + A\mathbf{x}_2$$
$$= \lambda_2(a\mathbf{y}_1 + \mathbf{x}_2) = \lambda_2\mathbf{y}_2 \tag{40}$$

where we have used $A\mathbf{y}_1 = \lambda_2\mathbf{y}_1$, as $\lambda_1 = \lambda_2$. Interestingly, $\mathbf{x}_3$ is orthogonal to both $\mathbf{y}_1$ and $\mathbf{y}_2$ and also to $\mathbf{x}_1$ and $\mathbf{x}_2$. That is,

$$\mathbf{x}_3^T\mathbf{y}_1 = 0 = \mathbf{x}_3^T\mathbf{y}_2$$
$$\mathbf{x}_3^T\mathbf{x}_1 = 0 = \mathbf{x}_3^T\mathbf{x}_2 \tag{41}$$

Is this unusual? No, this is true for normal matrices. Formally, if $\mathbf{x}_i$ and $\mathbf{x}_j$ are eigenvectors of a normal matrix A corresponding to the eigenvalues λ_i and λ_j, and $\lambda_i \neq \lambda_j$, then

$$\mathbf{x}_i^T\mathbf{x}_j = 0 \tag{42}$$

The vectors y_1, y_2, and $y_3 = x_3$ are orthogonal to each other. Normalizing these vectors, we have

$$z_1 = \frac{1}{\|y_1\|} y_1, \qquad z_2 = \frac{1}{\|y_2\|} y_2, \qquad z_3 = \frac{1}{\|y_3\|} y_3 \qquad (43)$$

where the lengths of the vectors $\|y_i\|$ are given by

$$\|y_1\| = \sqrt{5}, \qquad \|y_2\| = \sqrt{\tfrac{6}{5}}, \qquad \|y_3\| = \sqrt{6} \qquad (44)$$

Let us now form the matrix

$$Z = [z_1 \quad z_2 \quad z_3] = \begin{bmatrix} \dfrac{2}{\sqrt{5}} & \dfrac{1}{\sqrt{30}} & \dfrac{1}{\sqrt{6}} \\[2ex] -\dfrac{1}{\sqrt{5}} & \dfrac{2}{\sqrt{30}} & \dfrac{2}{\sqrt{6}} \\[2ex] 0 & -\dfrac{5}{\sqrt{30}} & \dfrac{1}{\sqrt{6}} \end{bmatrix} \qquad (45)$$

It is clear that Z is an orthogonal matrix. That is,

$$Z^T Z = Z Z^T = I \qquad (46)$$

Since the z_i are eigenvectors of A (note that the z_i are scalar multiples of y_i), we can write

$$A[z_1 \quad z_2 \quad z_3] = [z_1 \quad z_2 \quad z_3] \begin{bmatrix} \lambda_1 & & \\ & \lambda_2 & \\ & & \lambda_3 \end{bmatrix} \qquad (47)$$

or

$$AZ = Z\Lambda$$

where A was given in (26) and

$$\Lambda = \begin{bmatrix} 0 & & \\ & 0 & \\ & & 6 \end{bmatrix} \qquad (48)$$

Since Z is an orthogonal matrix, it follows that

$$Z^{-1} = Z^T \quad \text{and} \quad A = Z\Lambda Z^T \qquad (49)$$

This has a special form compared to (22) and is very useful for normal matrices. Formally, two matrices A and B are said to be unitarily (orthogonally) similar if

$$B = S^*AS \qquad (50)$$

where S^* is the conjugate transpose of S and S is assumed to be any unitary (orthogonal) matrix. From (48), we can say that A and Λ are orthogonally similar. We now state a basic theorem without proof for normal matrices that formalize some of the above statements (Perlis, 1952).

Theorem

Every normal matrix is unitarily similar to a diagonal matrix.

This theorem points out that for an $n \times n$ normal matrix there are n eigenvectors that are othogonal to each other corresponding to n eigenvalues of a given normal matrix. Note that normal matrices are a very important class of matrices because they include symmetric, orthogonal, and so on. The examples we have presented so far have the general property that the matrix A is similar to a diagonal matrix, and we identify such a matrix as a diagonable matrix. Next we need to ask the question: Are all matrices diagonable? The answer is no, and we consider this in the next section.

Problems

1. Given

$$A = \begin{bmatrix} 3 & -2 & -1 \\ -2 & 3 & -2 \\ -1 & -2 & 3 \end{bmatrix}$$

compute its eigenvalues and eigenvectors.

2. Let

$$S = \begin{bmatrix} 1 & 1 & 1 \\ 0 & 1 & -1 \\ -1 & 0 & 1 \end{bmatrix}$$

and $A = S^{-1}\Lambda S$, where Λ is a diagonal matrix. Is there a diagonal matrix Λ that is not equal to the identity matrix, for which A is symmetric?

3. Given

$$A = \frac{1}{\sqrt{2}} \begin{bmatrix} 1 & 1 \\ 1 & -1 \end{bmatrix}$$

compute its eigenvalues and eigenvectors.

6.4. Representations of Nondiagonable Matrices

Let us first consider a simple example to show that not every matrix is diagonable. Again, in this section, we assume that the field of operation is over the complex field and is not explicitly identified.

Example

Consider the matrix

$$A = \begin{bmatrix} \lambda_1 & 1 \\ 0 & \lambda_1 \end{bmatrix}$$

where λ_1 is any complex number. Since the matrix here is triangular, it follows that $|(\lambda I - A)| = (\lambda - \lambda_1)^2$. The eigenvalues are λ_1 and λ_1. If A is similar to a diagonal matrix, that is,

$$A = S^{-1}\Lambda S \tag{51a}$$

with

$$\Lambda = \begin{bmatrix} \lambda_1 & 0 \\ 0 & \lambda_1 \end{bmatrix} \tag{51b}$$

then

$$S^{-1}\Lambda S = \lambda_1 S^{-1} S = \lambda_1 I \neq A$$

Therefore, the decomposition in (51) is impossible with Λ given in (51b). This brings up the important point that we may have to consider triangular matrices in place of diagonal matrices for a representation of A. Note that the eigenvalues of a triangular matrix are its diagonal entries. We now consider a special class of triangular matrices.

6.4.1. Jordan Matrix

A Jordan matrix J is an upper triangular matrix with the following properties:

1. Equal eigenvalues appear next to each other along its diagonal.
2. The ij entries of J are 0 unless $j = i$ or $i + 1$. Its $(i, i + 1)$ entry may be 0 or 1 if the ith eigenvalue λ_i is equal to the $(i + 1)$th eigenvalue λ_{i+1}, but is 0 otherwise.

From these properties we can say that the matrix J can be expressed as a direct sum of Jordan matrices J_i each having eigenvalues λ_i on its diagonal,

1's immediately to the right of the diagonal (superdiagonal), and 0's else-where. That is,

$$
J = \begin{bmatrix} J_1 & 0 & \cdots & 0 \\ 0 & J_2 & \cdots & 0 \\ \vdots & \vdots & & \vdots \\ 0 & 0 & \cdots & J_k \end{bmatrix}, \quad
J_i = \begin{bmatrix} \lambda_i & 1 & 0 & \cdots & & 0 \\ 0 & \lambda_i & 1 & \cdots & & 0 \\ \vdots & \vdots & \vdots & & & \vdots \\ 0 & 0 & 0 & \cdots & \lambda_i & 1 \\ 0 & 0 & 0 & \cdots & 0 & \lambda_i \end{bmatrix} \tag{52}
$$

It is possible that the diagonal entries in J_i are the same as the diagonal entries in J_{i+1}. A limiting case of course is that J is a diagonal matrix. The diagonal matrix in (48) is an example of that. Another example of a Jordan matrix is

$$
J = \begin{bmatrix} \lambda_1 & 1 & 0 & 0 & 0 & 0 \\ 0 & \lambda_1 & 0 & 0 & 0 & 0 \\ 0 & 0 & \lambda_1 & 1 & 0 & 0 \\ 0 & 0 & 0 & \lambda_1 & 0 & 0 \\ 0 & 0 & 0 & 0 & \lambda_1 & 0 \\ 0 & 0 & 0 & 0 & 0 & \lambda_2 \end{bmatrix} \tag{53}
$$

We state an important theorem without proof.

Theorem

Every $n \times n$ matrix A over the complex field is similar to a Jordan matrix J, which is called the spectral matrix of A.

The above theorem states that given an $n \times n$ matrix A, there exists a nonsingular matrix X such that

$$
A = XJX^{-1} \quad \text{or} \quad J = X^{-1}AX \quad \text{or} \quad AX = XJ \tag{54}
$$

The matrices X and X^{-1} are usually referred to as a right modal matrix (or simply modal matrix) and left modal matrix, respectively. Finding the Jordan matrix J in (54) for arbitrary matrices is a rather difficult problem. Equation (54) can be symbolically written in terms of (52) as

$$
A[X_1 \quad X_2 \quad \cdots \quad X_k] = [X_1 \quad X_2 \quad \cdots \quad X_k]\begin{bmatrix} J_1 & & & \\ & J_2 & & \\ & & \ddots & \\ & & & J_k \end{bmatrix} \tag{55}
$$

where the X_i are now matrices with columns that are linearly independent. Now considering the ith Jordan block, we have

$$AX_i = X_iJ_i \tag{56}$$

Assuming that J_i is an $m \times m$ matrix, with the form in (52), we can express

$$A[\mathbf{x}_{i1}, \mathbf{x}_{i2}, \ldots, \mathbf{x}_{im}] = [\mathbf{x}_{i1}, \quad \mathbf{x}_{i2}, \quad \cdots \quad \mathbf{x}_{im}] \begin{bmatrix} \lambda_i & 1 & 0 & \cdots & & 0 \\ 0 & \lambda_i & 1 & \cdots & & 0 \\ \vdots & \vdots & \vdots & & \vdots & \vdots \\ 0 & 0 & 0 & \cdots & \lambda_i & 1 \\ 0 & 0 & 0 & \cdots & 0 & \lambda_i \end{bmatrix} \tag{57}$$

or

$$\begin{aligned} A\mathbf{x}_{i1} &= \lambda_i\mathbf{x}_{i1} \quad \text{or} \quad (A - \lambda_iI)\mathbf{x}_{i1} = 0 \\ A\mathbf{x}_{i2} &= \lambda_i\mathbf{x}_{i2} + \mathbf{x}_{i1} \quad \text{or} \quad (A - \lambda_iI)\mathbf{x}_{i2} = \mathbf{x}_{i1} \\ &\vdots \\ A\mathbf{x}_{im} &= \lambda_i\mathbf{x}_{im} + \mathbf{x}_{i(m-1)} \quad \text{or} \quad (A - \lambda_iI)\mathbf{x}_{im} = \mathbf{x}_{i(m-1)} \end{aligned} \tag{58}$$

It is clear that these expressions are applicable for each Jordan block. From (58), we see that if $\mathbf{x}_{i1}$ is known, then the remaining vectors can be found. What if there are other Jordan blocks that have the same eigenvalues? How do we divide the Jordan blocks with the same eigenvalues? These are discussed in the following.

First, let there be l Jordan blocks with the same eigenvalues λ_i. Then considering the first equation in (58), we can write

$$\begin{aligned} A\mathbf{x}_{i1} &= \lambda_i\mathbf{x}_{i1} \\ A\mathbf{x}_{(i+1)1} &= \lambda_i\mathbf{x}_{(i+1)1} \\ &\vdots \\ A\mathbf{x}_{(i+l-1)1} &= \lambda_i\mathbf{x}_{(i+l-1)1} \end{aligned} \tag{59}$$

or

$$A[\mathbf{x}_{i1} \quad \mathbf{x}_{(i+1)1} \quad \cdots \quad \mathbf{x}_{(i+l-1)1}]$$

$$= [\mathbf{x}_{i1} \quad \mathbf{x}_{(i+1)1} \quad \cdots \quad \mathbf{x}_{(i+l-1)1}] \begin{bmatrix} \lambda_i & & & \\ & \lambda_i & & \\ & & \ddots & \\ & & & \lambda_i \end{bmatrix} \tag{60}$$

or

$$AY = \lambda_iY \quad \text{or} \quad (A - \lambda_iI)Y = 0 \tag{61}$$

where Y is an $n \times l$ matrix. From (61), we see that the homogeneous equation

$$(A - \lambda_i I)\alpha = 0 \tag{62}$$

must have l independent solutions. In other words, the rank of $(A - \lambda_i I)$ must be of rank $(n - l)$ if l Jordan blocks exist. Therefore, to find the eigenvectors in (59), we need to solve for α in (62). This will be discussed later. The number of 1's associated with an eigenvalue λ_i in (52), that is, the number of 1's on the superdiagonal for all the blocks corresponding to the eigenvalue λ_i, is given by the order of λ_i in the characteristic equation minus the degeneracy of $(A - \lambda_i I)$. If n is the order of the matrix, n_i is the multiplicity of λ_i, and r is the rank of $(A - \lambda_i I)$, then the number of 1's is given by $n_i - (n - r)$. This can be visualized from the following example.

Example

Let $\lambda = \lambda_i$ be a root of the characteristic polynomial with multiplicity 6 and let the degeneracy of $(A - \lambda_i I)$ be 4. There are then 4 Jordan blocks associated with λ_i. The possibilities for the Jordan blocks (except for rearrangement of blocks) are

$$\begin{bmatrix} \lambda_i & 1 & & & & \\ & \lambda_i & 1 & & & \\ & & \lambda_i & & & \\ & & & \lambda_i & & \\ & & & & \lambda_i & \\ & & & & & \lambda_i \end{bmatrix} \qquad \begin{bmatrix} \lambda_i & 1 & & & & \\ & \lambda_i & & & & \\ & & \lambda_i & 1 & & \\ & & & \lambda_i & & \\ & & & & \lambda_i & \\ & & & & & \lambda_i \end{bmatrix} \tag{63}$$

The number of 1's is $6 - 4 = 2$, which verifies the earlier assertion. The selection of the correct possible Jordan matrix is a very difficult problem (Gantmacher, 1960) and is beyond our scope. We do not discuss the theory here. Let us now illustrate some of these ideas by examples.

Example

Given the matrix

$$A = \begin{bmatrix} 0 & 1 \\ -1 & 2 \end{bmatrix} \tag{64}$$

find the Jordan matrix and the corresponding modal matrices. The characteristic polynomial is given by

$$f(\lambda) = |\lambda I - A| = \lambda^2 - 2\lambda + 1 = (\lambda - 1)^2 \tag{65}$$

indicating that there is only one eigenvalue $\lambda_1 = 1$ and it is of multiplicity 2. The rank of

$$(A - \lambda_1 I) = (A - I) = \begin{bmatrix} -1 & 1 \\ -1 & 1 \end{bmatrix}$$

is 1. Therefore, there is only one Jordan block, given by

$$J = \begin{bmatrix} 1 & 1 \\ 0 & 1 \end{bmatrix} \qquad (66)$$

The eigenvector corresponding to $\lambda_1 = 1$ is given by

$$(A - I)\mathbf{x}_1 = 0$$

or

$$\begin{bmatrix} -1 & 1 \\ -1 & 1 \end{bmatrix} \begin{bmatrix} x_{11} \\ x_{21} \end{bmatrix} = 0$$

or

$$x_{11} = x_{21}$$

Selecting $x_{21} = 1$, we have the vector

$$\mathbf{x}_1 = \begin{bmatrix} 1 \\ 1 \end{bmatrix} \qquad (67)$$

The second column in the modal matrix is obtained from [see (58)]

$$A\mathbf{x}_2 = \lambda_1 \mathbf{x}_2 + \mathbf{x}_1$$

or

$$(A - I)\mathbf{x}_2 = \mathbf{x}_1$$

or

$$\begin{bmatrix} -1 & 1 \\ -1 & 1 \end{bmatrix} \begin{bmatrix} x_{12} \\ x_{22} \end{bmatrix} = \mathbf{x}_1 = \begin{bmatrix} 1 \\ 1 \end{bmatrix}$$

From this it follows that

$$x_{12} = x_{22} - 1$$

Using $x_{22} = 1$, we have $x_{12} = 0$. Therefore,

$$\mathbf{x}_2 = \begin{bmatrix} 0 \\ 1 \end{bmatrix} \tag{68}$$

The modal matrix is given by

$$X = \begin{bmatrix} 1 & 0 \\ 1 & 1 \end{bmatrix} \tag{69}$$

and the left modal matrix is

$$X^{-1} = \begin{bmatrix} 1 & 0 \\ -1 & 1 \end{bmatrix}$$

The result can be verified by expressing

$$X^{-1}AX = \begin{bmatrix} 1 & 0 \\ -1 & 1 \end{bmatrix}\begin{bmatrix} 0 & 1 \\ -1 & 2 \end{bmatrix}\begin{bmatrix} 1 & 0 \\ 1 & 1 \end{bmatrix} = \begin{bmatrix} 1 & 1 \\ 0 & 1 \end{bmatrix} = J$$

Let us now consider a more complicated example (DeRusso et al., 1965).

Example

Given

$$A = \begin{bmatrix} 0 & 0 & 0 & 0 \\ 0 & 0 & 0 & 1 \\ 1 & 0 & 0 & 0 \\ 0 & 0 & 0 & 0 \end{bmatrix} \tag{70}$$

find the decomposition in the form

$$A = XJX^{-1}$$

The characteristic polynomial

$$f(\lambda) = |\lambda I - A| = \lambda^4$$

indicating that $\lambda_1 = 0$ is an eigenvalue with a multiplicity of $n_1 = 4$. The degeneracy of $(\lambda_1 I - A)$ is obviously 2 as $(\lambda_1 I - A) = A$ and A has rank $r = 2$. Therefore, there are two Jordan blocks. The number of 1's on the superdiagonal is given by $n_1 - (n - r) = 4 - (4 - 2) = 2$. The two possibilities (except for the rearrangement of blocks) for the Jordan matrices are

$$
\begin{bmatrix} 0 & 1 & 0 & 0 \\ 0 & 0 & 1 & 0 \\ 0 & 0 & 0 & 0 \\ 0 & 0 & 0 & 0 \end{bmatrix} \text{ and } \begin{bmatrix} 0 & 1 & 0 & 0 \\ 0 & 0 & 0 & 0 \\ 0 & 0 & 0 & 1 \\ 0 & 0 & 0 & 0 \end{bmatrix} \tag{71}
$$

As pointed out earlier, the exact selection of the Jordan matrix is a difficult problem. We use trial and error here to identify the correct form. For explicit representation, we identify x_i, $i = 1, 2, 3, 4$, as the columns in the modal matrix X. Since there are two Jordan blocks, we need to solve for the two lead vectors, say x_{11} and x_{21}, from (59). If the first Jordan form in (71) is the right form, then $x_{11} = x_1$, $x_{12} = x_2$, $x_{13} = x_3$, and $x_{21} = x_4$. On the other hand, if the second form is the right one, then $x_{11} = x_1$, $x_{12} = x_2$, $x_{21} = x_3$, and $x_{22} = x_4$.

The next step is to solve for the vectors x_{11} and x_{21} in (59), corresponding to two Jordan blocks.

Expressing $(A - \lambda_1 I) x_1 = 0$, we have

$$
\begin{bmatrix} 0 & 0 & 0 & 0 \\ 0 & 0 & 0 & 1 \\ 1 & 0 & 0 & 0 \\ 0 & 0 & 0 & 0 \end{bmatrix} \begin{bmatrix} x_{11} \\ x_{21} \\ x_{31} \\ x_{41} \end{bmatrix} = 0 \tag{72}
$$

indicating that there are two independent solutions. From (72), we see that x_{21} and x_{31} are arbitrary, except that both can not be zero, and $x_{11} = 0$ and $x_{41} = 0$. The two independent vectors x_{11} and x_{21} can be generated from

$$
x_1 = \begin{bmatrix} 0 \\ x_{21} \\ x_{31} \\ 0 \end{bmatrix} \tag{73}
$$

with appropriate selections for x_{21} and x_{31}. We postpone this computation. Using the second equation in (58), we have

$$
Ax_2 = (0) \cdot x_2 + x_1 = x_1
$$

or

$$\begin{bmatrix} 0 & 0 & 0 & 0 \\ 0 & 0 & 0 & 1 \\ 1 & 0 & 0 & 0 \\ 0 & 0 & 0 & 0 \end{bmatrix} \begin{bmatrix} x_{12} \\ x_{22} \\ x_{32} \\ x_{42} \end{bmatrix} = \begin{bmatrix} 0 \\ x_{21} \\ x_{31} \\ 0 \end{bmatrix}$$

which indicates that

$$0 = 0 \qquad x_{42} = x_{21} \qquad x_{12} = x_{31} \qquad 0 = 0$$

and x_{22} and x_{32} are arbitrary. The vector $\mathbf{x}_2$ is

$$\mathbf{x}_2 = \begin{bmatrix} x_{31} \\ x_{22} \\ x_{32} \\ x_{21} \end{bmatrix} \qquad (74)$$

Since we do not know which Jordan matrix to consider in (71), we can continue with the first possibility to see whether it is the correct form. Using the third equation in (58), we have

$$A\mathbf{x}_3 = 0\mathbf{x}_3 + \mathbf{x}_2 = \mathbf{x}_2$$

or

$$\begin{bmatrix} 0 & 0 & 0 & 0 \\ 0 & 0 & 0 & 1 \\ 1 & 0 & 0 & 0 \\ 0 & 0 & 0 & 0 \end{bmatrix} \begin{bmatrix} x_{13} \\ x_{23} \\ x_{33} \\ x_{43} \end{bmatrix} = \begin{bmatrix} x_{31} \\ x_{22} \\ x_{32} \\ x_{21} \end{bmatrix} \qquad (75)$$

which results in

$$x_{31} = 0$$
$$x_{43} = x_{22}$$
$$x_{13} = x_{32} \qquad (76)$$
$$x_{21} = 0$$

The first and the last equation in (76) indicates that the first vector $\mathbf{x}_1$ in (73) is identically zero. This violates the condition that one of the columns in a modal matrix can not be identically zero. Therefore, the first possibility of the Jordan form in (71) is invalid and the second one is the correct form.

It is clear that for the second Jordan form in (71), we can compute the matrix X_1 for the first block such that

$$AX_1 = X_1 \begin{bmatrix} 0 & 1 \\ 0 & 0 \end{bmatrix}$$

where

$$X_1 = [\mathbf{x}_{11} \quad \mathbf{x}_{12}]$$

It is clear that $\mathbf{x}_{11}$ and $\mathbf{x}_{12}$ are

$$\mathbf{x}_{11} = \mathbf{x}_1 \quad \text{and} \quad \mathbf{x}_{12} = \mathbf{x}_2$$

For the second block, we have

$$AX_2 = X_2 \begin{bmatrix} 0 & 1 \\ 0 & 0 \end{bmatrix}$$

where

$$X_2 = [\mathbf{x}_{21} \quad \mathbf{x}_{22}]$$

We can repeat the above process and obtain

$$\mathbf{x}_3 = \mathbf{x}_{21} = \begin{bmatrix} 0 \\ x_{23} \\ x_{33} \\ 0 \end{bmatrix} \quad \text{and} \quad \mathbf{x}_4 = \mathbf{x}_{22} = \begin{bmatrix} x_{33} \\ x_{24} \\ x_{34} \\ x_{23} \end{bmatrix} \tag{77}$$

The entries must be selected such that $\mathbf{x}_{21}$ is independent of $\mathbf{x}_{11}$. The modal matrix

$$X = [\mathbf{x}_{11} \quad \mathbf{x}_{12} \quad \mathbf{x}_{21} \quad \mathbf{x}_{22}]$$

$$= \begin{bmatrix} 0 & x_{31} & 0 & x_{33} \\ x_{21} & x_{22} & x_{23} & x_{24} \\ x_{31} & x_{32} & x_{33} & x_{34} \\ 0 & x_{21} & 0 & x_{23} \end{bmatrix}$$

where the entries must be chosen such that the columns are independent. A solution is

$$X = \begin{bmatrix} 0 & 0 & 0 & 1 \\ 1 & 0 & 0 & 0 \\ 0 & 0 & 1 & 0 \\ 0 & 1 & 0 & 0 \end{bmatrix}$$

Therefore,

$$X^{-1} = \begin{bmatrix} 0 & 1 & 0 & 0 \\ 0 & 0 & 0 & 1 \\ 0 & 0 & 1 & 0 \\ 1 & 0 & 0 & 0 \end{bmatrix}$$

and

$$X^{-1}AX = \begin{bmatrix} 0 & 1 & 0 & 0 \\ 0 & 0 & 0 & 0 \\ 0 & 0 & 0 & 1 \\ 0 & 0 & 0 & 0 \end{bmatrix} = J$$

which gives the desired result.

As can be seen from the above discussion eigenvector decomposition is a difficult problem for arbitrary matrices. There are very few matrices that have analytical eigenvalue–eigenvector decomposition. There is one matrix that has wide interest, whose eigenvalues and eigenvectors can be analytically expressed in terms of the matrix entires. This is discussed in the next section.

Problem

Given

$$A = \frac{1}{5} \begin{bmatrix} 4 & 3 & -1 & 1 \\ -1 & 8 & -1 & 1 \\ -1 & 3 & 4 & 1 \\ 1 & -3 & 1 & 4 \end{bmatrix}$$

find the modal matrix X, the left modal matrix X^{-1}, and the Jordan matrix J such that

$$XJX^{-1} = A$$

Answer: One solution is

$$X = \begin{bmatrix} 1 & -5 & -3 & 1 \\ 1 & 0 & -1 & 0 \\ 1 & 0 & 0 & -1 \\ -1 & 0 & 0 & 0 \end{bmatrix}$$

6.5. Circulant Matrix and Its Eigenvectors

A matrix C is called a circulant matrix if we can write it in the form

$$C = \begin{bmatrix} c_0 & c_1 & \cdots & c_{N-1} \\ c_{N-1} & c_0 & \cdots & c_{N-2} \\ \vdots & \vdots & & \vdots \\ c_1 & c_2 & \cdots & c_0 \end{bmatrix} \tag{78}$$

From this we can see that C is known if we know the first column or the first row. An interesting result is that when c_i is over the real field, we have (Gray, 1972; Yarlagadda and Suresh Babu, 1980)

$$C = \frac{1}{N} A_{\mathrm{DFT}} D A_{\mathrm{DFT}}^* = \frac{1}{N} A_{\mathrm{DFT}}^* \Lambda A_{\mathrm{DFT}} \tag{79}$$

where A_{DFT} is the discrete Fourier transform matrix defined in Chapter 5, A_{DFT}^* is the complex conjugate transpose of A_{DFT}, and D and Λ are diagonal matrixes

$$D = \mathrm{dia}\,(d_{11}, d_{22}, \ldots, d_{NN}), \qquad \Lambda = \mathrm{dia}\,(\lambda_{11}, \lambda_{22}, \ldots, \lambda_{NN}) \tag{80}$$

and are given by

$$\begin{bmatrix} d_{11} \\ d_{22} \\ \vdots \\ d_{NN} \end{bmatrix} = A_{\mathrm{DFT}} \begin{bmatrix} c_0 \\ c_1 \\ \vdots \\ c_{N-1} \end{bmatrix}, \qquad \begin{bmatrix} \lambda_{11} \\ \lambda_{22} \\ \vdots \\ \lambda_{NN} \end{bmatrix} = A_{\mathrm{DFT}} \begin{bmatrix} c_0 \\ c_{N-1} \\ \vdots \\ c_1 \end{bmatrix}. \tag{81}$$

Equation (79) points out that C is unitarily similar to a diagonal matrix. The eigenvectors of a circulant can be obtained from the columns of the DFT matrix; and (81) indicates that the eigenvalues can be obtained by using the DFT matrix. Note that $(1/\sqrt{N})\,A_{\mathrm{DFT}}$ is a unitary matrix. Let us illustrate (79) by a simple example.

Example

Given a circulant matrix

$$C = \begin{bmatrix} 0 & 1 & 0 \\ 0 & 0 & 1 \\ 1 & 0 & 0 \end{bmatrix} \tag{82}$$

verify the first equality in (79).

The characteristic polynomial is given by

$$f(\lambda) = (\lambda^3 - 1)$$

and the roots are

$$\lambda_1 = 1 \qquad \lambda_2 = -\tfrac{1}{2} - j\frac{\sqrt{3}}{2} \qquad \lambda_3 = -\tfrac{1}{2} + j\frac{\sqrt{3}}{2} \tag{83}$$

By (81), we have

$$\begin{bmatrix} d_{11} \\ d_{22} \\ d_{33} \end{bmatrix} = \begin{bmatrix} 1 & 1 & 1 \\ 1 & e^{-j2\pi/3} & e^{-j4\pi/3} \\ 1 & e^{-j4\pi/3} & e^{-j2\pi/3} \end{bmatrix} \begin{bmatrix} 0 \\ 1 \\ 0 \end{bmatrix} = \begin{bmatrix} 1 \\ e^{-j2\pi/3} \\ e^{-j4\pi/3} \end{bmatrix} \tag{84}$$

It is clear from (83) and (84) that

$$\lambda_i = d_{ii}$$

The first eigenvector consists of all 1's corresponding to $\lambda_1 = 1$ and can be verified by inspection. For the second eigenvector, we have

$$C \begin{bmatrix} 1 \\ e^{-j2\pi/3} \\ e^{-j4\pi/3} \end{bmatrix} = \begin{bmatrix} e^{-j2\pi/3} \\ e^{-j4\pi/3} \\ 1 \end{bmatrix} = e^{-j2\pi/3} \begin{bmatrix} 1 \\ e^{-j2\pi/3} \\ e^{-j4\pi/3} \end{bmatrix} \tag{85}$$

which verifies the second eigenvector. Since the third eigenvalue is a conjugate of the second and C is real, it follows that the third eigenvector is a conjugate of the second eigenvector. This can be verified by inspection.

The decomposition in (79) is very useful and will be used in the next chapter. Next we consider some simple applications of the eigenvalue-eigenvector decompositions.

Problem

Consider the special case of a circulant given by

$$C = \begin{bmatrix} 3 & 2 & 1 & 2 \\ 2 & 3 & 2 & 1 \\ 1 & 2 & 3 & 2 \\ 2 & 1 & 2 & 3 \end{bmatrix}$$

Compute its eigenvalues and its eigenvectors. Since C is a symmetric matrix (also a Toeplitz matrix), the eigenvectors can be taken real. Can you find a real modal matrix? If so, give the result.

6.6. Simple Functions of Matrices

The eigenvalue-eigenvector decomposition discussed earlier for an arbitrary matrix, $A = (a_{ij})$, where the a_{ij} are over the complex field, in the form

$$A = XJX^{-1} \tag{86}$$

can be used as a basis in finding functions of matrices (Frame, 1964). For example,

$$A^2 = XJX^{-1}XJX^{-1} = XJ^2X^{-1}$$

and, in general,

$$A^m = XJ^mX^{-1} \tag{87}$$

indicating that finding any power of a matrix would be simplified once we know the Jordan matrix and the corresponding modal matrix X. From (52), we see that

$$J^m = \begin{bmatrix} J_1^m & & & \\ & J_2^m & & \\ & & \cdot & \\ & & & \cdot \\ & & & & J_k^m \end{bmatrix} \tag{88}$$

The Jordan submatrix J_i and its powers have some interesting properties. First, let J_i be an $l \times l$ matrix which can be written in the form

$$J_i = \lambda_i I + P \tag{89}$$

where

$$P = \begin{bmatrix} 0 & 1 & 0 & \cdots & 0 \\ 0 & 0 & 1 & \cdots & 0 \\ \vdots & \vdots & \vdots & & \vdots \\ 0 & 0 & 0 & \cdots & 1 \\ 0 & 0 & 0 & \cdots & 0 \end{bmatrix} \tag{90}$$

It can be seen that

$$p^m = 0 \quad \text{for all } m \geq l \tag{91}$$

Therefore, to compute J_i^m, expand the matrix polynomial $(\lambda_i I + P)^m$ and
delete all the terms that have powers of P higher than or equal to l. This
method of finding powers of A gets messy for large matrices. In Section
6.7, we consider another method.

The decomposition in (54) is particularly useful when A is an $n \times n$
diagonable matrix, that is, $J = D$, a diagonal matrix. For this case,

$$A^k = XD^kX^{-1} \tag{92}$$

where

$$D^k \doteq \text{dia}\,(d_{11}^k, \ldots, d_{nn}^k) \tag{93}$$

Finally, even though the eigenvalue–eigenvector decomposition is a
powerful tool, it is difficult to compute this decomposition for large matrices.
Therefore, for some applications, it is better to look for other decompositions. One such decomposition is given in the next section.

Problems

1. Find the kth power of the matrix A in (5) over the complex field.
2. Consider the matrix A in (26). Find A^2 over the complex field using its
modal matrix discussed in Section 6.3. If we assume that the entries of A

are over GF(5), can we use the A^2 computed above to find $A^2 = (c_{ij})$ where the c_{ij} are over GF(5)?

6.7. Singular Value Decomposition

Recently, there has been a good deal of interest in this decomposition because of its growing importance in numerical analysis, signal analysis, and signal estimation.

Let A be an $m \times n$ matrix over the complex field of rank r. Then there exists an $m \times m$ unitary matrix U, an $n \times n$ unitary matrix V, and an $r \times r$ diagonal matrix D with strictly positive elements such that (Bjorck and Anderson, 1969).

$$A = U\Sigma V^* \quad \text{and} \quad \Sigma = \begin{bmatrix} D & 0 \\ 0 & 0 \end{bmatrix} \tag{94}$$

The diagonal entries in D are called the singular values of A. If $r = m$ ($r = n$), then there will be no zero matrices in Σ at the bottom (at the right). If $r = m = n$, then $\Sigma = D$. The decomposition in (94) is called the singular value decomposition.

One of the reasons for interest in this decomposition is that the singular values are fairly insensitive to perturbations in the matrix elements compared to the eigenvalues, especially when considering nonsymmetric matrices.

The singular value decomposition can be reduced to the Hermitian matrix eigenvalue problem for a complex A (symmetric matrix problem for real A). Now

$$AA^* = U(\Sigma\Sigma^*)U^* \tag{95}$$

and

$$A^*A = V(\Sigma^*\Sigma)V^* \tag{96}$$

The modal matrices of (AA^*) and (A^*A), respectively, give U and V. We should point out that AA^* and A^*A have the same nonzero eigenvalues. Let us illustrate the decomposition now.

Example

Given

$$A = \begin{bmatrix} 5 & -5 \\ 1 & 7 \end{bmatrix} \tag{97}$$

find the decomposition in (94). Now

$$AA^T = \begin{bmatrix} 50 & -30 \\ -30 & 50 \end{bmatrix} \quad \text{and} \quad A^TA = \begin{bmatrix} 26 & -18 \\ -18 & 74 \end{bmatrix} \tag{98}$$

In this example, the characteristic polynomials are the same for (AA^T) and (A^TA) and

$$f(\lambda) = \lambda^2 - 100\lambda + 1600 = (\lambda - 80)(\lambda - 20) \tag{99}$$

The modal matrices for (AA^T) and (A^TA) can be computed using the procedure discussed earlier and they are given by

$$U = \frac{1}{\sqrt{2}} \begin{bmatrix} 1 & 1 \\ -1 & 1 \end{bmatrix} \quad \text{and} \quad V = \frac{1}{\sqrt{10}} \begin{bmatrix} 1 & 3 \\ -3 & 1 \end{bmatrix}$$

Using (94), we have

$$A = \frac{1}{2\sqrt{5}} \begin{bmatrix} 1 & 1 \\ -1 & 1 \end{bmatrix} \begin{bmatrix} \sqrt{80} & 0 \\ 0 & \sqrt{20} \end{bmatrix} \begin{bmatrix} 1 & -3 \\ 3 & 1 \end{bmatrix}$$

There are better methods to compute the decomposition (Wilkinson and Reinsch, 1971) than the method presented here. However, this method is intuitively simple to see.

Problems

1. Find a singular value decomposition for the matrix

$$A = \begin{bmatrix} 0 & 1 \\ -2 & -3 \end{bmatrix}$$

where the entries are assumed to be over the real field.

2. Give an example where the singular value decomposition is not unique. [Hint: Use a symmetric matrix with multiple eigenvalues.]

6.8. Characteristic Polynomials

When we first discussed the computation of characteristic polynomials, we pointed out that finding characteristic polynomials for large matrices is a tedious process. In this section, we consider an elegant method (Frame, 1964) for computing these polynomials over a field. The characteristic polynomial of an $n \times n$ matrix $A = (a_{ij})$ is given by

$$f(\lambda) = |\lambda I - A| = \lambda^n + d_1\lambda^{n-1} + \cdots + d_n \tag{100}$$

with the entries in A, a_{ij}, and the coefficients d_i belonging to the same field. We use the following example to illustrate various aspects first.

Example

Given

$$A = \begin{bmatrix} 1 & 0 & 1 \\ 1 & 1 & 0 \\ 0 & 1 & 1 \end{bmatrix} \tag{101}$$

over the real field, find the characteristic polynomial.
From (100) we have

$$f(\lambda) = |\lambda I - A| = \lambda^3 - 3\lambda^2 + 3\lambda - 2 \tag{102}$$

On the other hand, if we consider the entries A are over GF(2), then $f(\lambda)$ can be determined directly from the definition or from (102). In the latter case, we can use $-3 \equiv 1 \,(\mathrm{mod}\,2)$, $+3 \equiv 1 \,(\mathrm{mod}\,2)$, and $-2 \equiv 0 \,(\mathrm{mod}\,2)$. Therefore, over GF(2), the characteristic polynomial is given by

$$f(\lambda) = \lambda^3 + \lambda^2 + \lambda \tag{103}$$

Next, let us consider a method given by Fadeev, Frame, and others. To discuss this method, we first define a matrix $B(\lambda)$,

$$B(\lambda) = \mathrm{Adj}\,(\lambda I - A) = \sum_{i=0}^{n} B_i \lambda^{n-1-i}, \qquad B_0 = I, \quad B_n = 0 \tag{104}$$

usually referred to as the conjoint of A. For the example in (101), over the real field, we have

$$
B(\lambda) = \begin{bmatrix} (\lambda - 1)^2 & 1 & (\lambda - 1) \\ (\lambda - 1) & (\lambda - 1)^2 & +1 \\ 1 & (\lambda - 1) & (\lambda - 1)^2 \end{bmatrix}
$$

$$
= \begin{bmatrix} 1 & 0 & 0 \\ 0 & 1 & 0 \\ 0 & 0 & 1 \end{bmatrix} \lambda^2 + \begin{bmatrix} -2 & 0 & 1 \\ 1 & -2 & 0 \\ 0 & 1 & -2 \end{bmatrix} \lambda + \begin{bmatrix} 1 & 1 & -1 \\ -1 & 1 & 1 \\ 1 & -1 & 1 \end{bmatrix}
$$

$$
= B_0 \lambda^2 + B_1 \lambda + B_2 \tag{105}
$$

where the B_i are as identified.

The following theorem is useful in our development and is stated here without proof.

Theorem

The trace of the conjoint matrix $B(\lambda)$ over a field F equals the derivative of the characteristic polynomial.

For the example above, the trace of the conjoint matrix is

$$\operatorname{tr} B(\lambda) = 3\lambda^2 - 6\lambda + 3 \tag{106}$$

The derivative of the characteristic polynomial in (102) is given by

$$f'(\lambda) = 3\lambda^2 - 6\lambda + 3$$

which is the same as the polynomial derived in (106).

From the above theorem, we can state that

$$\operatorname{tr}(B_k) = (n - k)d_k \tag{107}$$

The proposed recursive algorithm for the computation of the characteristic polynomials and the conjoint matrix is based on the following theorem, which is given without proof (Frame, 1964).

Theorem

The scalar coefficients d_k in the characteristic polynomial $f(\lambda) = \sum_{k=0}^{n} d_k \lambda^{n-k}$ and the matrix coefficients B_k in $B(\lambda) = \sum_{k=0}^{n-1} B_k \lambda^{n-k-1}$ with the entries over the real field are related by

$$d_k = -\frac{1}{k}\operatorname{tr} AB_{k-1}, \qquad k = 1, 2, \ldots, n, \quad d_0 = 1 \tag{108}$$

$$B_k = AB_{k-1} + d_k I, \qquad k = 1, 2, \ldots, n, \quad B_0 = I, \quad B_n = 0 \tag{109}$$

where the operations are over the real field.

For the matrix A in (101), over the real field,

$$d_1 = -\operatorname{tr}(A) = -3$$

$$B_1 = A + d_1 I = \begin{bmatrix} -2 & 0 & 1 \\ 1 & -2 & 0 \\ 0 & 1 & -2 \end{bmatrix}$$

$$d_2 = -\tfrac{1}{2}\operatorname{tr}(AB_1) = -\tfrac{1}{2}\operatorname{tr}\begin{bmatrix} -2 & +1 & -1 \\ -1 & -2 & 1 \\ 1 & -1 & -2 \end{bmatrix} = 3$$

$$B_2 = \begin{bmatrix} 1 & 1 & -1 \\ -1 & 1 & 1 \\ 1 & -1 & 1 \end{bmatrix}$$

(110)

$$d_3 = -\tfrac{1}{3}\operatorname{tr}(AB_2) = -\tfrac{1}{3}\operatorname{tr}\begin{bmatrix} 2 & 0 & 0 \\ 0 & 2 & 0 \\ 0 & 0 & 2 \end{bmatrix} = -2$$

$$B_3 = 0$$

These results check with the earlier computation. Note that $B_n = 0$ can be used as a check.

The equations in (108) and (109) are not valid for all field representations. For example, over GF(2), the expression in (108) is given in terms of $1/k$. This is not defined when k is even. However, the equations can still be used for finite fields provided the characteristic polynomial and the conjoint matrix are found first in terms of real field representation and then using the finite field representation. For the matrix A in (101), over GF(2), we have

$$d_1 = 1, \qquad d_2 = 1, \qquad d_3 = 0$$

$$B_0 = I, \qquad B_1 = \begin{bmatrix} 0 & 0 & 1 \\ 1 & 0 & 0 \\ 0 & 1 & 0 \end{bmatrix}$$

$$B_2 = \begin{bmatrix} 1 & 1 & 1 \\ 1 & 1 & 1 \\ 1 & 1 & 1 \end{bmatrix}, \qquad B_3 = 0$$

An interesting result follows from (109) by setting $k = n$:

$$A^{-1} = -\frac{1}{d_n}B_{n-1}, \qquad \text{if } |-A| = d_n \neq 0 \tag{111}$$

It is clear that the inverse of A in (101) exists if the entries in A are over the real field and the inverse does not exist if they are over GF(2).

Another important result follows from (109) by successively finding

$$B_k = A^k + d_1 A^{k-1} + \cdots + d_{k-1}A + d_k I, \qquad k = 1, 2, \ldots, n \quad (112)$$

and noting that $B_n = 0$.

Theorem

Every square matrix over a field satisfies its characteristic polynomial. That is, if

$$f(\lambda) = |\lambda I - A| = 0 \quad \text{then} \quad f(A) = 0 \tag{113}$$

This theorem is usually referred to as the Cayley–Hamilton theorem. This is a powerful theorem and has some very interesting applications. First, from (100) and (113)

$$f(A) = A^n + d_1 A^{n-1} + \cdots + d_n I = 0 \tag{114}$$

which implies that any power of a matrix can be expressed in terms of powers of A^i, $i \leq n - 1$, using the characteristic polynomial. Also, when $d_n \neq 0$, we can write from (114)

$$I = -\frac{1}{d_n}(A^{n-1} + d_1 A^{n-2} + \cdots + d_{n-1}I)A \tag{115}$$

indicating that

$$A^{-1} = -\frac{1}{d_n}(A^{n-1} + d_1 A^{n-2} + \cdots + d_{n-1}I) \tag{116}$$

Note that the inverse exists only if $(-d_n)$, the determinant of A, is nonzero.

Second, we need to ask an important question: Given a characteristic polynomial with coefficients over a given field, can we find a matrix with entries over a given field that has the given characteristic polynomial? The answer is yes, and there are a few that can be given by inspection. Let us define a companion matrix

$$C = \begin{bmatrix} 0 & 1 & 0 & \cdots & 0 \\ 0 & 0 & 1 & \cdots & 0 \\ \vdots & \vdots & \vdots & & \vdots \\ 0 & 0 & 0 & \cdots & 1 \\ -a_n & -a_{n-1} & -a_{n-2} & \cdots & -a_1 \end{bmatrix} \tag{117}$$

that has the characteristic polynomial

$$|\lambda I - C| = \lambda^n + a_1\lambda^{n-1} + \cdots + a_{n-1}\lambda + a_n \tag{118}$$

Note the entries in the last row of (117) and the corresponding coefficients in (118). Let us verify this for $n = 3$.

Example

  Given

$$C = \begin{bmatrix} 0 & 1 & 0 \\ 0 & 0 & 1 \\ -a_3 & -a_2 & -a_1 \end{bmatrix}$$

find

$$|\lambda I - C| = \begin{vmatrix} \lambda & -1 & 0 \\ 0 & \lambda & -1 \\ a_3 & a_2 & \lambda + a_1 \end{vmatrix}$$

Expanding the determinant using the last row, we have

$$|\lambda I - C| = (\lambda + a_1)\begin{vmatrix} \lambda & -1 \\ 0 & \lambda \end{vmatrix} - a_2\begin{vmatrix} \lambda & 0 \\ 0 & -1 \end{vmatrix} + a_3\begin{vmatrix} -1 & 0 \\ 0 & -1 \end{vmatrix}$$
$$= (\lambda + a_1)\lambda^2 + a_2\lambda + a_3 \tag{119}$$

Note that successive terms in (119) have successive powers of λ. This can be generalized.

  There are other matrices that can be given for a given characteristic polynomial. Since

$$|\lambda I - C^T| = |(\lambda I - C)|$$

we have the matrix

$$C_1 = \begin{bmatrix} 0 & 0 & \cdots & \cdot & -a_n \\ 1 & 0 & \cdots & \cdot & -a_{n-1} \\ \vdots & \vdots & & \vdots & \vdots \\ 0 & 0 & \cdots & 1 & -a_1 \end{bmatrix} \tag{120}$$

that has the same characteristic polynomial as in (118). There are obviously other variations of C that can be used in this context. It should be pointed out that the correspondence between (117) and (118) is valid over any field, provided the entries in C are interpreted in an appropriate fashion. For example, over GF(2),

$$-a_i = a_i \tag{121}$$

which is either 0 or 1.

Over the complex field, the characteristic polynomial can be factored in the form

$$f(\lambda) = \prod_{i=1}^{n} (\lambda - \lambda_i) \tag{122}$$

where the λ_i are the characteristic roots. Noting that the determinant of a diagonal matrix is equal to the product of the diagonal entries, we can write

$$J = \text{dia}(\lambda_1, \lambda_2, \ldots, \lambda_n) \tag{123}$$

and

$$f(\lambda) = |\lambda I - J|$$

The only problem here is that some λ_i may be complex. If so, J will be complex. This can be circumvented, if the characteristic polynomial is a real polynomial, by using the following procedure. Let

$$f_1(\lambda) = (\lambda - \alpha + j\beta)(\lambda - \alpha - j\beta) \tag{124}$$

The matrix

$$B = \begin{bmatrix} \alpha & \beta \\ -\beta & \alpha \end{bmatrix} \tag{125}$$

has the characteristic polynomial in (124). The structure in (125) or a companion form for (124) can be used as blocks in place of the complex conjugate eigenvalues in (123).

Example

Use the form in (125) to find a matrix with real entries that has the characteristic polynomial

$$f(\lambda) = (\lambda^2 + \lambda + 1)(\lambda + 1)$$

The corresponding matrix is given by

$$B_1 = \begin{bmatrix} -1 & 0 \\ 0 & B \end{bmatrix}$$

where the entries in B from (125) are $\alpha = -\frac{1}{2}$ and $\beta = \sqrt{3}/2$.

Before we finish this section, we should point out that matrices of the form in (125) do not exist for finite field representation.

Problems

1. We are given the characteristic polynomial

$$f(\lambda) = (\lambda^2 + \lambda + \tfrac{1}{2}) = \prod_{i=1}^{2} (\lambda + \lambda_i)$$

with coefficients in the polynomial over the real field. For second-order polynomials, with $|\lambda_i| \leq 1$, an interesting representation is $f(\lambda) = |\lambda I - K|$, where (Ledbetter and Yarlagadda, 1974)

$$K = \begin{bmatrix} k_1 & 1 \\ -1 & k_2 \end{bmatrix}$$

Find k_1 and k_2 for the above polynomial.

2. Find the inverses of the following matrices using the Cayley–Hamilton theorem:

(a) $\begin{bmatrix} 1 & 1 & 2 \\ 0 & 1 & 1 \\ 0 & 0 & 1 \end{bmatrix}$ GF(3)

(b) $\begin{bmatrix} 1 & 1 \\ -1 & 0 \end{bmatrix}$ Real field

6.9. Minimal Polynomial

In Section 6.7 it was stated that every $n \times n$ matrix A over a field F satisfies its characteristic polynomial. However, it is not necessarily the least degree polynomial that the matrix A satisfies. The least degree polynomial that satisfies A is called the minimal polynomial. Let the minimal polynomial, $m(\lambda)$, of an $n \times n$ matrix A over field F be given by

$$m(\lambda) = \lambda^b + a_1 \lambda^{b-1} + \cdots + a_b, \qquad b \leq n \tag{126}$$

such that

$$m(A) = A^b + a_1 A^{b-1} + \cdots + a_b = 0 \tag{127}$$

It is clear that the entries a_{ij} in the matrix $A = (a_{ij})$ and the coefficients a_i in (127) must be over the same field.

The minimal polynomial can be determined from the Adj $(\lambda I - A)$ by using the following procedure (Ogata, 1967).

1. Compute $B(\lambda) = \text{Adj} (\lambda I - A)$ and the characteristic polynomial, $f(\lambda)$, by using the cofactors or by using the methods discussed in Section 6.7.
2. Write the elements of $B(\lambda)$ in factored form. That is, write the element $B_{ij}(\lambda)$, the ijth element of $B(\lambda)$, in a factored form

$$b_{ij}(\lambda) = \prod_{l=1}^{k} b_{ij}^{(l)}(\lambda)$$

where $b_{ij}^{(l)}(\lambda)$ are polynomials in an appropriate field.
3. Determine $d(\lambda)$ as the greatest common devisor (GCD) of all the elements of $B(\lambda)$. We can also say that $d(\lambda)$ is the GCD of all the $(n-1)$st-order minors of $(\lambda I - A)$.
4. The minimal polynomial $m(\lambda)$ is given by

$$m(\lambda) = \frac{f(\lambda)}{d(\lambda)} \tag{128}$$

It is clear that the minimal polynomial equals the characteristic polynomial when $d(\lambda) = 1$. Let us illustrate this by a simple example.

Example

Given the matrix

$$A = \begin{bmatrix} 1 & a & 0 \\ 0 & 1 & 1 \\ 0 & 0 & 1 \end{bmatrix} \tag{129}$$

where a is some number to be assigned later, compute the minimal polynomial over the real field.

The characteristic polynomial $f(\lambda)$ is given by

$$f(\lambda) = (\lambda - 1)^3 \tag{130}$$

The conjoint of A, $B(\lambda) = \text{Adj}(\lambda I - A)$ is given by

$$B(\lambda) = \begin{bmatrix} (\lambda - 1)^2 & a(\lambda - 1) & a \\ 0 & (\lambda - 1)^2 & (\lambda - 1) \\ 0 & 0 & (\lambda - 1)^2 \end{bmatrix} \tag{131}$$

It is clear that we have two cases:

1. When $a = 0$, $d(\lambda) = (\lambda - 1)$, and therefore

$$m(\lambda) = \frac{f(\lambda)}{d(\lambda)} = (\lambda - 1)^2 \tag{132}$$

2. When $a \neq 0$, $d(\lambda) = 1$, and therefore

$$m(\lambda) = f(\lambda) \tag{133}$$

As a check, consider the minimal polynomial in (132), $m(\lambda) = \lambda^2 - 2\lambda + 1$, and its associated matrix polynomial

$$m(A) = A^2 - 2A + I \tag{134}$$

Using (129) in (134) for $a = 0$, we have

$$m(A) = \begin{bmatrix} 1 & 0 & 0 \\ 0 & 1 & 2 \\ 0 & 0 & 1 \end{bmatrix} - 2 \begin{bmatrix} 1 & 0 & 0 \\ 0 & 1 & 1 \\ 0 & 0 & 1 \end{bmatrix} + \begin{bmatrix} 1 & 0 & 0 \\ 0 & 1 & 0 \\ 0 & 0 & 1 \end{bmatrix} = 0$$

which validates the earlier assertion.

The above example illustrates the idea that if the multiple eigenvalues of A are not linked, that is, if a pair of equal eigenvalue Jordan blocks are not linked, then the minimal polynomial is of lower degree than the characteristic polynomial. Obviously, it is not necessary to find the Jordan form to find the minimal polynomial, except that it does illustrate the concepts. The problem with the above method is finding the GCD of all the elements in $B(\lambda)$. In the above example, we picked a simple example to illustrate the concepts. One can visualize that for large matrices finding the GCD for all $b_{ij}(\lambda)$ is a difficult task. Next we consider a method over the real field that does not require factoring and is based on the Gram–Schmidt orthogonalization process (Gelbaum, 1983). For complex field applications, only a few minor changes need be made.

The idea is based on first forming a row vector from the matrix A^k, $k = 0, 1, 2, \ldots, n$, in the form

$$\boldsymbol{\phi}^T(A^k) = [a_{11}^{(k)} \quad a_{12}^{(k)} \quad \cdots \quad a_{1n}^{(k)} \quad a_{21}^{(k)} \quad \cdots \quad a_{2n}^{(k)} \quad \cdots \quad a_{n1}^{(k)} \quad \cdots \quad a_{nn}^{(k)}]$$

$$(135)$$

where $a_{ij}^{(k)}$ corresponds to the ijth entry of A^k. Now we can treat these as vectors in an n^2-dimensional vector space. Apply the Gram–Schmidt orthogonalization to compute a set of b vectors that are independent and finally obtain a relation

$$\boldsymbol{\phi}^T(A^b) + \sum_{k=1}^{b} a_k \boldsymbol{\phi}^T(A^{b-k}) = 0 \qquad (136)$$

Let y_k^T be the normalized row vectors obtained from the Gram–Schmidt orthogonalization. To compute these, let

$$\mathbf{y}_0^T = \frac{\boldsymbol{\phi}^T(I)}{\sqrt{n}} \qquad (137)$$

and let

$$\mathbf{x}_{p+1}^T = \boldsymbol{\phi}^T(A^{p+1}) - \sum_{k=0}^{p} (\boldsymbol{\phi}^T(A^{p+1})\mathbf{y}_k)\mathbf{y}_k^T \qquad (138)$$

Furthermore, let

$$\mathbf{y}_{p+1}^T = \frac{\mathbf{x}_{p+1}^T}{\|\mathbf{x}_{p+1}\|} \qquad (139)$$

where $\|\mathbf{x}_{p+1}\|$ is the length of the vector. Since the process stops at the first power of A that is linearly dependent on the preceding powers of A, it follows that

$$\mathbf{x}_{p+1}^T = 0$$

gives the matrix polynomial that corresponds to the minimum polynomial. Let us illustrate this procedure by an example.

Example

Given the matrix A in (129) with $a = 0$, find the minimal polynomial using the above procedure.

From (129) and (135), we can write

$$\phi^T(A^0) = [1 \quad 0 \quad 0 \quad 0 \quad 1 \quad 0 \quad 0 \quad 0 \quad 1]$$

$$\phi^T(A^1) = [1 \quad 0 \quad 0 \quad 0 \quad 1 \quad 1 \quad 0 \quad 0 \quad 1]$$

$$\phi^T(A^2) = [1 \quad 0 \quad 0 \quad 0 \quad 1 \quad 2 \quad 0 \quad 0 \quad 1]$$

$$\phi^T(A^3) = [1 \quad 0 \quad 0 \quad 0 \quad 1 \quad 3 \quad 0 \quad 0 \quad 1]$$

Now using the procedure discussed above, we have

$$\mathbf{y}_0^T = \frac{1}{\sqrt{3}}[1 \quad 0 \quad 0 \quad 0 \quad 1 \quad 0 \quad 0 \quad 0 \quad 1]$$

$$\mathbf{x}_1^T = [1 \quad 0 \quad 0 \quad 0 \quad 1 \quad 1 \quad 0 \quad 0 \quad 1]$$
$$\qquad - \tfrac{3}{3}[1 \quad 0 \quad 0 \quad 0 \quad 1 \quad 0 \quad 0 \quad 0 \quad 1]$$

$$\mathbf{y}_1^T = [0 \quad 0 \quad 0 \quad 0 \quad 0 \quad 1 \quad 0 \quad 0 \quad 0]$$

$$\mathbf{x}_2^T = [1 \quad 0 \quad 0 \quad 0 \quad 1 \quad 2 \quad 0 \quad 0 \quad 1]$$
$$\qquad - \tfrac{3}{3}[1 \quad 0 \quad 0 \quad 0 \quad 1 \quad 0 \quad 0 \quad 0 \quad 1]$$
$$\qquad - 2[0 \quad 0 \quad 0 \quad 0 \quad 0 \quad 1 \quad 0 \quad 0 \quad 0]$$
$$\qquad = 0$$

Therefore, we have

$$\phi^T(A^2) - (\phi^T(A^2)\mathbf{y}_0)\mathbf{y}_0^T - (\phi^T(A^2)\mathbf{y}_1)\mathbf{y}_1^T = 0$$

where

$$\phi^T(A^2)\mathbf{y}_0 = \sqrt{3}, \qquad \phi^T(A^2)\mathbf{y}_1 = 2 \qquad \phi^T(A^1)\mathbf{y}_0 = \sqrt{3}$$

and

$$\mathbf{y}_0^T = \frac{1}{\sqrt{3}}\phi^T(A^0)$$

$$\|\mathbf{x}_1\|\mathbf{y}_1^T = \phi^T(A^1) - (\phi^T(A^1)\mathbf{y}_0)\mathbf{y}_0^T$$

$$= \phi^T(A^1) - \frac{\sqrt{3}}{\sqrt{3}}\phi^T(A^0)$$

$$\phi^T(A^2) = \phi^T(A^0) + 2(\phi^T(A^1) - \phi^T(A^0)) = 0$$

or

$$\phi^T(A^2) - 2\phi^T(A) + \phi^T(A^0) = 0$$

giving the minimal polynomial in terms of $\boldsymbol{\phi}^T(A^k)$. From this we can write the minimal polynomial as obtained before.

Finally, we want to point out that search procedures can be used to solve for the minimal polynomial. This can be achieved by investigating the dependency of vectors starting from $\boldsymbol{\phi}^T(A^0), \ldots, \boldsymbol{\phi}^T(A^{n-1})$. If they are independent, then the characteristic polynomial is equal to the minimal polynomial; if not, continue with $\boldsymbol{\phi}^T(A^0), \ldots, \boldsymbol{\phi}^T(A^{n-2})$. Once a state is reached where $\boldsymbol{\phi}^T(A^0), \ldots, \boldsymbol{\phi}^T(A^{b-1})$ are independent vectors, then we solve for the coefficients a_i from $\boldsymbol{\phi}^T(A^b) = -a_1\boldsymbol{\phi}^T(A^{b-1}) - \cdots - a_b$ from the previous step. This is not too bad a procedure since b, the order of the minimal polynomial, is usually close to n, the order of the characteristic polynomial. This method is particularly attractive for computer implementation for finite field representations.

Problems

1. Given the matrix A in (26), where the entries are assumed to be over the real field, find the minimal polynomial using the method discussed in this section.

2. Assuming that the entries in A in (26) are over GF(5), compute the minimal polynomial using the iterative technique discussed above.

6.10. Powers of Some Special Matrices

Earlier, we discussed powers of matrices and how to compute them. There are some matrices that can be characterized on the basis of powers of matrices. These have important applications in various areas and we briefly touch on some of these.

6.10.1. Idempotent Matrices

A matrix E over a field F is idempotent if $E = E^2$. An example of an idempotent matrix over any field is

$$E_1 = \begin{bmatrix} 1 & 0 \\ 0 & 1 \end{bmatrix}$$

Examples over the real field are

$$E_2 = \begin{bmatrix} -\frac{1}{2} & -\frac{1}{2} \\ \frac{3}{2} & \frac{3}{2} \end{bmatrix} \quad \text{and} \quad E_3 = \begin{bmatrix} \frac{3}{2} & \frac{1}{2} \\ -\frac{3}{2} & -\frac{1}{2} \end{bmatrix} \tag{140}$$

Idempotent matrices play an important role in functions of diagonable matrices. For example, the eigenvector decomposition of A in (22) and (23) can be written in terms of $\mathbf{x}_1$, $\mathbf{x}_2$ and $(\bar{\mathbf{x}}^{-1})^T = [\mathbf{y}_1 \quad \mathbf{y}_2]$:

$$\lambda_1(\mathbf{x}_1\mathbf{y}_1^T) + \lambda_2\mathbf{x}_2\mathbf{y}_2^T = \lambda_1 E_2 + \lambda_2 E_3$$

where E_2 and E_3 are given in (140) and are sometimes referred to as constituent idempotent matrices (Frame, 1964). The necessary and sufficient condition for E being an idempotent matrix is that it must satisfy one of the following conditions: (1) $E = 0$; (2) $E = I$; (3) E has a minimal polynomial $(\lambda^2 - \lambda)$. It is clear that most idempotents come under the last category because the first two are trivial cases.

6.10.2. Nilpotent Matrices

A nilpotent matrix N over a field F is a square matrix for which N^t, for some integer t, is the null matrix. An example of a nilpotent matrix is

$$N = \begin{bmatrix} 0 & 1 & 0 \\ 0 & 0 & 1 \\ 0 & 0 & 0 \end{bmatrix} \tag{141}$$

as $N^3 = 0$, a null matrix. This class of matrices is an interesting class over the complex field since for any Jordan block J_i corresponding to an eigenvalue λ_i, we have $(\lambda_i I - J_i)$, a nilpotent matrix.

6.10.3. Involutory Matrices

A matrix A_I over a field F is called an involutory matrix if $A_I^2 = I$, an identity matrix. An example of such a matrix over the real field is

$$A_I = \begin{bmatrix} \cos\theta & \sin\theta \\ \sin\theta & -\cos\theta \end{bmatrix} \tag{142}$$

where $0 \le \theta \le 2\pi$. From the definition, we see that A_I is its own inverse and we can classify it as a symmetric orthogonal matrix. These matrices were used in early algebraic cryptography work several decades ago. These matrices have important applications for matrices over the real and complex fields.

Generation of symmetric orthogonal matrices has been of interest in many areas of numerical analysis. For example, the matrix $A_s = (I - 2r\mathbf{v}\mathbf{v}^T)$ is a symmetric orthogonal matrix for a nonzero vector $\mathbf{v}$ with $\|\mathbf{v}\|^2 = 1/r$. These have been used in tridiagonalizing symmetric matrices. A matrix is tridiagonal if it has nonzero entries on the diagonal, subdiagonal, and superdiagonal and zeros everywhere else. Another matrix that is of interest is the unitary Hermitian matrix,

$$A_H = \begin{bmatrix} a & \mathbf{x}^* \\ \mathbf{x} & -I + \mathbf{x}\mathbf{x}^*/(1+a) \end{bmatrix} \tag{143}$$

where a is a nonnegative real number, and the vector

$$\mathbf{v}_1 = \begin{bmatrix} a \\ \mathbf{x} \end{bmatrix} \tag{144}$$

is a of length1, that is, $\|\mathbf{v}_1\| = 1$. When the vector $\mathbf{v}_1$ is real, A_H is a symmetric orthogonal matrix. Let us illustrate this.

Example

Given below the vector $\mathbf{v}_1$, where the entries are over the real field, find A_H in (143):

$$\mathbf{v}_1 = \frac{1}{\sqrt{2}} \begin{bmatrix} 1 \\ 1 \end{bmatrix} = \begin{bmatrix} a \\ \mathbf{x} \end{bmatrix}$$

First, from (143), we have $-I + (\mathbf{x}\mathbf{x}^*/(1+a)) = -1/\sqrt{2}$. Therefore,

$$A_H = \frac{1}{\sqrt{2}} \begin{bmatrix} 1 & 1 \\ 1 & -1 \end{bmatrix}$$

The matrices in (143) are widely used in transforming an arbitrary matrix to a triangular matrix over the complex field. For example, if $\mathbf{v}_1$ in (144) is the eigenvector of an $n \times n$ matrix B corresponding to an eigenvalue λ_1 and $A_H^{(1)}$ is generated from $\mathbf{v}_1$ via (143), then (Frame, 1964)

$$BA_H^{(1)} = A_H^{(1)} \begin{bmatrix} \lambda_1 & \mathbf{c}_{12}^T \\ 0 & B_1(n-1) \end{bmatrix} \tag{145}$$

where $\mathbf{c}_{12}^T$ is an $(n-1)$-dimensional row vector and $B_1(n-1)$ is an $(n-1) \times (n-1)$ matrix. Continuing with the procedure, we can see that the matrix B can be unitarily transformed to a triangular matrix. That is, let

$$S = A_H^{(1)} \cdots A_H^{(n-1)}$$

with

$$A_H^{(k+1)} = \begin{bmatrix} I_k & 0 \\ 0 & A_{Hk} \end{bmatrix}$$

where A_{Hk} is the unitary Hermitian matrix generated for $B_k(n-k)$ corresponding to the $(k+1)$th eigenvalue λ_{k+1}. From (145), it follows that

$$B = SJ_A S^*$$

where J_A is a triangular matrix, not necessarily a Jordan matrix, with all the eigenvalues of B located on the diagonal. A special case, but an important one, is when B is normal. Then S corresponds to a unitary matrix that transforms B into a diagonal matrix. That is, when B is normal, $J_A = D$, a diagonal matrix.

6.10.4. Roots of Identity Matrices

The mathematical implications are numerous when we talk about roots. Here we consider a simple definition. An $n \times n$ matrix A over a field F is a root of the identity matrix, I, if

$$A^n = I \tag{146}$$

for some n. Some examples were given earlier for $n = 2$. These matrices have important applications in signal analysis and finite field applications, such as coding, cryptography, and many other areas. Let us consider some examples first.

Example

The discrete Fourier transform matrix defined earlier over the complex field

$$A_{\text{DFT}} = (a_{ik})$$

where $a_{ik} = e^{-j(2\pi/N)ik}$, $i, k = 0, 1, \ldots, N-1$, satisfies the equation

$$\left(\frac{1}{\sqrt{N}} A_{\text{DFT}}\right)^4 = I$$

for any N. It is clear that this implies that the eigenvalues of $(1/\sqrt{N})A_{\text{DFT}}$ can only be from the set $1, -1, j, -j$.

For real and complex field applications, we can generate a matrix A in (146) with A being unitarily similar to a diagonal matrix. That is,

$$A = PDP^*$$

where D is a diagonal matrix with the diagonal entries d_{ii} satisfying the condition that $d_{ii}^n = 1$. For finite field matrices, we can not use eigenvector decomposition. However, we can use the Cayley–Hamilton theorem. Let us consider an example first.

Example

Given the matrix

$$A = \begin{bmatrix} 1 & 0 & 1 \\ 1 & 0 & 0 \\ 0 & 1 & 0 \end{bmatrix}$$

with entries over GF(2), we have the characteristic polynomial

$$f(\lambda) = \lambda^3 + \lambda^2 + 1 \tag{147a}$$

By the Cayley–Hamilton theorem, we have

$$A^3 = A^2 + I \tag{147b}$$

It is clear that since $|A| = 1$, A is nonsingular. Successively computing A^k, we have

$$A^2 = \begin{bmatrix} 1 & 1 & 1 \\ 1 & 0 & 1 \\ 1 & 0 & 0 \end{bmatrix}, \quad A^3 = \begin{bmatrix} 0 & 1 & 1 \\ 1 & 1 & 1 \\ 1 & 0 & 1 \end{bmatrix}, \quad A^4 = \begin{bmatrix} 1 & 1 & 0 \\ 0 & 1 & 1 \\ 1 & 1 & 1 \end{bmatrix}$$
$$A^5 = \begin{bmatrix} 0 & 0 & 1 \\ 1 & 1 & 0 \\ 0 & 1 & 1 \end{bmatrix}, \quad A^6 = \begin{bmatrix} 0 & 1 & 0 \\ 0 & 0 & 1 \\ 1 & 1 & 0 \end{bmatrix}, \quad A^7 = \begin{bmatrix} 1 & 0 & 0 \\ 0 & 1 & 0 \\ 0 & 0 & 1 \end{bmatrix} \tag{148}$$

indicating that 7th power of A is the identity matrix. The inverse of A can be computed from (147b):

$$A(A + A^2) = I \tag{149}$$

and therefore

$$A^{-1} = A + A^2 = \begin{bmatrix} 0 & 1 & 0 \\ 0 & 0 & 1 \\ 1 & 1 & 0 \end{bmatrix} = A^6 \tag{150}$$

indicating that the inverse of a finite field matrix is a power of A when the inverse of A exists. Let us consider another example before we consider the justification for the above statement.

Example

Given the matrix

$$A = \begin{bmatrix} 1 & 2 \\ 1 & 0 \end{bmatrix}$$

with entries over GF(3), the characteristic polynomial is

$$f(\lambda) = \lambda^2 + 2\lambda + 1$$

By the Cayley–Hamilton theorem, we have

$$A^2 = A + 2I$$

Successively computing A^k, we have

$$A^2 = \begin{bmatrix} 0 & 2 \\ 1 & 2 \end{bmatrix}, \quad A^3 = \begin{bmatrix} 2 & 0 \\ 0 & 2 \end{bmatrix}, \quad A^4 = \begin{bmatrix} 2 & 1 \\ 2 & 0 \end{bmatrix}$$

$$A^5 = \begin{bmatrix} 0 & 1 \\ 2 & 1 \end{bmatrix}, \quad A^6 = \begin{bmatrix} 1 & 0 \\ 0 & 1 \end{bmatrix} \tag{151}$$

and

$$A^{-1} = A^5 \tag{152}$$

Are the results in (148), (150), (151), and (152) unusual? The answer is no. First, we are considering this discussion only finite field nonsingular matrices. Second, there are only a finite number of nth-order matrices that are distinct in a given finite field. For example, there are only $2^4 = 16$ possible matrices of dimension 2×2 over GF(2). Third, A^k in the above example is a nonsingular matrix for any k. Therefore, there exists an integer m such that $A^m = A$. That is, the matrices will start repeating. Since A is nonsingular, it follows that

$$A^t = I, \quad \text{with } t = m - 1 \tag{153}$$

or

$$A^{t-1} = A^{-1} \tag{154}$$

thus establishing the assertion.

Next, we need to ask, how do we find the value of t? The direct method of course is to use the Cayley-Hamilton theorem for functions starting with λ^{n+1} until we get $\lambda^t + 1 = 0$. For the polynomial in (147a), we have $\lambda^3 = \lambda^2 + 1$, $\lambda^4 = \lambda^3 + \lambda$, $\lambda^5 = \lambda + 1$, $\lambda^6 = \lambda^2 + \lambda$, and $\lambda^7 = 1$, thus establishing the fact. From this we can see that finding t in (153) is rather cumbersome. Finding t for arbitrary finite fields is a very difficult problem, sometimes referred to as the discrete logarithmic problem. For GF(2), if the characteristic polynomial is primitive, then the smallest integer that satisfies (153) is $t = 2^n - 1$. We discuss some of these concepts in a later chapter.

6.11. Matrix Norms

It is useful to describe a vector or a matrix by a number which gives an overall description of the size of a vector or a matrix. The concept is similar to the one used to describe the modulus of a complex number. For vectors and matrices, certain functions of the elements, called norms, are used to describe the size. For a good reference on this topic, see Wilkinson (1965). It is clear that matrix norms are defined for matrices over real or complex fields. Therefore, the discussion here is for these fields.

In an earlier chapter we discussed the length of a vector. Let us formalize this and extend it so that we can use it for matrices.

The norm of a vector $\mathbf{x}$ is usually denoted by $\|\mathbf{x}\|$ and it satisfies the relations (1) $\|\mathbf{x}\| > 0$ unless $\mathbf{x} = 0$; (2) $\|k\mathbf{x}\| = |k|\,\|\mathbf{x}\|$ for any complex scalar k; and (3) $\|\mathbf{x} + \mathbf{y}\| \leq \|\mathbf{x}\| + \|\mathbf{y}\|$. Normally a subscript p on $\|\mathbf{x}\|$, $\|\mathbf{x}\|_p$ denotes

$$\|\mathbf{x}\|_p = [|x_1|^p + |x_2|^p + \cdots + |x_n|^p]^{1/p} \tag{155}$$

where we assumed that $\mathbf{x}$ is a column vector (or a row vector) with entries $x_1, x_2, \ldots, x_n$. Normally p takes the values of 1, 2, or ∞. Other values of p can be considered also. For $p = \infty$,

$$\|\mathbf{x}\|_\infty = \max |x_i| \tag{156}$$

Similar to a vector norm, we can also define a matrix norm $\|A\|$ for a matrix A. This is defined by

$$\|A\| = \max_{\mathbf{x} \neq 0} \frac{\|A\mathbf{x}\|}{\|\mathbf{x}\|} \tag{157}$$

This is called the spectral norm or the bound norm and is the smallest norm that satisfy the inequality

$$\|A\mathbf{x}\| \leq \|A\|\,\|\mathbf{x}\| \tag{158}$$

for all $\mathbf{x}$. The $\|A\|_p$ norms are given by

$$\|A\|_1 = \max_j \sum_i |a_{ij}| \tag{159}$$

$$\|A\|_\infty = \max_i \sum_j |a_{ij}| \tag{160}$$

$$\|A\|_2 = (\text{maximum eigenvalue of } A^*A)^{1/2} \tag{161}$$

Another norm which satisfies (158) and is a simple function of the matrix element is the Schur or Euclidean norm:

$$\|A\|_E = \left(\sum_{i,j} |a_{ij}|^2 \right)^{1/2} \tag{162}$$

The spectral radius is defined by

$$\rho(A) = \max \left[|\text{eigenvalue of } A| \right]$$

and it can be shown that

$$\rho(A) \le \|A\| \tag{163}$$

The norms are used in many applications wherein some iterative methods are used in solving matrix equations. These allow us to validate a method or actually use a parameter, based on a norm, in an algorithm. Finding the spectral radius is particularly important in many applications.

6.11.1. A Simple Iterative Method for Finding $\rho(A)$

It is clear that we are interested in only one eigenvalue—the one with largest magnitude— and perhaps the corresponding eigenvector. For this case, it is unnecessary to go through the complete eigenvalue-eigenvector analysis. In the following, we discuss a simple iterative method, called the power method (Wilkinson, 1965) that gives good results.

We assume first that the eigenvalues of the $n \times n$ matrix A, λ_i, $i = 1, 2, \ldots, n$, can be ordered in the form

$$|\lambda_1| > |\lambda_2| \ge |\lambda_3| > \cdots \ge |\lambda_n| \tag{164}$$

We are obviously assuming that there is a unique eigenvalue of maximum magnitude. Let $\mathbf{x}_0$ be an arbitrary nonzero vector and let the sequence of vectors $\mathbf{y}_i$ and $\mathbf{x}_j$ be defined by

$$\mathbf{y}_{j+1} = A\mathbf{x}_j \tag{165}$$

with

$$\mathbf{x}_{j+1} = \frac{\mathbf{y}_{j+1}}{\max (\mathbf{y}_{j+1})} \tag{166}$$

where max $(\mathbf{y}_{j+1})$ is the element in $\mathbf{y}_{j+1}$ that has the maximum magnitude. It can be shown that under some mild conditions $\mathbf{x}_j$ approaches the eigenvector $\mathbf{e}_1$ corresponding to the eigenvalue λ_1 and max (y_j) approaches λ_1. Let us illustrate these concepts by a simple example.

Example

Given

$$A = \begin{bmatrix} 0 & 1 \\ -2 & -3 \end{bmatrix}$$

find $\rho(A)$ and the corresponding eigenvector.

Assuming

$$\mathbf{x}_0 = \begin{bmatrix} 1 \\ 0 \end{bmatrix}$$

we can successively compute $\mathbf{y}_{j+1}$, $\mathbf{x}_j$, and max $(\mathbf{y}_{j+1})$. The vectors $\mathbf{x}_j$ and the corresponding max $(\mathbf{y}_{j+1})$ are listed in Table 1 for $j = 0, 1, \ldots, 7$. From the table it is clear that max $(\mathbf{y}_{j+1}) \to \lambda_1 = -2$, and $\mathbf{x}_j$ approaches the corresponding eigenvector

$$\mathbf{e}_1 = \begin{bmatrix} -0.5 \\ 1 \end{bmatrix}$$

Therefore, $\rho(A) = 2$.

The rate of convergence depends on the ratio $|\lambda_1/\lambda_2|$. When the ratio is close to 1, the convergence is slow. If λ_1 is a multiple eigenvalue with several independent eigenvectors, say $\mathbf{e}_1, \mathbf{e}_2, \ldots, \mathbf{e}_r$, then the iterative process still converges and the resultant eigenvector lies in the subspace spanned by the eigenvectors $\mathbf{e}_1, \mathbf{e}_2, \ldots, \mathbf{e}_r$. Earlier we mentioned that under some mild conditions the solution converges. These conditions can be overcome by selecting different starting vectors.

Table 1. Vectors $\mathbf{x}_j$ and Corresponding Max $(\mathbf{y}_{j+1})$

	$\mathbf{x}_0$	$\mathbf{x}_1$	$\mathbf{x}_2$	$\mathbf{x}_3$	$\mathbf{x}_4$	$\mathbf{x}_5$	$\mathbf{x}_6$	$\mathbf{x}_7$
	1	0	$-\frac{1}{3}$	-0.4285	-0.467	-0.483	-0.492	-0.496
	0	1	1	1	1	1	1	1
max $(\mathbf{y}_{j+1})$	-2	-3	-2	-2.143	-2.06	-2.03	-2.01	-2.007

Problem

Given the matrix A below over the complex field (a) compute the norms in (159)–(162) and (b) compute the spectral radius and the corresponding eigenvector:

$$A = \begin{bmatrix} 0 & 1 \\ -1 & -1 \end{bmatrix}$$

In this chapter we discussed various concepts associated with matrix representations. In Chapter 7 we use some of these results.

References

Dahlquist, G. and A. Bjorck (1969), *Numerical Methods*, Prentice-Hall, Englewood Cliffs, N.J.

DeRusso, P. M., R. J. Roy, and C. M. Close (1965), *State Variables for Engineers*, Wiley, New York.

Frame, J. S. (1964), Matrix Functions and Applications, *IEEE Spectrum*, Part I, March, 208–220; Part II, April, 102–108; Part III, May, 100–109; Part IV, June, 123–131; Part V, July, 103–109.

Gantmacher, F. R. (1960), *Theory of Matrices*, Vols. I and II, Chelsea Publishing, New York.

Gelbaum, B. R. (1983), An Algorithm for the Minimal Polynomial of a Matrix, *Amer. Math. Monthly*, Vol. 90, 43–44.

Gray, R. M. (1972), On the Asymptotic Eigenvalue Distribution of Toeplitz Matrices, *IEEE Transactions on Information Theory*, Vol. IT-18, 725–730.

Ledbetter, J. and R. Yarlagadda (1974), A Tridiagonal Approach to Digital Filter Synthesis, *IEEE Transactions on Circuit Theory*, Vol. CT-20, 322–324.

Ogata, K. (1967), *State Space Analysis of Control Systems*, Prentice-Hall, Englewood Cliffs, NJ.

Perlis, S. (1952), *Theory of Matrices*, Addison-Wesley, Reading, MA.

Wilkinson, J. H. (1965), *The Algebraic Eigenvalue Problem*, Clarendon Press, Oxford.

Wilkinson, J. H. and C. Reinsch (1971), *Handbook for Automatic Computation*, Vol. 2, *Linear Algebra*, Springer, Berlin.

Yarlagadda, R., and B. N. Suresh Babu (1980), A Note on the Application of FFT to the Solution of a System of Toeplitz Normal Equations, *IEEE Transactions on Circuits and Systems*, Vol. CAS-27, 151–154.

Applications of Matrices to Discrete Data System Analysis

7.1. Introduction

In the last three chapters we discussed various aspects of matrix theory. In this chapter we apply some of these concepts to discrete data systems. In the process we show that matrix analysis is a logical tool to use for solving discrete data problems. The major topics of this chapter are convolution, deconvolution, and difference equations. In an earlier chapter we pointed out that matrix transformations can be visualized as converting one set of numbers to another set of numbers. For discrete data systems, we need to identify the order of these numbers, that is, the time these numbers appear before we process. In the next section we discuss discrete signals in terms of some special functions.

7.2. Discrete Systems

In discrete system analysis, we are concerned about processing signals that are represented by sequences. A sequence of numbers x is usually represented by

$$x = \{x(n)\}, \qquad -\infty < n < \infty \tag{1}$$

where $x(n)$ is the nth number in the sequence. It is convenient to write $x(n) = x_n$ and we use this notation when there is no notational confusion. A simple, but important, sequence is the unit-sample sequence, $\delta(n)$, defined by

$$\delta(n) = \begin{cases} 0, & n \neq 0 \\ 1, & n = 0 \end{cases} \tag{2}$$

Some authors use terms such as a delta function, a discrete impulse, or simply an impulse for $\delta(n)$.

In the analysis of discrete systems, sequences are manipulated by the following simple rules.

1. The product of two sequences x and y is defined as

$$x \cdot y = \{x(n)y(n)\} \tag{3}$$

2. The sum of two sequences x and y is defined as

$$x + y = \{x(n) + y(n)\} \tag{4}$$

3. Multiplication of a sequence x by a constant α is defined by

$$\alpha \cdot x = \{\alpha x(n)\} \tag{5}$$

4. A sequence y is a delayed or shifted version of a sequence x if y has values

$$y(n) = x(n - n_0) \tag{6}$$

where n_0 is an integer.

A simple example of a shifted sequence is the delayed unit-sample sequence,

$$\delta(n - n_0) = \begin{cases} 0, & n \neq n_0 \\ 1, & n = n_0 \end{cases} \tag{7}$$

With these simple rules, we can express any arbitrary sequence x by

$$x(n) = \sum_{k=-\infty}^{\infty} x(k)\delta(n - k) \tag{8}$$

Example

Given that

$$x(-1) = 1, \qquad x(0) = 1, \qquad x(1) = 2$$

and the remaining $x(n)$ are zero, express the sequence in the form of (8).

$$x(n) = 1\delta(n + 1) + 1\delta(n) + 2\delta(n - 1) \tag{9}$$

Next we consider the important concept of linear shift variance.

7.2.1. Linear Shift-Invariant Systems

In our approach, we assume that a system is defined as a unique transformation that maps an input sequence $x(n)$ into an output sequence $y(n)$. This is denoted by

$$y(n) = T[x(n)] \tag{10}$$

Classes of discrete systems are defined by putting constraints on the transformation $T[\ ]$. There are three concepts that are of interest to us: linearity, shift invariance, and causality.

Linearity: Linearity is defined by using the principle of superposition. Given the responses $y_1(n)$ and $y_2(n)$ corresponding to the respective inputs $x_1(n)$ and $x_2(n)$, a system is linear if and only if

$$T[ax_1(n) + bx_2(n)] = aT[x_1(n)] + bT[x_2(n)]$$
$$= ay_1(n) + by_2(n) \tag{11}$$

for arbitrary constants a and b.

Example

The delay operation

$$r(n) = T[x(n)] = x(n - n_0)$$

is a linear operation since

$$T[a_1 x_1(n) + a_2 x_2(n)] = a_1 x_1(n - n_0) + a_2 x_2(n - n_0)$$

Example

A system characterized by

$$r(n) = T[x(n)] = nx(n - n_0)$$

is linear because

$$T[a_1 x_1(n) + a_2 x_2(n)] = a_1 n x_1(n - n_0) + a_2 n x_2(n - n_0)$$

Example

A system characterized by

$$r(n) = T[x(n)] = x^2(n)$$

is nonlinear because

$$T[a_1x_1(n) + a_2x_2(n)] = (a_1x_1(n) + a_2x_2(n))^2$$
$$\neq a_1^2x_1^2(n) + a_2^2x_2^2(n)$$

Shift Invariance: It is characterized by the property that if $y(n)$ is the response to the input $x(n)$, then $y(n - k)$ is the response to $x(n - k)$.

Example

A system characterized by

$$r(n) = T[x(n)] = x^2(n)$$

is shift invariant because

$$T[x(n - k)] = x^2(n - k)$$

Example

A system characterized by

$$r(n) = nx(n - n_0)$$

is shift variant because if the input is $x(n - k)$, then the response is $nx(n - k - n_0)$ and not $(n - k)x(n - k - n_0)$.

Causality: A causal system is a system for which the output for any time n_0 depends on the inputs for $n \leq n_0$ only. That is, the response does not depend on the future inputs but relies only on the present and past inputs. It is clear that this is an important constraint because physically realizable systems can not predict what will happen in the future.

7.2.2. Characterization of a Linear System

The representation of an arbitrary signal given in (8) allows us to examine the fact that a linear system can be characterized by its unit-sample response. If $h_k(n)$ is the unit-sample response of a linear system for the input $\delta(n - k)$, that is,

$$h_k(n) = T[\delta(n - k)] \tag{12}$$

then the response for the input $x(n)$ is given by

$$r(n) = T[x(n)] = T\left[\sum_{k=-\infty}^{\infty} x(k)\delta(n-k)\right]$$

$$= \sum_{k=-\infty}^{\infty} x(k)T[\delta(n-k)]$$

$$= \sum_{k=-\infty}^{\infty} x(k)h_k(n) \tag{13}$$

In addition to the linearity, let us add the additional constraint of shift invariance. For a shift-invariant system, we know that if $h(n)$ is the unit-sample response, then $h(n-k)$ is the response to the input $\delta(n-k)$. Therefore, for a shift-invariant system, we can write

$$h_k(n) = h(n-k) \tag{14}$$

Using this in (13), we have

$$r(n) = \sum_{k=-\infty}^{\infty} x(k)h(n-k) \tag{15}$$

or

$$r(n) = \sum_{m=-\infty}^{\infty} h(m)x(n-m) \tag{16}$$

where the last equation can be derived from (15) by defining $m = n - k$. The equations in (15) and (16) can be symbolically represented by

$$r(n) = x(n) * h(n) = h(n) * x(n) \tag{17}$$

which is called the convolution and the symbol $*$ represents convolution.

Since our aim in this book is not digital signal processing, we do not consider any more theory associated with these systems. However, we would like to illustrate matrix concepts as an application tool. In the next section, we consider a matrix approach to the computation of the discrete convolution.

Problem

Test the following equations for linearity and shift invariance:

(a) $x(n+1) + 2x(n) = y(n)$.
(b) $x(n+1) + nx(n) = y(n)$.
(c) $x(n+1)x(n) = y(n)$.

7.3. Discrete Convolution

In engineering two of the most important and widely used concepts are convolution and correlation. Before we go into the details, let us identify the notations we plan to use in the following discussion. Let Figure 7.1 represent a linear shift-invariant system and let the input and the corresponding output of this system be represented by $x(n)$ and $r(n)$, respectively. If $x(n) = \delta(n)$, a unit sample, then we assume $r(n) = h(n)$, the unit-sample response. For ease of notation, we use

$$x_n = x(n), \qquad r_n = r(n), \qquad \delta_n = \delta(n), \qquad h_n = h(n) \tag{18}$$

in the following. Now from (8), (15), and (16), we can write

$$x_n = \sum_{k=-\infty}^{\infty} x_k \delta_{n-k} \tag{19}$$

$$r_n = \sum_{k=-\infty}^{\infty} x_k h_{n-k}$$

$$= \sum_{k=-\infty}^{\infty} h_k x_{n-k} \tag{20}$$

For physical systems, we know that $h_n = 0$, $n < 0$, that is, they are causal, and we assume for practical reasons the input x_n and the unit-sample response are time-limited functions. Therefore, in this section, we assume that

$$x_n = 0, \quad \text{for } n < 0 \text{ and for } n \geq L \tag{21}$$

$$h_n = 0, \quad \text{for } n < 0 \text{ and for } n \geq N \tag{22}$$

For these, we can write (20) in the form

$$r_n = \sum_{k=0}^{L-1} x_k h_{n-k} \tag{23a}$$

$$= \sum_{k=0}^{N-1} h_k x_{n-k} \tag{23b}$$

δ(n)   h(n)

x(n)   r(n)   Figure 7.1. A discrete system.

or, symbolically,

$$r_n = x_n * h_n = h_n * x_n \tag{23c}$$

Let us illustrate these ideas by a simple example.

Example

Given

$$x_0 = 1, \quad x_1 = 1, \quad \text{and} \quad x_n = 0, \quad \text{for } n < 0 \text{ and } n > 1$$
$$h_0 = 1, \quad h_1 = 2, \quad \text{and} \quad h_n = 0, \quad \text{for } n < 0 \text{ and } n > 1$$

find r_n using (23).

From (23a) we have

$$\begin{aligned}
r_n &= 0, \qquad n < 0 \\
r_0 &= x_0 h_0 + x_1 h_{-1} = x_0 h_0 \\
r_1 &= x_0 h_1 + x_1 h_0 \\
r_2 &= x_0 h_2 + x_1 h_1 = x_1 h_1 \\
r_n &= 0, \qquad n > 2
\end{aligned} \tag{24}$$

Note that (23b) gives the same result. The equations in (24) for r_n, $0 \leq n \leq 2$, can be written in a matrix form:

$$\begin{bmatrix} r_0 \\ r_1 \\ r_2 \end{bmatrix} = \begin{bmatrix} h_0 & 0 \\ h_1 & h_0 \\ 0 & h_1 \end{bmatrix} \begin{bmatrix} x_0 \\ x_1 \end{bmatrix} \tag{25}$$

If we assume that the data x_i and the unit-sample responses are over the real field, then

$$\begin{bmatrix} r_0 \\ r_1 \\ r_2 \end{bmatrix} = \begin{bmatrix} 1 & 0 \\ 2 & 1 \\ 0 & 2 \end{bmatrix} \begin{bmatrix} 1 \\ 1 \end{bmatrix} = \begin{bmatrix} 1 \\ 3 \\ 2 \end{bmatrix} \tag{26}$$

If the entries are over GF(3), we have

$$\begin{bmatrix} r_0 \\ r_1 \\ r_2 \end{bmatrix} = \begin{bmatrix} 1 \\ 0 \\ 2 \end{bmatrix} \tag{27}$$

From this example, we see that the response of a discrete linear system can be obtained by matrix multiplication. Let us now generalize these results. Using (23a), we can write

$$
\begin{bmatrix}
r_0 \\
r_1 \\
r_2 \\
\cdot \\
\cdot \\
\cdot \\
r_{N-1} \\
r_N \\
\cdot \\
\cdot \\
\cdot \\
r_{N+L-2}
\end{bmatrix}
=
\begin{bmatrix}
h_0 & 0 & \cdot & \cdot & \cdot & 0 \\
h_1 & h_0 & 0 & \cdot & \cdot & 0 \\
h_2 & h_1 & h_0 & \cdot & \cdot & 0 \\
\cdot & \cdot & \cdot & & & \\
\cdot & \cdot & \cdot & & & \\
\cdot & \cdot & \cdot & & & \\
h_{N-1} & h_{N-2} & h_{N-3} & \cdot & \cdot & \cdot \\
0 & h_{N-1} & h_{N-2} & \cdot & \cdot & \cdot \\
\cdot & 0 & h_{N-1} & \cdot & \cdot & \cdot \\
\cdot & \cdot & \cdot & \cdot & \cdot & \\
\cdot & \cdot & \cdot & \cdot & \cdot & h_{N-2} \\
0 & 0 & \cdot & \cdot & 0 & h_{N-1}
\end{bmatrix}
\begin{bmatrix}
x_0 \\
x_1 \\
\cdot \\
\cdot \\
\cdot \\
x_{L-1}
\end{bmatrix}
\qquad (28)
$$

or, in symbolic form,

$$\mathbf{r} = H\mathbf{x} \qquad (29)$$

where $\mathbf{x}$ and $\mathbf{r}$ are column vectors of dimension L and $N + L - 1$, respectively, and H is a rectangular matrix of dimension $(N + L - 1) \times L$. It is clear from this and the above example that the length of the response of a convolution of two sequences of length L and N is $(N + L - 1)$. The coefficient matrix H in (29) has an interesting pattern. The jth column in H is given by

$(j - 1)$ zeros

N coefficients written in the order of $h_0, h_1, \ldots, h_{N-1}$

$(L - j)$ zeros

where $1 \le j \le L$.

The matrix form in (28) is obtained by using (23a). Clearly, we can use (23b) to write the convolution in matrix form

$$\mathbf{r} = X\mathbf{h} \qquad (30)$$

where **h** and **r** are column vectors of dimension N and $(N + L - 1)$, respectively, and X is a rectangular matrix of dimension $(N + L - 1) \times N$, where the jth column in X is given by

$(j - 1)$ zeros

L points of data written in the order $x_0, x_1, \ldots, x_{L-1}$ (31)

$(N - j)$ zeros

where $1 \leq j \leq N$ and the vector $\mathbf{h} = \text{col}\,(h_0, h_1, \ldots, h_{N-1})$.

7.3.1. Discrete Correlation

The discrete correlation of the two signals x_n and h_n given earlier is defined by

$$y_n = \sum_{m=0}^{N-1} h_m x_{n+m} = \sum_{m=0}^{L-1} x_m h_{n+m} \qquad (32)$$

and is symbolically represented by

$$y_n = x_n \circ h_n = h_n \circ x_n \qquad (33)$$

where $\circ$ represents correlation. The value n represents the shift of the second signal with respect to the first. The concept of correlation is widely used in communication theory. For example, the autocorrelation of a periodic signal x_n, $y_n = x_n \circ x_n$, will have its largest amplitude when n is a multiple of the period of the signal. We illustrate this with a periodic sequence example.

Example

Compute the autocorrelation of the following periodic binary sequence with period 7. One period of this sequence is given by

$$x = \{-1, 1, -1, -1, -1, 1, 1\}$$

Note that $x_{n+7m} = x_n$ for any integer m. The correlation expression is given by

$$y_n = \sum_{k=0}^{6} x_k x_{k+n} \qquad (34)$$

Since the sequences x_n and y_n are periodic with the same period, we can compute y_n for $0 \leqslant n \leqslant 6$.

$$
\begin{bmatrix} y_0 \\ y_1 \\ y_2 \\ y_3 \\ y_4 \\ y_5 \\ y_6 \end{bmatrix} = \begin{bmatrix} x_0 & x_1 & x_2 & x_3 & x_4 & x_5 & x_6 \\ x_6 & x_0 & x_1 & x_2 & x_3 & x_4 & x_5 \\ x_5 & x_6 & x_0 & x_1 & x_2 & x_3 & x_4 \\ x_4 & x_5 & x_6 & x_0 & x_1 & x_2 & x_3 \\ x_3 & x_4 & x_5 & x_6 & x_0 & x_1 & x_2 \\ x_2 & x_3 & x_4 & x_5 & x_6 & x_0 & x_1 \\ x_1 & x_2 & x_3 & x_4 & x_5 & x_6 & x_0 \end{bmatrix} \begin{bmatrix} x_0 \\ x_1 \\ x_2 \\ x_3 \\ x_4 \\ x_5 \\ x_6 \end{bmatrix} \tag{35}
$$

where the coefficient matrix is a circulant as discussed in Chapter 6. Substituting the appropriate values for x_n, we have

$$
y_0 = 7, \quad y_1 = -1, \quad y_2 = -1, \quad y_3 = -1, \quad y_4 = -1, \quad y_5 = -1, \quad y_6 = -1
$$

indicating the maximum value when the sequences match. These types of approach are used in synchronization schemes. Binary periodic sequences play an important role in communication theory. We discuss binary sequences in Chapter 8. Now let us consider a matrix representation of (32):

$$
\begin{bmatrix} y_{-(N-1)} \\ y_{-(N-2)} \\ \cdot \\ \cdot \\ \cdot \\ y_0 \\ y_1 \\ \cdot \\ \cdot \\ \cdot \\ y_{(L-1)} \end{bmatrix} = \begin{bmatrix} h_{N-1} & 0 & \cdot & \cdot & \cdot & 0 \\ h_{N-2} & h_{N-1} & \cdot & \cdot & \cdot & 0 \\ \cdot & \cdot & \cdot & \cdot & \cdot & \vdots \\ \cdot & \cdot & \cdot & \cdot & \cdot & h_{N-1} \\ \cdot & \cdot & \cdot & \cdot & \cdot & h_{N-2} \\ h_0 & h_1 & \cdot & \cdot & \cdot & \cdot \\ 0 & h_0 & \cdot & \cdot & \cdot & \cdot \\ \cdot & 0 & \cdot & \cdot & \cdot & \cdot \\ \cdot & \cdot & \cdot & \cdot & \cdot & \cdot \\ \cdot & \cdot & \cdot & \cdot & \cdot & \cdot \\ 0 & 0 & \cdot & \cdot & \cdot & h_0 \end{bmatrix} \begin{bmatrix} x_0 \\ x_1 \\ \cdot \\ \cdot \\ x_{(L-1)} \end{bmatrix} \tag{36}
$$

or, in symbolic form,

$$
\mathbf{y} = H_1 \mathbf{x} \tag{37}
$$

It is clear that (28) and (36) have the same general form and, for all practical purposes, the computational procedure is identical.

Next we consider the computational aspects of convolutions using matrix transformations.

7.3.2. FFT Computation of Convolutions

Direct computation of convolutions and correlations over the real field is computationally intensive. For example, the matrix multiplication in (28) takes approximately on the order of $(N + L - 1)(L)$ multiplications and additions, that is,

$$N_1 = O[(N + L - 1)(L)] \tag{38}$$

multiplications and/or additions. It is clear that when N and L are large, this number could also be large. Some reductions can be achieved by noting the zeros in the matrix H in (29). A common approach is to use fast Fourier transforms to compute convolutions. The clue is to use the relation between the circulants and the discrete Fourier transforms discussed in Chapter 6.

The idea is to rewrite (29) in the form

$$\mathbf{r}_a = H_c \mathbf{x}_a \tag{39}$$

where $\mathbf{r}_a$ and $\mathbf{x}_a$ are augmented vectors obtained from

$$\mathbf{r}_a^T = [\mathbf{r}^T \quad \mathbf{0}_{M-(N+L-1)}^T] \tag{40}$$

and

$$\mathbf{x}_a^T = [\mathbf{x}^T \quad \mathbf{0}_{(M-L)}^T] \tag{41}$$

where $\mathbf{0}_b^T$ is a null vector of dimension b. The matrix H_c is a circulant matrix generated from H by using the following procedure. We pointed out in Chapter 6 that the first column in a circulant matrix defines the circulant. The first column in H_c is obtained from the first column of H by augmenting it by $M - (N + L - 1)$ zeros. That is, the first column of H_c, say $\mathbf{h}_a$, is

$$\mathbf{h}_a = \begin{bmatrix} \text{First column of } H \\ \text{Column of } M - (N + L - 1) \text{ zeros} \end{bmatrix}$$

where M is taken as a power of 2 for efficient FFT implementation. Now

(39) can be written explicitly as

$$
\begin{bmatrix}
r_0 \\
r_1 \\
\cdot \\
\cdot \\
\cdot \\
r_{N-1} \\
r_N \\
\cdot \\
\cdot \\
r_{N+L-2} \\
0 \\
\cdot \\
\cdot \\
0
\end{bmatrix}
=
\begin{bmatrix}
h_0 & 0 & \cdot & \cdot & \cdot & 0 & \cdot & \cdot & 0 & h_{N-1} & \cdot & \cdot & h_1 \\
h_1 & h_0 & & & & 0 & & & 0 & 0 & & & \\
 & \cdot & & & & & & & & & & & h_{N-1} \\
 & & & & h_0 & & & & & & & & \\
h_{N-1} & & & & & & & & & & & & \\
0 & & h_{N-1} & & & & & & & & & & \\
\cdot & & & & & & & & & & & & \\
\cdot & & & & & & & & & & & & \\
\cdot & \cdot & \cdot & \cdot & h_{N-1} & & & & & & & & \\
 & \cdot & & & 0 & & & & & & & & \\
 & \cdot & & & & & & & & & & & \\
 & \cdot & & & & & & & & & & & \\
0 & & 0 & 0 & 0 & \cdot & 0 & h_{N-1} & h_{N-2} & \cdot & \cdot & h_0
\end{bmatrix}
\begin{bmatrix}
x_0 \\
x_1 \\
\cdot \\
\cdot \\
x_{L-1} \\
0 \\
0 \\
\cdot \\
\cdot \\
\cdot \\
\cdot \\
\vdots \\
0
\end{bmatrix}
\tag{42}
$$

which is sometimes referred to as cyclic or period convolution because the extended sequence vectors $\mathbf{r}_a$, $\mathbf{h}_a$, and $\mathbf{x}_a$ can be considered as vectors obtained from periodic sequences with period M.

Since the circulant H_c can be expressed by (see (79) in Chapter 6)

$$
H_c = \frac{1}{M} A_{\text{DFT}}^* \Lambda A_{\text{DFT}} \tag{43}
$$

we can write (39) in the form

$$
(A_{\text{DFT}} \mathbf{r}_a) = \Lambda A_{\text{DFT}} \mathbf{x}_a \tag{44}
$$

Convolution via FFT consists of computing Λ, $A_{\text{DFT}} \mathbf{x}_a$, and taking the inverse DFT of $(\Lambda A_{\text{DFT}} \mathbf{x}_a)$. The last $(M - (N + L - 1))$ values in (42) are 0 and can be discarded. Similar steps can be taken for the correlation.

Let us illustrate some of these ideas using the example in (26). Writing (26) in the form of (42), we have

$$
\begin{bmatrix}
r_0 \\
r_1 \\
r_2 \\
r_3
\end{bmatrix}
=
\begin{bmatrix}
1 & 0 & 0 & 2 \\
2 & 1 & 0 & 0 \\
0 & 2 & 1 & 0 \\
0 & 0 & 2 & 1
\end{bmatrix}
\begin{bmatrix}
1 \\
1 \\
0 \\
0
\end{bmatrix}
\tag{45}
$$

$$
\mathbf{r}_a \quad = \quad H_c \quad \mathbf{x}_a
$$

To use (44), we need to compute [see Section 6.5 on circulants in Chapter 6, Equations (79)-(80)]

$$\begin{bmatrix} \lambda_{11} \\ \lambda_{22} \\ \lambda_{33} \\ \lambda_{44} \end{bmatrix} = A_{DFT} \begin{bmatrix} 1 \\ 2 \\ 0 \\ 0 \end{bmatrix} \quad \text{and} \quad A_{DFT} = \begin{bmatrix} 1 & 1 & 1 & 1 \\ 1 & -j & -1 & j \\ 1 & -1 & 1 & -1 \\ 1 & j & -1 & -j \end{bmatrix}$$

from which we have

$$\lambda_{11} = 3, \qquad \lambda_{22} = 1 - 2j, \qquad \lambda_{33} = -1, \qquad \lambda_{44} = 1 + 2j$$

$$A_{DFT}\mathbf{x}_a = \begin{bmatrix} 2 \\ 1 - j \\ 0 \\ 1 + j \end{bmatrix}$$

and

$$\Lambda A_{DFT}\mathbf{x}_a = \begin{bmatrix} 6 \\ -1 - 3j \\ 0 \\ -1 + 3j \end{bmatrix}$$

Now

$$\begin{bmatrix} r_0 \\ r_1 \\ r_2 \\ r_3 \end{bmatrix} = \frac{1}{M} A_{DFT}^* (\Lambda A_{DFT}\mathbf{x}_a) = \frac{1}{4} \begin{bmatrix} 4 \\ 12 \\ 8 \\ 0 \end{bmatrix}$$

which coincides with the results obtained in (26). From this example, it appears that the number of computations is excessive when compared to the simple matrix multiplication in (26). This is valid when $(N + L - 1)$ is small. For $(N + L - 1)$ large, say larger than 64, convolution via FFT is efficient.

Even though convolution via FFT is efficient, there are computational errors, such as round-off. To circumvent some of the problems, number theoretic transforms have been used. This area is rather wide so we present only some basics and an example. Interested readers should consult the references.

7.3.3. Applications of Number Theoretic Transforms to Convolution

In Chapter 5 we introduced a Number Theoretic Transform that had the general structure of a DFT. The transform pair is given below for ease [see (146) and (147) in Chapter 5]

$$F(k) = \sum_{n=0}^{N_1-1} x(n)r^{nk}, \qquad k = 0, 1, \ldots, N_1 - 1$$

$$x(n) = N_1^{-1} \sum_{k=0}^{N_1-1} F(k)r^{-nk}, \qquad n = 0, 1, \ldots, N_1 - 1 \qquad (46)$$

and r is called a root of unity of order N_1 if N_1 is the least positive integer such that $r^{N_1} = 1 \pmod{M}$. Agarwal and Burrus (1975) gave an important result which we state below.

A length-N_1 Number Theoretic Transform having the DFT structure will implement cyclic convolution if and only if there exists an inverse of N_1 and an element r which is a root of unity of order N_1.

The property of circular convolution enables us to find the convolution of two sequences x_n and h_n as the inverse transform of the product of the transforms.

That is, we define

$$X_k = \sum_{n=0}^{N_1-1} x_n r^{nk}, \qquad k = 0, 1, \ldots, N_1 - 1 \qquad (47a)$$

$$H_k = \sum_{n=0}^{N_1-1} h_n r^{nk}, \qquad k = 0, 1, \ldots, N_1 - 1 \qquad (47b)$$

and

$$R_k = H_k x_k \qquad (48)$$

The inverse transform of R_k is

$$r_n = N_1^{-1} \sum_{k=0}^{N_1-1} R_k r^{-nk}, \qquad n = 0, 1, \ldots, N_1 - 1 \qquad (49)$$

which gives the desired values.

Example

Illustrate the use of the Number Theoretic Transform (NTT) discussed above to the cyclic convolution in (45).

The following steps can be used to compute r_n. First, find NTT coefficients in (47a) by using

$$\begin{bmatrix} x_0 \\ x_1 \\ x_2 \\ x_3 \end{bmatrix} = T \begin{bmatrix} 1 \\ 1 \\ 0 \\ 0 \end{bmatrix} \qquad (50)$$

where T is given in (148) in Chapter 5 and is given below for convenience:

$$T = \begin{bmatrix} 1 & 1 & 1 & 1 \\ 1 & 4 & 16 & 13 \\ 1 & 16 & 1 & 16 \\ 1 & 13 & 16 & 4 \end{bmatrix} \qquad (51)$$

Second, computing (50), we have

$$\begin{bmatrix} x_0 \\ x_1 \\ x_2 \\ x_3 \end{bmatrix} \equiv \begin{bmatrix} 2 \\ 5 \\ 0 \\ 14 \end{bmatrix} \pmod{17} \qquad (52)$$

Third, from (47b), we have

$$\begin{bmatrix} H_0 \\ H_1 \\ H_2 \\ H_3 \end{bmatrix} = T \begin{bmatrix} h_0 \\ h_1 \\ h_2 \\ h_3 \end{bmatrix} = T \begin{bmatrix} 1 \\ 2 \\ 0 \\ 0 \end{bmatrix} \equiv \begin{bmatrix} 3 \\ 9 \\ 16 \\ 10 \end{bmatrix} \pmod{17} \qquad (53)$$

Fourth, multiplying the transform coefficients using (48) we have

$$R_0 = 6, \qquad R_1 = 11, \qquad R_2 = 0, \qquad R_3 = 4 \qquad (54)$$

The inverse transform is given by

$$\begin{bmatrix} r_0 \\ r_1 \\ r_2 \\ r_3 \end{bmatrix} = T^{-1} \begin{bmatrix} 6 \\ 11 \\ 0 \\ 4 \end{bmatrix} \qquad (55)$$

where [see (150) in Chapter 5]

$$T^{-1} = 13 \begin{bmatrix} 1 & 1 & 1 & 1 \\ 1 & 13 & 16 & 4 \\ 1 & 16 & 1 & 16 \\ 1 & 4 & 16 & 13 \end{bmatrix} \qquad (56)$$

Finally, using (56) in (55), we have

$$r_0 = 13 \ (21) \equiv 1 \pmod{17}$$

$$r_1 = 13 \ (165) \equiv 3 \pmod{17}$$

$$r_2 = 13 \ (246) \equiv 2 \pmod{17}$$

$$r_3 = 13\,(102) \equiv 0 \quad (\mathrm{mod}\ 17)$$

validating the earlier results. We can omit the last data point as we did before.

Problems

1. Given

$$x = \{1, 0, -1, 1, 0\} \quad \text{and} \quad h = \{1, -1, 0, 1\}$$

find

$$y = x * h \quad \text{and} \quad y = x \circ h$$

2. Set the equations up for solving Problem 1 using fast Fourier transforms.

7.4. Discrete Deconvolution

In Section 7.3, we were interested in the convolution of two signals, namely, the input data, x_n, and the unit-sample response of the system, h_n. In symbolic form, the result of convolution was expressed by

$$r_n = x_n * h_n \tag{57}$$

The system under consideration might be a communication channel, a measurement device, and so on. In most field applications, we know h_n. The problem is to find x_n, given r_n and h_n. In real field applications, we call this deconvolution and in finite field applications, we call it a decoding. Finite field decoding is a major problem by itself and therefore we do not discuss it further. In the following we consider the deconvolution problem for real field applications.

   The problem can be cast as a matrix problem. Given the following system of equations [see Equations (28) and (29)]

$$\mathbf{r} = H\mathbf{x} \tag{58}$$

where $\mathbf{r}$ is an $(N + L - 1)$-dimensional vector, H is a rectangular matrix of dimension $(N + L - 1) \times L$ and of rank L, and $\mathbf{x}$ is an L-dimensional vector, find $\mathbf{x}$ with the assumption that $\mathbf{r}$ and H are known and that there is a unique solution for $\mathbf{x}$. If these are the only aspects we need to consider, all we need to do is pick L independent rows from the matrix H and solve for the unknowns. This is possible since the matrix H is assumed to be of rank L. This seems to be a simple problem; however, the problem is really complicated by the fact that the received vector is not the same as r_n since it is generally corrupted by noise. That is, the received vector, say $\mathbf{d}$, is

$$\mathbf{d} = \mathbf{r} + \mathbf{n} \tag{59}$$

where **n** is some noise vector. In most applications, we assume the noise is additive. Now the problem is, given

$$\mathbf{d} = H\mathbf{x} \tag{60}$$

where H is known, solve for **x**. This problem does not have a solution, in general, because there are more equations (constraints) than unknowns. This is an overdetermined system. Therefore, for real field applications, we need to resort to a solution such as a least squares solution, discussed in Chapter 5. For matrices such as H in (29), fast algorithms are available to solve this problem.

The least squares solution of (60) over the real field is given by [see Equations (57) and (58) in Chapter 5]

$$\mathbf{x} = (H^T H)^{-1} H^T \mathbf{d} \tag{61}$$

Example

Given

$$H = \begin{bmatrix} h_0 & 0 & 0 \\ h_1 & h_0 & 0 \\ h_2 & h_1 & h_0 \\ 0 & h_2 & h_1 \\ 0 & 0 & h_2 \end{bmatrix}$$

find $(H^T H)$.

Now

$$H^T H = \begin{bmatrix} h_0^2 + h_1^2 + h_2^2 & h_0 h_1 + h_1 h_2 & h_0 h_2 \\ h_0 h_1 + h_1 h_2 & h_0^2 + h_1^2 + h_2^2 & h_0 h_1 + h_1 h_2 \\ h_0 h_2 & h_0 h_1 + h_1 h_2 & h_0^2 + h_1^2 + h_2^2 \end{bmatrix}$$

which is a symmetric Toeplitz matrix discussed earlier. This is true for any size H in (60). Therefore, fast algorithms developed in Chapter 5 for inverses of such matrices can be used to solve for the least squares solution.

Let us complete this section using the example in (26).

Example

We are given

$$\begin{bmatrix} 1 + \varepsilon \\ 3 - \varepsilon \\ 2 + \varepsilon \end{bmatrix} = \begin{bmatrix} 1 & 0 \\ 2 & 1 \\ 0 & 2 \end{bmatrix} \begin{bmatrix} x_0 \\ x_1 \end{bmatrix}$$

and

$$\mathbf{d} = H\mathbf{x}$$

where ε is assumed to be some noise term and is explicitly shown here for illustrative purposes. Find the least squares solution.

The symmetric Toeplitz matrix, (H^TH), and its inverse are given by

$$H^TH = \begin{bmatrix} 5 & 2 \\ 2 & 5 \end{bmatrix} \quad \text{and} \quad (H^TH)^{-1} = \frac{1}{21}\begin{bmatrix} 5 & -2 \\ -2 & 5 \end{bmatrix}$$

The vector $H^T\mathbf{d}$ is given by

$$H^T\mathbf{d} = \begin{bmatrix} 1 & 2 & 0 \\ 0 & 1 & 2 \end{bmatrix}\begin{bmatrix} 1+\varepsilon \\ 3-\varepsilon \\ 2+\varepsilon \end{bmatrix} = \begin{bmatrix} 7-\varepsilon \\ 7+\varepsilon \end{bmatrix}$$

and the least squares solution by

$$\mathbf{x} = (H^TH)^{-1}H^T\mathbf{d} = \frac{1}{21}\begin{bmatrix} 5 & -2 \\ -2 & 5 \end{bmatrix}\begin{bmatrix} 7-\varepsilon \\ 7+\varepsilon \end{bmatrix}$$

$$= \begin{bmatrix} 1-\frac{1}{3}\varepsilon \\ 1+\frac{1}{3}\varepsilon \end{bmatrix}$$

It is clear then when the data are noise-free, that is, $\varepsilon = 0$ in our example, the solution gives the actual input, whereas when noise is present, the desired input signal is off by $(H^TH)^{-1}H^T\mathbf{n}$ where $\mathbf{n}$ corresponds to the noise vector. Another problem one would encounter is that the unit-sample response of the system may not be known exactly and (H^TH) may be singular. In such cases, we can use $(H^TH) + \varepsilon I$, with a small $\varepsilon > 0$, instead of (H^TH).

The method presented here is based on the least squares algorithm. That is, the procedure is based on minimizing the squared errors, $E = \sum_i (d_i - r_i)^2$. Deconvolution based on minimizing $\sum_i |d_i - r_i|^p$ for p other than 2 requires the use of iterative techniques discussed in Chapter 5 (see Yarlagadda *et al.* 1985). The above discussion is obviously applicable for real field applications. Unfortunately, the least squares method is not transferable to other fields. Since finite field decoding is a vast area, it can not be dealt with in any detail here.

So far in this chapter we have been concerned with systems that have finite unit-sample responses, that is, $h_n = 0$ for $n < 0$, $n > L$, a finite number. There are cases where we need to generalize this. The next section deals with matrix applications to such systems.

Problem

Consider

$$
\begin{bmatrix} y_0 \\ y_1 \\ y_2 \\ y_3 \end{bmatrix} = \begin{bmatrix} 1 & 0 \\ 2 & 1 \\ 1 & 2 \\ 0 & 1 \end{bmatrix} \begin{bmatrix} x_1 \\ x_2 \end{bmatrix}
$$

for the cases

(a) $y_0 = 1, y_1 = 0, y_2 = 0, y_3 = 0$.
(b) $y_0 = 2, y_1 = 1, y_2 = 0, y_3 = 0$.
(c) $y_0 = 1, y_1 = 2, y_2 = 1, y_3 = 1$.

Solve for x_i using the deconvolution procedure discussed in this section for each case. Each one of these cases has special significance. If the convolution is expressed by $y = h * x$, then case (a) corresponds to finding x such that $h * x$ approximates a unit sample. Case (b) corresponds to the idea of one-step prediction. Durbin's Algorithm discussed in an earlier chapter is applicable for this case. Case (c) corresponds to an arbitrary deconvolution problem.

7.5. Linear Constant-Coefficient Difference Equations

In an earlier section, we considered sequences $x_i = x_i(n)$ that represent some variables at discrete instants of time identified by n. Dynamic discrete systems can be described by difference equations. Before we discuss this in detail, let us consider a simple differential equation, over the real field,

$$
\frac{d}{dt} p(t) = z(t) \tag{62}
$$

and see how we could implement this on a computer. The derivative of the function $p(t)$, say at $t = (n + 1)T$, can be approximated by

$$
\left[\frac{dp}{dt} \right]_{t=(n+1)T} \simeq \left[\frac{p(t) - p(t - T)}{T} \right]_{t=(n+1)T}
$$

$$
\simeq \frac{p((n + 1)T) - p(nT)}{T} \tag{63}
$$

Using this, we can simulate (62) in terms of

$$
p((n + 1)T) = p(nT) + Tz((n + 1)T) \tag{64}
$$

which says that the function p at time $(n + 1)T$ is obtained from the function p at time (nT) plus T times the function z at time $(n + 1)T$. If we know the initial condition on p, that is, the starting value, and the input function, then we can compute $p(nT)$. Since T in $p(nT)$ is a constant, it can be omitted and (64) is usually expressed by

$$p(n + 1) = p(n) + Tz(n + 1) \tag{65}$$

We can redefine the time variable and rewrite the difference equation in (65) in the form

$$p(n) = p(n - 1) + Tz(n) \tag{66}$$

In an earlier chapter, we used the concept of approximating an integral by a digital sum. These point out the fact that difference equations play a major role in the analysis and design of any type of a discrete dynamic system. Therefore, we study some of the concepts associated with difference equations in the framework of matrices.

We limit our discussion to the linear constant-coefficient difference equations and omit the nonlinear and time-varying cases (DeRusso *et al.*, 1965). In Chapter 2, we gave a thumbnail sketch of the solutions of difference equations. In the following we use matrix concepts.

An Nth-order linear constant-coefficient difference equation is defined by

$$a_N p(n + N) + a_{N-1} p(n + N - 1) + \cdots + a_0 p(n)$$

$$= b_m z(n + M) + b_{m-1} z(n + M - 1) + \cdots + b_0 z(n) \tag{67}$$

where we assume that $z(n)$ is the input and $p(n)$ is the response. For obvious reasons, we assume that $M \leq N$. In compact form (67) can be written

$$\sum_{j=0}^{N} a_j p(n + j) = \sum_{j=0}^{M} b_j z(n + j) \tag{68}$$

The equation in (67) [or in (68)] is called an Nth-order inhomogeneous difference equation, and the equation

$$\sum_{j=0}^{N} a_j p(n + j) = 0 \tag{69}$$

is called an Nth-order homogeneous difference equation. It is clear that (68) can be written

$$p(n + N) = -\frac{1}{a_N} \sum_{j=0}^{N-1} a_j p(n + j) + \sum_{j=0}^{M} b_j z(n + j) \left(\frac{1}{a_N}\right) \tag{70}$$

which is a recurrence expression. If $p(0), p(1), \ldots, p(N-1)$ are known and, of course, the inputs are assumed to be known, then $p(n)$ can be computed for all values of $n \geq N$ by repeated application of the recurrence formula in (70). While this method is not elegant, it still allows us to compute $p(n)$. Direct computation by (70) is not usually considered a solution. We discuss the solutions later. From (70), it is clear that the solution is unique if $p(0), p(1), \ldots, p(N-1)$ are specified. It is clear that the initial conditions, $p(n), 0 \leq n \leq N-1$, are arbitrary. Note that we did not specify any constraints, such as causality, on the system.

In general, the solution of (67) can be expressed in the form

$$p(n) = p_H(n) + p_p(n) \tag{71}$$

where $p_H(n)$ is the solution of the homogeneous equation in (69) and $p_p(n)$ is any one solution that satisfies (68). The components p_H and p_p are sometimes referred to as complementary and particular solutions. It is not our intention here to discuss these concepts using standard techniques to obtain (71). However, we are interested in developing matrix techniques to solve an Nth-order difference equation. We do this in the next section by representing an Nth-order difference equation by replacing it by N first-order difference equations (Kailath, 1980). There are two concepts that we use. These are linearity and shift invariance. The difference equation in (68) is a linear shift-invariant system. Therefore, we can use these properties.

Problem

Solve the following differential equation

$$\frac{dp}{dt} = 0.5p(t)$$

with $p(0) = 1$ using the discrete approximation discussed in this section. Pick two values for T and compare the results with the actual solution.

7.6. Matrix Representation of an Nth-Order Constant-Coefficient Difference Equation

We first consider the case $N = 2$. Let

$$p(n+2) + a_1 p(n+1) + a_0 p(n) = z(n) \tag{72}$$

be the second-order equation representing a system. We want to develop a system of two first-order difference equations that are equivalent to (72) in the sense that both sets give the same solution.

We claim that the difference equation in (72) and the following system of equations

$$q_1(n + 1) = q_2(n) \tag{73a}$$

$$q_2(n + 1) = -(a_0 q_1(n) + a_1 q_2(n)) + z(n) \tag{73b}$$

$$p(n) = q_1(n) \tag{73c}$$

will yield the same solution. This can be shown by successively computing

$$a_0 p(n) = a_0 q_1(n)$$
$$a_1 p(n + 1) = a_1 q_2(n) \tag{74}$$
$$p(n + 2) = -(a_0 q_1(n) + a_1 q_2(n)) + z(n)$$

Adding these equations will result in (72).

Next let us change the input $z(n)$ to $z(n + 1)$ in (72). Since the difference equation in (72) represents a linear shift-invariant system, all we need do is replace (73c) by

$$p(n) = q_1(n + 1) = q_2(n) \tag{75}$$

Similarly, if the right-hand side of (72) is changed to $z(n + 2)$, then Equation (73c) should be changed to

$$p(n) = q_1(n + 2) = q_2(n + 1)$$
$$= -(a_0 q_1(n) + a_1 q_2(n)) + z(n) \tag{76}$$

Finally, we can use the linearity property to derive a pair of first-order difference equations for the following second-order difference equation. Let

$$p(n + 2) + a_1 p(n + 1) + a_0 p(n) = d_2 z(n + 2) + d_1 z(n + 1) + d_0 z(n) \tag{77}$$

Then from (73), (75), and (76) we have

$$q_1(n + 1) = q_2(n) \tag{78a}$$

$$q_2(n + 1) = -(a_0 q_1(n) + a_1 q_2(n)) + z(n) \tag{78b}$$

$$p(n) = (d_0 - d_2 a_0) q_1(n) + (d_1 - d_2 a_1) q_2(n) + d_2 z(n) \tag{78c}$$

It is clear that (77) and (78) will yield the same solution. Equation (78) can be conveniently written in a matrix form,

$$\begin{bmatrix} q_1(n+1) \\ q_2(n+1) \end{bmatrix} = \begin{bmatrix} 0 & 1 \\ -a_0 & -a_1 \end{bmatrix} \begin{bmatrix} q_1(n) \\ q_2(n) \end{bmatrix} + \begin{bmatrix} 0 \\ 1 \end{bmatrix} z(n) \qquad (79a)$$

$$p(n) = [d_0 - d_2 a_0 \quad d_1 - d_2 a_1] \begin{bmatrix} q_1(n) \\ q_2(n) \end{bmatrix} + d_2 z(n) \qquad (79b)$$

or in symbolic form,

$$\begin{aligned} \mathbf{q}(n+1) &= A\mathbf{q}(n) + \mathbf{B}z(n) \\ p(n) &= \mathbf{C}^T\mathbf{q}(n) + Dz(n) \end{aligned} \qquad (80)$$

where

$$\mathbf{q}(n) = \begin{bmatrix} q_1(n) \\ q_2(n) \end{bmatrix}$$

The matrix representation in (80) is called a discrete state model and the vector $\mathbf{q}(n)$ is a state vector, and $q_1(n)$ and $q_2(n)$ are state variables. Note that the matrix A in (79a) has the form of a companion matrix. The results in (79) can be generalized for an Nth-order difference equation

$$p(n+N) + a_{N-1}p(n+N-1) + \cdots + a_0 p(n)$$
$$= d_N z(n+N) + d_{N-1}z(n+N-1) + \cdots + d_0 z(n) \qquad (81)$$

The state model representation of (81) is given by (80), where $\mathbf{q}(n)$ is now an N-dimensional vector, and B and C are N-dimensional vectors. One set of values for A, $\mathbf{B}$, $\mathbf{C}$, and D are given by

$$A = \begin{bmatrix} 0 & 1 & 0 & \cdots & 0 \\ 0 & 0 & 1 & \cdots & 0 \\ \vdots & \vdots & \vdots & & \vdots \\ 0 & 0 & 0 & \cdots & 1 \\ -a_0 & -a_1 & -a_2 & \cdots & -a_{N-1} \end{bmatrix}, \quad \mathbf{B} = \begin{bmatrix} 0 \\ 0 \\ \vdots \\ 0 \\ 1 \end{bmatrix} \qquad (82a)$$

$$\mathbf{C}^T = [d_0 - d_N a_0 \quad d_1 - d_N a_1 \quad \cdots \quad d_{N-1} - d_N a_{N-1}], \quad D = d_N \qquad (82b)$$

The set A, $\mathbf{B}$, $\mathbf{C}^T$, D in (80) is not unique for the given difference equation. This can be seen from the following. Since the vector $\mathbf{q}(n)$ is an intermediate vector, we can define a new N-dimensional vector, say $\mathbf{g}(n)$, by

$$\mathbf{g}(n) = T\mathbf{q}(n) \tag{83}$$

where T is any arbitrary nonsingular matrix. Then (80) can be rewritten

$$\begin{aligned} \mathbf{g}(n+1) &= TAT^{-1}\mathbf{g}(n) + (T\mathbf{B})z(n) \\ p(n) &= (\mathbf{C}^T T^{-1})\mathbf{g}(n) + Dz(n) \end{aligned} \tag{84}$$

which is also a state model representation of (81). Note that the input $z(n)$ and the output $p(n)$ are not modified by the transformation in (83).

Problem

Find a state model representation for the following difference equations:

(a) $y(n+3) + 2y(n+2) + y(n+1) = z(n) + z(n+1) + z(n+2)$.
(b) $y(n+2) + y(n) = z(n+2) + z(n)$.

7.7. Solutions of an *N*th-Order Difference Equation Using a State Model Representation

State models give an effective way of computing solutions of difference equations. It is clear that from (80) we have

$$\mathbf{q}(1) = A\mathbf{q}(0) + \mathbf{B}z(0)$$

$$\begin{aligned} \mathbf{q}(2) &= A\mathbf{q}(1) + \mathbf{B}z(1) \\ &= A^2\mathbf{q}(0) + A\mathbf{B}z(0) + \mathbf{B}z(1) \end{aligned}$$

$$\mathbf{q}(3) = A^3\mathbf{q}(0) + A^2\mathbf{B}z(0) + A\mathbf{B}z(1) + \mathbf{B}z(2)$$

In general,

$$\mathbf{q}(n) = A^n\mathbf{q}(0) + \sum_{i=0}^{n-1} A^{n-1-i}\mathbf{B}z(i) \tag{85}$$

and

$$p(n) = C^T\left(A^n\mathbf{q}(0) + \sum_{i=0}^{n-1} A^{n-1-i}\mathbf{B}z(i) \right) + Dz(n) \tag{86}$$

From this we can see that all the matrix techniques we developed earlier can be used to compute $p(n)$ in closed form.

We have not said anything about initial conditions yet. How do we use initial conditions $p(0), p(1), \ldots, p(N-1)$? Let us discuss this next.

7.7.1. Computation of q(0)

Let us illustrate this first for $N = 2$. From (80) and (86) we have

$$p(0) = \mathbf{C}^T \mathbf{q}(0) + Dz(0)$$

and

$$p(1) = \mathbf{C}^T A \mathbf{q}(0) + (\mathbf{C}^T \mathbf{B}) z(0) + Dz(1)$$

or in matrix form

$$\begin{bmatrix} \mathbf{C}^T \\ \mathbf{C}^T A \end{bmatrix} \mathbf{q}(0) = \begin{bmatrix} p(0) \\ p(1) \end{bmatrix} - \begin{bmatrix} Dz(0) \\ Dz(1) + \mathbf{C}^T \mathbf{B} z(0) \end{bmatrix} \tag{87}$$

or

$$F\mathbf{q}(0) = \mathbf{p} - \mathbf{e} \tag{88}$$

For the general case, the matrix F and the vectors $\mathbf{p}$ and $\mathbf{e}$ are given by

$$F = \begin{bmatrix} C^T \\ C^T A \\ \vdots \\ C^T A^{N-1} \end{bmatrix}, \quad \mathbf{p} = \begin{bmatrix} p(0) \\ p(1) \\ \vdots \\ p(N-1) \end{bmatrix}, \quad \mathbf{E} = \begin{bmatrix} e_0 \\ e_1 \\ \vdots \\ e_{N-1} \end{bmatrix} \tag{89}$$

where

$$e_0 = Dz(0)$$

$$e_n = C^T \sum_{i=0}^{n-1} A^{n-1-i} \mathbf{B} z(i) + Dz(n), \quad 1 \leq n \leq N-1 \tag{90}$$

For our present situation, the inverse of F does exist since the state variables have to be independent and the above selection of the state variables allows for the solution. However, for an arbitrary state model the observation of the outputs $p(0), p(1), \ldots, p(N-1)$ may not yield a solution for the vector $\mathbf{q}(0)$ (Kailath, 1980). This problem is discussed in a later section.

Example

Solve the following difference equation using the above method:

$$p(n + 2) + 3p(n + 1) + 2p(n) = z(n + 2) + z(n + 1) + z(n) \qquad (91)$$

with

$$z(n) = \delta(n)$$

For illustrative purposes, we pick two sets of initial conditions, say p_A and p_B, and obtain solutions for these cases.
Set A:

$$p_A(0) = 1 \quad \text{and} \quad p_A(1) = -2 \qquad (92a)$$

Set B:

$$p_B(0) = 1 \quad \text{and} \quad p_B(1) = 1 \qquad (92b)$$

A state model representation of (91) is given by

$$\begin{bmatrix} q_1(n + 1) \\ q_2(n + 1) \end{bmatrix} = \underbrace{\begin{bmatrix} 0 & 1 \\ -2 & -3 \end{bmatrix}}_{A} \begin{bmatrix} q_1(n) \\ q_2(n) \end{bmatrix} + \underbrace{\begin{bmatrix} 0 \\ 1 \end{bmatrix}}_{B} z(n) \qquad (93a)$$

$$p(n) = \underbrace{[-1 \quad -2]}_{C^T} \begin{bmatrix} q_1(n) \\ q_2(n) \end{bmatrix} + \underbrace{(1)z(n)}_{D} \qquad (93b)$$

7.7.2. Initial Conditions

For Set *A*, we have

$$F = \begin{bmatrix} -1 & -2 \\ 4 & 5 \end{bmatrix}, \qquad \mathbf{p} = \begin{bmatrix} 1 \\ -2 \end{bmatrix}, \qquad \mathbf{e} = \begin{bmatrix} e_0 \\ e_1 \end{bmatrix}$$

where

$$e_0 = Dz(0) = 1 \quad \text{and} \quad e_1 = C^T Bz(0) + Dz(1) = -2$$

Therefore, for Set *A*,

$$\mathbf{q}(0) = F^{-1}(\mathbf{p} - e) = \begin{bmatrix} 0 \\ 0 \end{bmatrix}_A \qquad (94a)$$

Similarly, for Set B, we have

$$\mathbf{q}(0) = \begin{bmatrix} 2 \\ -1 \end{bmatrix}_B \tag{94b}$$

To compute the general solution, it is better to express the matrix A in (93a) in its Jordan form. The characteristic polynomial

$$f(\lambda) = \lambda^2 + 3\lambda + 2 = (\lambda + 2)(\lambda + 1)$$

and the modal matrix and its inverse are given by

$$M = \begin{bmatrix} 1 & 1 \\ -1 & -2 \end{bmatrix} \quad \text{and} \quad M^{-1} = \begin{bmatrix} 2 & 1 \\ -1 & -1 \end{bmatrix}$$

Noting that

$$A = M\Lambda M^{-1}, \qquad \Lambda = \begin{bmatrix} -1 & 0 \\ 0 & -2 \end{bmatrix}$$

and using these in (93) with

$$\mathbf{g}(n) = M^{-1}\mathbf{q}(n) \tag{95}$$

we have

$$\mathbf{g}(n+1) = \Lambda\mathbf{g}(n) + (M^{-1}\mathbf{B})z(n)$$
$$p(n) = (\mathbf{C}^T M)\mathbf{g}(n) + Dz(n) \tag{96}$$

where

$$M^{-1}B = \begin{bmatrix} 2 & 1 \\ -1 & -1 \end{bmatrix}\begin{bmatrix} 0 \\ 1 \end{bmatrix} = \begin{bmatrix} 1 \\ -1 \end{bmatrix}$$

$$C^T M = [-1 \quad -2]\begin{bmatrix} 1 & 1 \\ -1 & -2 \end{bmatrix} = [1 \quad 3]$$

Explicitly, we have

$$\begin{bmatrix} g_1(n+1) \\ g_2(n+1) \end{bmatrix} = \begin{bmatrix} -1 & 0 \\ 0 & -2 \end{bmatrix}\begin{bmatrix} g_1(n) \\ g_2(n) \end{bmatrix} + \begin{bmatrix} 1 \\ -1 \end{bmatrix}z(n) \tag{97a}$$

$$p(n) = [1 \quad 3]\begin{bmatrix} g_1(n) \\ g_2(n) \end{bmatrix} + z(n) \tag{97b}$$

Note that the variables $g_1(n)$ and $g_2(n)$ are not coupled in (97a). The initial conditions on $g_i(n)$ can be computed from (94) and (95). From these, we have $\mathbf{g}(0)$ for Sets A and B, respectively,

$$\mathbf{g}(0) = \begin{bmatrix} 0 \\ 0 \end{bmatrix}_A, \quad \begin{bmatrix} 3 \\ -1 \end{bmatrix}_B$$

In our example, $z(n) = \delta(n) = 1$, $n = 0$, and $z(n) = 0$, $n \neq 0$. Therefore, from (96), we have

$$p(0) = (1)_A, \quad (1)_B \tag{98}$$

which validates the original assumption [see Eq. (92)]. For $n \geq 1$, we have from (85)

$$\mathbf{g}(n) = \Lambda^n \mathbf{g}(0) + \sum_{i=0}^{n-1} \Lambda^{n-1-i} M^{-1} \mathbf{B} z(i)$$

$$= \begin{bmatrix} -(-1)^n \\ \frac{1}{2}(-2)^n \end{bmatrix}_A, \quad \begin{bmatrix} 2(-1)^n \\ -\frac{1}{2}(-2)^n \end{bmatrix}_B, \quad n \geq 1$$

Now, from (96), $p(n)$ is given by

$$p(n) = (-(-1)^n + \tfrac{3}{2}(-2)^n)_A, \quad (2(-1)^n - \tfrac{3}{2}(-2)^n)_B, \quad n \geq 1$$

We can incorporate (98) with the above and write

$$p(n) = (-(-1)^n + \tfrac{3}{2}(-2)^n + \tfrac{1}{2}\delta(n))_A, \quad n \geq 0 \tag{99a}$$

$$= (2(-1)^n - \tfrac{3}{2}(-2)^n + \tfrac{1}{2}\delta(n))_B, \quad n \geq 0 \tag{99b}$$

From these, we can compute the first few values of $p(n)$:

$$p(n) = (1, -2, 5, 11, \ldots)_A, \quad n \geq 0 \tag{100a}$$

$$= (1, 1, -4, 10, \ldots)_B, \quad n \geq 0 \tag{100b}$$

Do these satisfy the difference equation in (91)? By substituting these in (91), we can show that they satisfy for all values of $n \geq 0$. What about $p(n)$, $n < 0$. Let us express (91) in the form

$$p(n) = -\tfrac{1}{2}(p(n+2) + 3p(n+1)) + \tfrac{1}{2}(\delta(n+2) + \delta(n+1) + \delta(n)) \tag{101}$$

Evaluating (101) for the two cases, we have

$$p(-1) = (0)_A, \quad (-\tfrac{3}{2})_B$$
$$p(-2) = (0)_A, \quad (\tfrac{9}{4})_B$$

indicating $p(n) = 0$, $n < 0$ for case A and $p(n) \neq 0$ for case B. We consider this example again in a later section.

The detailed example presented here is to illustrate various concepts we have learned in the last few chapters. The state model representation is obviously valid for all field representations. However, eigenvalue–eigenvector decomposition is not valid for finite field applications. We illustrate the finite field case later. Let us now review and generalize some of the concepts we have learned about state models.

7.7.3. State Equations—Some Generalizations

The analysis of systems via state equations is very important in many areas including system theory, communication theory, and many others. Because of the availability of computers, more and more analysis and design techniques are used in the time domain rather than in the conventional operator domain.

Earlier we identified the following model as a state model,

$$\mathbf{q}(n+1) = A\mathbf{q}(n) + \mathbf{B}z(n)$$
$$p(n) = \mathbf{C}^T\mathbf{q}(n) + Dz(n) \tag{102}$$

corresponding to a single input–single output case. Obviously, we can generalize this to the multi-input–multi-output case by writing

$$\mathbf{q}(n+1) = A\mathbf{q}(n) + B\mathbf{z}(n) \tag{103a}$$

$$\mathbf{p}(n) = C^T\mathbf{q}(n) + D\mathbf{z}(n) \tag{103b}$$

where $\mathbf{p}(n)$ and $\mathbf{z}(n)$ are now vectors of dimension, say M and L, respectively. Let the vector $\mathbf{q}$ be an N-dimensional vector. Then A is an $N \times N$ matrix, B is an $N \times L$ matrix, C^T is an $M \times N$ matrix, and D is an $M \times L$ matrix. So far, we have been interested in shift-invariant systems. That is, matrices A, B, C, and D have constants as entries. For a shift-variant case, of course, these matrices are time dependent. State models are particularly useful for handling shift-variant systems. We do not discuss the shift-variant case here, however, because it is beyond our scope. Next we discuss the meaning of the state, state vector, and state model.

The state (or condition), in common terms, contains all the necessary information concerning the past history to determine the response of the system. The state vector specifies the state (or condition) of the system at that time. Equation (103a) describes the relation between the state variables, that is, the variables that describe the state, and the inputs. This equation describes the dynamics of the system. Equations (103b), usually referred to as the output equations, describe the relations between the inputs, state variables, and responses. Let us illustrate some of these concepts by an example.

Example

In Chapter 8, we will be interested in generating maximum-length shift register sequences. We postpone the complete discussion to that time; however, these systems illustrate the concepts and so we use it here as a system. Consider the sequence generator shown in Figure 7.2, where there are three registers identified by s_0, s_1, and s_2, and a summer (mod 2 adder). The contents in the registers are identified by $s_i(n)$ corresponding to the time n. The output is identified by $p(n)$. At every clock cycle (not shown in the figure) the contents of the registers are shifted toward the left. Initially, say at instant 0, the three registers have the data $s_0(0)$, $s_1(0)$, and $s_2(0)$. After the first shift, the registers s_1 and s_2 have the contents of the registers s_0 and s_1, respectively, before the shift, and the register s_0 has the contents that is the mod 2 sum of the contents in the registers s_1 and s_2. These can be generalized as

$$s_1(n) = s_0(n - 1)$$
$$s_2(n) = s_1(n - 1)$$
$$s_0(n) = s_1(n - 1) + s_2(n - 1)$$

or

$$s_2(n + 1) = s_1(n)$$
$$s_1(n + 1) = s_0(n)$$
$$s_0(n + 1) = s_2(n) + s_1(n)$$
$$p(n) = s_2(n)$$

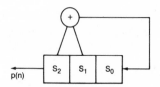

Figure 7.2. A binary sequence generator.

which can be conveniently expressed in a matrix form

$$\begin{bmatrix} s_2(n+1) \\ s_1(n+1) \\ s_0(n+1) \end{bmatrix} = \begin{bmatrix} 0 & 1 & 0 \\ 0 & 0 & 1 \\ 1 & 1 & 0 \end{bmatrix} \begin{bmatrix} s_2(n) \\ s_1(n) \\ s_0(n) \end{bmatrix} \tag{104a}$$

$$p(n) = \begin{bmatrix} 1 & 0 & 0 \end{bmatrix} \begin{bmatrix} s_2(n) \\ s_1(n) \\ s_0(n) \end{bmatrix} \tag{104b}$$

Symbolically, we have

$$\mathbf{s}(n+1) = A\mathbf{s}(n)$$
$$p(n) = C^T\mathbf{s}(n) \tag{105}$$

It is clear from (104) that $s_0(n)$, $s_1(n)$, and $s_2(n)$ represent the states and (104) gives the status of the system at any time. That is, knowing the initial state, we can compute $p(n)$ for any n. Knowing the state vector $\mathbf{s}(n)$, we can compute the output $p(n)$ by using (104b).

Using the concepts discussed earlier, we have

$$\mathbf{s}(n) = A^n\mathbf{s}(0)$$
$$p(n) = C^T A^n \mathbf{s}(0) \tag{106}$$

For computing A^n, for finite field applications, it is best to use the Cayley-Hamilton Theorem. The characteristic polynomial for the system in (105) is given by

$$f(\lambda) = \lambda^3 + \lambda + 1 \tag{107}$$

Later on we will use a different version of this polynomial,

$$g(x) = x^3 + x^2 + 1$$

usually referred to as the generator polynomial, obtained from $f(\lambda)$ by

$$g(x) = (f(\lambda)|_{\lambda=1/x})x^3 \tag{108}$$

The reason for using it is that x represents the delay shift, whereas λ represents the advance shift. The latter representation is based on physical aspects.

For illustrative purposes, let us compute $p(3)$ with the assumption that the initial states are $s_0(0) = 1$, $s_1(0) = 1$, and $s_2(0) = 1$. From (106), we have

$$p(3) = C^T A^3 s(0)$$

where

$$A^3 = A + I = \begin{bmatrix} 1 & 1 & 0 \\ 0 & 1 & 1 \\ 1 & 1 & 1 \end{bmatrix} \tag{109}$$

and

$$p(3) = \begin{bmatrix} 1 & 0 & 0 \end{bmatrix} \begin{bmatrix} 1 & 1 & 0 \\ 0 & 1 & 1 \\ 1 & 1 & 1 \end{bmatrix} \begin{bmatrix} 1 \\ 1 \\ 1 \end{bmatrix} = 0$$

This completes a simple example illustrating the state model concepts for finite field applications.

Finally, for real field applications, we want to briefly comment on the transfer functions. We realize that the operator theory is a vast area, but a simple introduction to this subject is necessary and is presented in the next section.

Problems

1. Solve the following difference equation using state models:

$$p(n + 2) + p(n + 1) + p(n) = \delta(n)$$

Assume $p(0) = 1$ and $p(1) = 1$.

2. Solve the following state equations

$$\begin{bmatrix} q_1(n + 1) \\ q_2(n + 1) \end{bmatrix} = \begin{bmatrix} \frac{1}{2} & 1 \\ 0 & \frac{1}{2} \end{bmatrix} \begin{bmatrix} q_1(n) \\ q_1(n) \end{bmatrix}$$

$$p(n) = q_1(n)$$

with $q_1(0) = 1$ and $q_2(0) = 1$.

3. Find $p(7)$ in (104b), assuming the initial conditions $s_0(0) = 0$, $s_1(0) = 0$, and $s_2(0) = 1$. What can you see about the periodicity of $p(n)$?

7.8. Transfer Functions: An Introduction to Z Transforms

Transfer function analysis is based on the Z-transform operation. If $x(n)$ is a sequence, then the one-sided Z transform of $x(n)$ is defined by

$$X(z) = \sum_{n=0}^{\infty} x(n)z^{-n} \tag{110}$$

and its existence depends on the absolute summability of the sequence in a region of convergence. Notationwise, we write $X(z) = Z[x(n)]$. We should note that the transform operation is a linear operation.

Example

The Z transform of the sequence

$$x(n) = a^n U(n)$$

where $U(n)$ is a unit step defined by

$$U(n) = 1, \qquad n \geq 0$$
$$= 0, \qquad n < 0$$

is given by

$$X(z) = \sum_{n=0}^{\infty} a^n z^{-n} = \sum_{n=0}^{\infty} (az^{-1})^n = \frac{1}{1 - az^{-1}} \quad \text{for } |z| > a \tag{111}$$

The region of convergence here includes the entire z plane for $|z| > a$.

In our analysis we are interested in rational functions of z. That is, $X(z)$ can be expressed by a ratio of two polynomials

$$X(z) = \frac{A(z)}{D(z)}$$

In factored form, $A(z)$ and $D(z)$ can be expressed as

$$A(z) = \prod_{i=1}^{M} (z - a_i)$$

and

$$D(z) = \prod_{i=1}^{N} (z - d_i)$$

where the d_i are called the poles and the a_i are called the zeros. In (111), $X(z)$ has a pole at $z = a$ and a zero at $z = 0$.

Example

The Z transform of a unit-sample sequence $\delta(n)$ is given by

$$\sum_{n=0}^{\infty} \delta(n)z^{-n} = 1 \tag{112}$$

Example

Noting that $Z[x(n)] = X(z)$, the Z transforms of $x(n + k)$ and $x(n - k)$ for some k are given by

$$Z[x(n + k)] = z^k \left[X(z) - \sum_{m=0}^{k-1} x(m)z^{-m} \right] \tag{113}$$

and

$$Z[x(n - k)] = z^{-k} \left[X(z) + \sum_{m=1}^{k} x(-m)z^{m} \right] \tag{114}$$

Let us prove (113). By definition

$$Z[x(n + k)] = \sum_{n=0}^{\infty} x(n + k)z^{-n}$$

$$= z^k \sum_{n=0}^{\infty} x(n + k)z^{-(n+k)}$$

$$= z^k \left[X(z) - \sum_{m=0}^{k-1} x(m)z^{-m} \right]$$

In a similar manner, (114) can be proved.

The inverse Z transform is defined by

$$x(n) = \frac{1}{2\pi j} C \oint X(z) z^{n-1}\, dz \qquad (115a)$$

where C is a counterclockwise closed contour in the region of convergence of $X(z)$ encircling the origin of the z plane. The inverse transform can be efficiently computed using the residue theorem for rational Z-transform functions (Oppenheim and Schafer, 1975). That is,

$$x(n) = \sum [\text{Residues of } x(z)z^{n-1} \text{ at the poles inside } C] \qquad (115b)$$

Let

$$X(z)z^{n-1} = \frac{b(z)}{(z - z_0)^n}$$

where z_0 is a pole of $X(z)z^{n-1}$ of multiplicity n and z_0 is not a pole or a zero of $b(z)$. The residue at this pole is given by

$$\text{Res}\,[X(z)z^{n-1} \text{ at } z = z_0] = \frac{1}{(n-1)!} \left. \frac{d^{n-1}b(z)}{dz^{n-1}} \right|_{z=z_0} \qquad (115c)$$

Example

The inverse transform of

$$X(z) = \frac{z}{z - a}$$

is

$$x(n) = \text{Res}\,\frac{z}{z - a} z^{n-1} \bigg|_{z=a} = a^n$$

With these simple ideas we are ready to find the transfer functions for the state model

$$\mathbf{q}(n + 1) = A\mathbf{q}(n) + \mathbf{B}e(n)$$
$$p(n) = \mathbf{C}^T\mathbf{q}(n) + De(n) \qquad (116)$$

where $e(n)$ is the input function, $p(n)$ is the output function, and $\mathbf{q}(n)$ is the state vector.

Using (113) and the linearity property of Z transforms, we have

$$Z[\mathbf{q}(n+1)] = z(\mathbf{Q}(z) - \mathbf{q}(0))$$

$$Z[A\mathbf{q}(n)] = A\mathbf{Q}(z)$$

$$Z[e(n)] = E(z)$$

$$Z[p(n)] = P(z) \tag{117}$$

$$Z[C^T\mathbf{q}(n)] = \mathbf{C}^T\mathbf{Q}(z)$$

$$Z[De(n)] = DE(z)$$

where A, B, C^T, and D are assumed to be constant matrices. Using (117) in (116), we have

$$z\mathbf{Q}(z) = A\mathbf{Q}(z) + \mathbf{B}E(z) + z\mathbf{q}(0)$$

$$P(z) = \mathbf{C}^T Q(z) + DE(z) \tag{118}$$

From the first equation, we have

$$(zI - A)Q(z) = \mathbf{B}E(z) + z\mathbf{q}(0)$$

or

$$Q(z) = (zI - A)^{-1}\mathbf{B}E(z) + (zI - A)^{-1}z\mathbf{q}(0)$$

Substituting this in the second equation in (118), we have

$$P(z) = [D + \mathbf{C}^T(zI - A)^{-1}\mathbf{B}]E(z) + \mathbf{C}^T(zI - A)^{-1}z\mathbf{q}(0) \tag{119a}$$

The first term with $E(z) = 1$ (that is, $e(n) = \delta(n)$)

$$H(z) = D + \mathbf{C}^T(zI - A)^{-1}\mathbf{B} \tag{119b}$$

is called a transfer function and is obtained from $P(z)$ by setting the initial conditions to zero with $E(z) = 1$. Computation of $H(z)$ from a state model is conceptually simple. It involves the computation of

$$(zI - A)^{-1} = \frac{1}{|zI - A|} \text{Adj}\,(zI - A) \tag{120}$$

In Section 6.7, we discussed an efficient method, given by Faddeev, Frame, and others, for computing (120). Let us illustrate these ideas by a simple example.

Example

Given the state model in (93), compute the transfer function.
Using the results in Section 6.7 or by straight computation, we have

$$\text{Adj}\,(zI - A) = \begin{bmatrix} z + 3 & 1 \\ -2 & z \end{bmatrix}$$

and

$$|zI - A| = z^2 + 3z + 2$$

Therefore,

$$H(z) = 1 + [-1 \quad -2]\frac{1}{z^2 + 3z + 2}\begin{bmatrix} z + 3 & 1 \\ -2 & z \end{bmatrix}\begin{bmatrix} 0 \\ 1 \end{bmatrix} \tag{121a}$$

$$= 1 + \frac{-1 - 2z}{z^2 + 3z + 2} = D + \frac{n_1(z)}{f(z)} \tag{121b}$$

$$= \frac{z^2 + z + 1}{z^2 + 3z + 2} = \frac{N(z)}{f(z)} \tag{121c}$$

Note the correspondence between the following:

Polynomial	Matrix or a Vector
$f(z)$ in (121b)	A in (93)
$n_1(z)$ in (121b)	C^T in (93)

This correspondence indicates that a state model can be obtained by inspection from a transfer function. This involves two steps. First, let

$$H(z) = D + \frac{c_{N-1}z^{N-1} + \cdots + c_0}{z^N + d_{N-1}z^{N-1} + \cdots + d_0} \tag{122}$$

Second, the entries in the state model are given by

$$
A = \begin{bmatrix}
0 & 1 & 0 & \cdot & \cdot & \cdot & 0 \\
0 & 0 & 1 & \cdot & \cdot & \cdot & 0 \\
\cdot & \cdot & \cdot & \cdot & \cdot & \cdot & \cdot \\
\cdot & \cdot & \cdot & \cdot & \cdot & \cdot & \cdot \\
0 & 0 & 0 & \cdot & \cdot & \cdot & 1 \\
-d_0 & -d_1 & -d_2 & \cdot & \cdot & \cdot & -d_{N-1}
\end{bmatrix}, \quad
B = \begin{bmatrix}
0 \\ 0 \\ \cdot \\ \vdots \\ \cdot \\ 1
\end{bmatrix} \quad (123)
$$

$$
C^T = \begin{bmatrix} c_0 & c_1 & c_2 & \cdots & c_{N-1} \end{bmatrix}
$$

and D is known from (122). The reason for this simplicity is that for a companion matrix A given in (123), $(\text{Adj}\,(zI - A))$ has a special form:

$$
(\text{Adj}\,(zI - A)) = \begin{bmatrix}
\times & \times & \times & \times & 1 \\
\times & \times & \times & \times & z \\
\times & \times & \times & \times & z^2 \\
\cdot & \cdot & \cdot & \cdot & \cdot \\
\cdot & \cdot & \cdot & \cdot & \cdot \\
\cdot & \cdot & \cdot & \cdot & \cdot \\
\times & \times & \times & \times & z^{N-1}
\end{bmatrix} \quad (124)
$$

where the $\times$'s are not important. The entries in the last column are $1, z, \ldots, z^{N-1}$.

Finally, let us illustrate the use of Z transforms in solving difference equations.

Example

Find the solution of the difference equation in (91) with the input $z(n) = \delta(n)$ using Z transforms.

The difference equation in (91) is rewritten here for ready reference:

$$
p(n + 2) + 3p(n + 1) + 2p(n) = \delta(n + 2) + \delta(n + 1) + \delta(n) \quad (125)
$$

Using (112) and (113) we have

$$
z^2 P(z) - p(0)z^2 - p(1)z + 3[zP(z) - p(0)z] + 2P(z) = 1
$$

or

$$
P(z) = \frac{p(0)z^2 + z(p(1) + 3p(0)) + 1}{(z + 1)(z + 2)}
$$

For the two sets of initial conditions in (92), we have

$$P(z) = \left(\frac{z^2 + z + 1}{(z+1)(z+2)}\right)_A, \quad \left(\frac{z^2 + 4z + 1}{(z+1)(z+2)}\right)_B$$

For case B, we have

$$p(n) = \delta(n)\left(\text{Res of } \frac{P(z)}{z} \text{ at } z = 0\right) + \text{Res of } P(z)z^{n-1} \text{ at } z = -1$$

$$+ \text{Res of } P(z)z^{n-1} \text{ at } z = -2 \tag{126}$$

$$= \delta(n)\left.\frac{z^2 + 4z + 1}{(z+1)(z+2)}\right|_{z=0} + \left.\frac{z^2 + 4z + 1}{z + 2} z^{n-1}\right|_{z=-1}$$

$$+ \left.\frac{z^2 + 4z + 1}{(z+1)} z^{n-1}\right|_{z=-2}$$

$$= \tfrac{1}{2}\delta(n) + (-2)(-1)^{n-1} + (3)(-2)^{n-1}$$

$$= \tfrac{1}{2}\delta(n) + 2(-1)^n - \tfrac{3}{2}(-2)^n \tag{127}$$

which is what we obtained in (99b). Similar results can be obtained for case A. Note that the first term appeared in (126) because $p(z)z^{n-1}$ has a pole at the origin only when $n = 0$.

Finally, we want to make some comments about stability. Without going into the theory of stability, we want to state that the system described by a state model is stable, if the eigenvalues of the A matrix are located inside the unit circle. That is, if the characteristic polynomial of the system is

$$f(z) = |zI - A| = \prod_{i=1}^{N} (z - z_i) \tag{128}$$

then

$$|z_i| < 1 \tag{129}$$

in order for the system to be stable. $|z_i| = 1$ corresponds to the oscillatory case. The justification is that, for a source-free system, the response [see (85)] in terms of the initial conditions is given by

$$q(n) = A^n q(0) \tag{130}$$

which clearly indicates that if A has eigenvalues outside the unit circle, then the entries in A^n will be unbounded as $n \to \infty$.

Problems

1. Solve the following difference equations using Z transforms:

   (a) $p(n + 2) + 0.5p(n + 1) + 0.25 = \delta(n)$; $p(0) = 0$, $p(1) = 0$.
   (b) $p(n + 1) + 0.5p(n) = U(n)$; $p(0) = 1$ and

$$U(n) = \begin{cases} 1, & n \geq 0 \\ 0, & n < 0 \end{cases}$$

2. Find the inverse Z transforms of the following functions:

   (a) $\dfrac{z}{(z + 1)^2}$

   (b) $\dfrac{z}{(z + 0.5)(z + 0.3)}$

7.9. Observability Problem

We touched on this problem earlier. Formally, we can say that the discrete system over a field F (Kailath, 1980)

$$\mathbf{q}(n + 1) = A\mathbf{q}(n)$$
$$p(n) = \mathbf{C}^T\mathbf{q}(n) \tag{131}$$

is completely observable if there is a finite number N such that the knowledge of the outputs $p(n)$, $n = 0, 1, \ldots, N - 1$, is sufficient to determine the values of the initial state.

The above statement implies that the system is completely observable if we can solve the equations

$$\begin{bmatrix} p(0) \\ p(1) \\ \vdots \\ p(N - 1) \end{bmatrix} = \begin{bmatrix} \mathbf{C}^T \\ \mathbf{C}^T A \\ \vdots \\ \mathbf{C}^T A^{N-1} \end{bmatrix} \mathbf{q}(0) \tag{132}$$

or

$$\mathbf{p} = F\mathbf{q}(0) \tag{133}$$

for $\mathbf{q}(0)$. For the single-output case, that is, when $p(n)$ is a scalar variable, we can state that the system in (131) with N state variables is completely observable if and only if F in (133) is nonsingular. The formal proof of this is omitted.

Example

Check the observability of the following system over the real field:

$$\begin{bmatrix} q_1(n+1) \\ q_2(n+1) \end{bmatrix} = \begin{bmatrix} 1 & 1 \\ 0 & 2 \end{bmatrix} \begin{bmatrix} q_1(n) \\ q_2(n) \end{bmatrix}$$

$$p(n) = \begin{bmatrix} 0 & 1 \end{bmatrix} \begin{bmatrix} q_1(n) \\ q_2(n) \end{bmatrix}$$

(134)

The observability matrix

$$F = \begin{bmatrix} C^T \\ C^T A \end{bmatrix} = \begin{bmatrix} 0 & 1 \\ 0 & 2 \end{bmatrix}$$

indicates that F is singular. Therefore, the system is not observable. It is clear that there is no way we can observe the state $q_1(n)$. We use the observability concept in Chapter 8.

Similar to the observability problem, we can state the concept of controllability of discrete systems. This concept is obviously of considerable importance in control theory. However, we are less interested in this topic. For completeness, we state the concept. The single input system

$$\mathbf{q}(n+1) = A\mathbf{q}(n) + \mathbf{B}e(n)$$

(135)

is completely state controllable if for $\mathbf{q}(0) = 0$ and any given vector $\mathbf{q}_1$, there exists a set of inputs $e(0), e(1), \ldots, e(N-1)$, for N finite, such that $\mathbf{q}(N) = \mathbf{q}_1$.

Again, the test for controllability can be derived by constructing

$$\mathbf{q}(1) = \mathbf{B}e(0)$$

$$\mathbf{q}(2) = A\mathbf{q}(1) + \mathbf{B}e(1)$$

$$= AB e(0) + \mathbf{B}e(1)$$

$$\vdots$$

$$\mathbf{q}(N) = A^{N-1}\mathbf{B}e(0) + \cdots + \mathbf{B}e(N-1)$$

and solving for $e(i)$. For the single-input case, let

$$\mathbf{q}_1 = \mathbf{q}(N) = \begin{bmatrix} \mathbf{B} & AB & \cdots & A^{N-1}\mathbf{B} \end{bmatrix} \begin{bmatrix} e(N-1) \\ e(N-2) \\ \vdots \\ e(0) \end{bmatrix}$$

Then the discrete system described in (135) with N state variables is completely state controllable if and only if the matrix

$$C = [\mathbf{B} \quad \mathbf{AB} \quad \cdots \quad \mathbf{A}^{N-1}\mathbf{B}]$$

is a nonsingular matrix.

Problem

Test the controllability and observability of the following system:

$$\begin{bmatrix} q_1(n+1) \\ q_1(n+1) \end{bmatrix} = \begin{bmatrix} 0 & 1 \\ -1 & 0 \end{bmatrix} \begin{bmatrix} q_1(n) \\ q_2(n) \end{bmatrix} + \begin{bmatrix} 0 \\ 1 \end{bmatrix} e(n)$$

$$p(n) = [1 \quad 0] \begin{bmatrix} q_1(n) \\ q_2(n) \end{bmatrix}$$

In this chapter we have discussed various concepts that interest communications engineers, signal processors, control theorists, and others. The concepts presented here are developed in terms of examples. Interested readers can pursue the theoretical aspects by studying the references.

References

Agarwal, R. C. and C. S. Burrus (1975), Number Theoretic Transforms to Implement Fast Digital Convolutions, *Proceedings of the IEEE*, Vol. 63, pp. 550–560.

DeRusso, P. M., R. J. Roy, and C. M. Close (1965), *State Variables for Engineers*, Wiley, New York.

Kailath, T. (1980), *Linear Systems*, Prentice-Hall, Englewood Cliffs, NJ.

Oppenheim, A. and R. W. Schafer (1975), *Digital Signal Processing*, Prentice-Hall, Englewood Cliffs, NJ.

Yarlagadda, R., J. B. Bednar, and T. Watt (1985), Fast Algorithms for L_p Deconvolution, *IEEE Transactions on Acoustics, Speech, and Signal Processing*, Vol. 33, pp. 174–182.

Random and Pseudorandom Sequences

8.1. Introduction

The study of random sequences is required by many aspects of data transmission. Synchronization and privacy are but two of these. In this chapter we examine the behavior of random sequences and conclude with a study of the m-sequence—an often used approximation to a random sequence.

One of the most commonly encountered problems looks something like the following. We observe the output of a balanced Bernoulli source until it *first* produces a specified sequence, or one of a specified set of sequences, of length k. We want to know two things. First, what is the mean time until observation of a specified sequence? Second, for *each* specified sequence, what is the probability that it will be the *first* of the specified sequences produced and what is the average waiting time until one of the specified sequences is produced? These are traditionally "sticky" questions. They, and others, are discussed and unified in a very fine paper by Blom and Thorburn (1982) and we use their results.

But first, as an introduction, and as motivation, we look at an unsettling problem that has been bouncing around in the mathematical community for the last few decades. It is a truly marvelous example of nontransitive relationships and was well cast by Gardner (1979) in his famous and entertaining series in *Scientific American* magazine. We draw heavily on his presentation.

Suppose we observe a balanced Bernoulli source for three consecutive outputs. We would all agree that any one of the eight possible triplets $000, 001, \ldots, 110, 111$ is equally likely to occur. But suppose we ask a slightly different question. Suppose we observe the output of a balanced Bernoulli source until a particular triple occurs. Are some triples more likely to occur *before* others? The first surprise is that the answer is yes. To see this immediately, consider the two triples 000 and 100. The *only* way the triple 000 could occur before the triple 100 is if it occurred as the first three bits out of the source. (Why?) Thus, the probability that the triple 000 will precede the triple 100 is only 1/8.

The second surprise which may be devastating to the intuition, occurs when we consider pairs of triples. Table 1, taken from Gardner's article and modified slightly, gives the probability that triple B will occur before triple A. Note that each column has an entry which is greater than $\frac{1}{2}$. Thus, for any specified triple, there is always a different triple which is more likely to occur before it! This nontransitivity also obtains for all higher-order n-tuples.

Problem

(Wolk's Rule via Gardner, 1974.) A good memory aid to determine the triple most likely to occur before a given triple $t_3 t_2 t_1$ was discovered by Wolk. Consider $t_3 t_2$ as a normal binary number (0 to 3), multiply it by 5 and add 4. The three least significant bits of the result expressed in normal binary is the triple most likely to occur before the triple $t_3 t_2 t_1$. Show that this rule works for all triples in Table 1.

The key to calculating the answers to the questions cited at the introduction is to develop what are called "leading numbers," denoted by $e_r(i, j)$, their transforms, the elements of a matrix $F = (f_{ij})$, and some ancillary numbers q_i. We let the specified sequences of interest, the k-tuples, be denoted by $S_1, S_2, \ldots, S_n$. We define the leading numbers as functions of two, not necessarily distinct, k-tuples, S_i and S_j, as $e_r(i, j) = 1$ if and only if the *last* r bits of S_i are the same as the *first* r bits of s_j; otherwise, $e_r(i, j) = 0$. [When does $e_r(i, j) = e_r(j, i)$?] We define the transform of the leading numbers by $f_{ij} = \sum_{r=1}^{k} e_r(i, j) 2^{r-k}$. We term the $\{f_{ij}\}$ as transforms because the $\{e_r\}$ can be derived from the $\{f_{ij}\}$. We form the auxiliary numbers, $\{q_i\}$, according to the formula $q_r(i, j) = e_{k-r}(i, j) 2^{-r}$ for $r = 0, 1, \ldots, k - 1$.

Table 1. Probability that Triple B Occurs Before Triple A[a]

Triple B	Triple A							
	000	001	010	011	100	101	110	111
000	—	$\frac{1}{2}$	$\frac{2}{5}$	$\frac{2}{5}$	$\frac{1}{8}$	$\frac{5}{12}$	$\frac{3}{10}$	$\frac{1}{2}$
001	$\frac{1}{2}$	—	$\frac{2}{3}$	$\frac{2}{3}$	$\frac{1}{4}$	$\frac{5}{8}$	$\frac{1}{2}$	$\frac{7}{10}$
010	$\frac{3}{5}$	$\frac{1}{3}$	—	$\frac{1}{2}$	$\frac{1}{2}$	$\frac{1}{2}$	$\frac{3}{8}$	$\frac{7}{12}$
011	$\frac{3}{5}$	$\frac{1}{3}$	$\frac{1}{2}$	—	$\frac{1}{2}$	$\frac{1}{2}$	$\frac{3}{4}$	$\frac{7}{8}$
100	$\frac{7}{8}$	$\frac{3}{4}$	$\frac{1}{2}$	$\frac{1}{2}$	—	$\frac{1}{2}$	$\frac{1}{3}$	$\frac{3}{5}$
101	$\frac{7}{12}$	$\frac{3}{8}$	$\frac{1}{2}$	$\frac{1}{2}$	$\frac{1}{2}$	—	$\frac{1}{3}$	$\frac{3}{5}$
110	$\frac{7}{10}$	$\frac{1}{2}$	$\frac{5}{8}$	$\frac{1}{4}$	$\frac{2}{3}$	$\frac{2}{3}$	—	$\frac{1}{2}$
111	$\frac{1}{2}$	$\frac{3}{10}$	$\frac{5}{12}$	$\frac{1}{8}$	$\frac{2}{5}$	$\frac{2}{5}$	$\frac{1}{2}$	—

The first results that we state concern only a single specified sequence of interest; let it be S_1. It can be shown that the probability, p_i, that the k-tuple of interest occurs for the first time after exactly i bits have left the source ($i \geq k$) is determined recursively by

$$p_i = 2^{-k} - 2^{-k} \sum_{j=k}^{i-k} p_j - \sum_{j=i-k+1}^{i-1} p_j q_{i-j}(1, 1) \tag{1}$$

where we use the convention that the sum is zero if the upper limit on the sum is less than the lower limit. $p_j = 0$ for $j < k$. In determining $q_r(i, j)$, we need to consider the specified k-tuple sequence.

Problem

Work out $\{e_r(1, 1)\}$ for $r = 1, 2$, and 3 for the $k = 3$ tuple 101. Also work out f_{11}. Finally, derive $\{p_i\}$ for $i = 3, 4, 5, 6, 7, 8, 9$, and 10. What is the probability that 101 does not occur within the first 10 bit outputs from the balanced Bernoulli source?

The probability generating function, $E(t^X)$, is as follows:

$$E(t^X) = \left[1 + (1 - t)d\left(\frac{2}{t}\right) \right]^{-1} \tag{2}$$

where

$$d(x) = \sum_{r=1}^{k} e_r(1, 1)x^r \tag{3}$$

Problems

1. Show that the mean waiting time for a single specified sequence is given by $d(2)$. Show that the variance of the waiting time for a single specified sequence is given by $d^2(2) + d(2) - 4d'(2)$. $d'(2)$ is the derivative of $d(x)$ with respect to x and subsequently evaluated at $x = 2$.

2. Assume we start "listening to" a balanced Bernoulli source that produces one bit at every time interval. We start listening at $t = 0$. What are the means and variances of the waiting times until (a) the sequence 0000011001010 (known as a Barker sequence) appears and (b) the sequence 0101010101010 appears?

3. (Wolk's Observation via Gardner, 1974.) The following observation may also prove to be an assault on the intuition. Show by direct calculations and comparisons that the quadruplet 1010 is more likely to occur before the quadruplet 0100 *but* the mean waiting time for the quadruplet 1010 is *greater* than the mean waiting time for the quadruplet 0100.

We now look at some results pertaining to n specified k-length sequences of interest, $S_1 - S_n$. It can be shown that the mean time to wait until one of the $\{S_i\}$ appears is

$$\left(\sum_{i=1}^{n} x_i \right)^{-1} \tag{4}$$

and the probability that sequence S_j will be the first of the S_i observed is

$$\pi_j = x_j \left(\sum_{i=1}^{n} x_i \right)^{-1} \tag{5}$$

where the $x_1, x_2, \ldots, x_n$ are solutions to the set of n linear equations

$$\sum_{i=1}^{n} f_{ij} x_i = 2^{-k}, \qquad j = 1, 2, \ldots, n \tag{6}$$

As an example, let us calculate the probability that the triple $S_1 = 010$ will be produced before the triple $S_2 = 111$. We first determine the elementary quantities:

$$e_1(1, 1) = 1 \qquad e_1(2, 1) = 0 \qquad f_{11} = \tfrac{5}{4}$$

$$e_2(1, 1) = 0 \qquad e_2(2, 1) = 0 \qquad f_{12} = 0$$

$$e_3(1, 1) = 1 \qquad e_3(2, 1) = 0 \qquad f_{21} = 0$$

$$e_1(1, 2) = 0 \qquad e_1(2, 2) = 1 \qquad f_{22} = \tfrac{7}{4}$$

$$e_2(1, 2) = 0 \qquad e_2(2, 2) = 1$$

$$e_3(1, 2) = 0 \qquad e_3(2, 2) = 1$$

We then solve the two equations

$$\tfrac{5}{4} x_1 = \tfrac{1}{8}$$

$$\tfrac{7}{4} x_2 = \tfrac{1}{8} \tag{7}$$

which yields

$$x_1 = \tfrac{1}{10}$$

$$x_2 = \tfrac{1}{14} \tag{8}$$

and thus

$$\pi_1 = \frac{\frac{1}{10}}{\frac{1}{10} + \frac{1}{14}} = \frac{7}{12}$$

$$\pi_2 = \frac{\frac{1}{14}}{\frac{1}{10} + \frac{1}{14}} = \frac{5}{12}$$

(9)

which agrees with the appropriate entry in Table 1.

Problem

Verify the remainder of the entries in the third row of Table 1.

8.2. Markov Chains

We now turn our attention to Markov chains. These are very useful models of the activities of many discrete systems that can be characterized by a countable set of states.† The system may change states only at allowed transition times, which occur periodically, according to a transition probability matrix, T. The ijth entry of T is interpreted as the probability that the system will transition to state number j at allowed transition time $t + 1$ if it is in state i at time t. This implies a key defining property of Markov chains, that is, the probability that the chain will transition to a particular state at the next allowable transition time depends *only* on the state it is presently in.

There are four types of states:

1. Initial State—the state in which the Markov chain is started.
2. Recurring State—a state which, if the chain enters it once, it will eventually enter again an infinite number of times as $t \to \infty$,
3. Transient State—a state which, if entered once, will be entered again only a finite number of times as $t \to \infty$,
4. Absorbing State—a state which, if entered by the chain, will never be left.

We review some very important results that pertain to chains incorporating some of these types of state. But first an example is in order. Figure 8.1 depicts a three-state Markov chain. The states are represented by circles or nodes. The numbers inside the circles are the state numbers. They may be arbitrary, or they may represent a particular variable; but they *must* be

---

† The word "chain" should not connote a necessarily connected set of states.

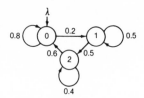

Figure 8.1. A three-state Markov chain.

unique. The arrows leaving a state represent allowed transitions from that state to other states. If the arrow returns to the state from where it came, it indicates that the chain, if in that state, may remain in that state at the next transition interval. A number above a state transition arrow represents the ijth entry of the transition probability matrix, T, that is, the probability the chain will transition from the state at the arrow's tail to the state at the arrow's head. The lambda with an arrow identifies the initial state—the state the chain is in at $t = 0$.

What might the Markov chain of Figure 8.1 be modeling? Consider that we have a modulo 3 counter which we start at zero. As it increments by unity, the counter moves through the states $0, 1, 2$ and then returns to 0. Assume we have three Bernoulli processes:

Process number 0, with $p_0(1) = 0.2$

Process number 1, with $p_1(1) = 0.5$

Process number 2, with $p_2(1) = 0.6$

The rule for the Markov chain at an allowed transition time is as follows: if the chain is in state i, $i = 0, 1$, or 2, the ith Bernoulli process is sampled. The counter is incremented if and only if the output of the Bernoulli process is a 1.

Let us look at the transition probability matrix, T, for this process. By inspection it is seen to be

$$T = \begin{bmatrix} 0.8 & 0.2 & 0 \\ 0 & 0.5 & 0.5 \\ 0.6 & 0 & 0.4 \end{bmatrix} \tag{10}$$

The value of representing transition probabilities by a matrix form is readily apparent when we ask the question, What is the probability that the process will be in a particular state after the next transition interval? We let $\mathbf{S}^T = [s_0 \quad s_1 \quad s_2]$ be the row vector of probabilities that the chain is in states s_0, s_1, or s_2. By forming the product $\mathbf{S}^T T$ we have a compact form for our answer. For the example of Figure 8.1, $S^T T =$ $[0.8s_0 + 0.6s_2 \quad 0.2s_0 + 0.5s_1 \quad 0.5s_1 + 0.4s_2]$. Thus, for example, the probability that the chain will be in state s_0 after a transition interval is 0.8 times the probability that it was in state s_0 prior to the transition interval *plus* 0.6 times the probability it was in state s_2 prior to the transition interval.

A basic property of the matrix T is that it is a "row-wise stochastic" matrix. This means that each element of t_{ij} satisfies $0 \leq t_{ij} \leq 1$ and that the elements in each row of T sum to unity.

Problem

Prove that two row-wise stochastic matrices, A and B, produce a row-wise stochastic matrix product.

One of the nicest things that happens when you are able to characterize a process as a Markov chain is the extreme easiness of calculating the probabilities of long, conditional events. Suppose, for example, we wanted to know the probability of being in state 2 (of Figure 8.1) after ten allowed transition times. Just for this simple example, consider that each state has two possible successors and therefore, by "brute force," we would have to lay out 2^{10} or over a thousand possible "chains." But because Figure 8.1 represents a Markov chain, the Markov property allows us to perform these seemingly involved calculations in a very compact and efficient manner. All we need to do to calculate the transition probability matrix for m allowed transition times, versus a single allowed transition time, is to compute T^m.

Problem

Compute T^2 for the example of Figure 8.1 and then derive the elements of the transition matrix by a more brute force method, that is, compute expressions of the form:

probability (of a transition from state i to state j)
   $= \sum_k$ probability (of a transition from state i to state k in one transition) · probability (of a transition from state k to state j in one transition)

Note that this preceding form is the canonical form for the computation of matrix element ij in the multiplication of two matrices. Compare the results of these brute force computations with the elements in T^2.

If the Markov chain passes through at least one recurring state which has at least two nonzero transition probabilities, then we can calculate what we call a stationary or steady-state distribution of state probabilities. This is a very useful concept and can be intuitively grasped by considering the question, If we do not observe the Markov chain for a great many allowed transition times, what are the probabilities that we will find the chain in state 0, state 1, ... and so on? We can answer this question by a very simple argument. Let $\mathbf{S}^{T'}$ be the vector of steady-state probabilities, that is, $\mathbf{S}^{T'} = [s'_0 \quad s'_1 \quad \cdots]$ where s'_i is the probability of finding the chain in state i after a great many steps (formally the limit as the number of steps goes to infinity). The argument now proceeds as follows. Because the transition probabilities are steady state, it should not matter if, in carrying out an ensemble of measurements, we calculate the probabilities by averaging the outcomes after m steps ($m \gg 1$) or $m + 1$ steps; therefore we may write

$$\mathbf{S}^{T'} = \mathbf{S}^{T'} T \tag{11}$$

Writing (11) in terms of $\mathbf{S}^{T'}$ and T we get

$$[s'_0 \quad s'_1 \quad s'_2] = [s'_0 \quad s'_1 \quad s'_2] \begin{bmatrix} 0.8 & 0.2 & 0 \\ 0 & 0.5 & 0.5 \\ 0.6 & 0 & 0.4 \end{bmatrix} \tag{12}$$

In attempting to solve (12), and also in general, we find that we need one additional equation because the equations are not independent. The additional equation is provided by the observation that

$$\sum_i s'_i = 1 \tag{13}$$

Solving (12) and (13) we find that

$$\mathbf{S}^{T'} = [s'_0 \quad s'_1 \quad s'_2] = [\tfrac{15}{26} \quad \tfrac{6}{26} \quad \tfrac{5}{26}] \tag{14}$$

Another way of looking at (14) is to say that, on the average, the Markov chain of Figure 8.1 "spends" $\tfrac{15}{26}$ of its time in state 0, $\tfrac{6}{26}$ of its time in state 1, and $\tfrac{5}{26}$ of its time in state 2. Note that the eigenvector of T^T corresponding to the eigenvalue 1 is given by s' in (14).

Problems

1. Calculate T^m for a number of "octaves," T^2, $T^4 = (T^2)^2$, $T^8 = (T^4)^2$, $T^{16} = (T^8)^2$, and observe how all the entries in any particular column approach the same values. Why is this?

2. There are some Markov chains of special type which are very useful in modeling many situations of interest and for which convenient, special form solutions exist. One of these is the birth/death chain. This process concerns a chain of $n + 1$ states as shown in Figure 8.2.

The following rules apply at each allowable transition time:

1. For state i, $0 < i < n$, the probability of transition to state $i + 1$ is p_i; the probability of transition to state $i - 1$ is q_i.
2. For state 0, the probability of remaining in state 0 is $1 - p_0$; the probability of transition to state 1 is p_0.
3. For state n, the probability of remaining in state n is $1 - q_n$; the probability of transition to state $n - 1$ is q_n.

(Note that states 0 and n are the only states in which the chain may remain at an allowed transition time.) Define

$$S_i = \begin{cases} 1, & i = 0 \\ \dfrac{p_0 \cdots p_{i-1}}{q_1 \cdots q_i}, & i \geq 1 \end{cases}$$

Show that the steady-state probability vector, S', has components

$$S'_i = \frac{S_i}{\sum_{j=0}^{n} S_j}$$

Now let all the p_i and q_i be $\frac{1}{2}$. Note that this situation corresponds to an "up/down counter," whose bottom range is 0 and whose top range is n. The counter is driven by a balanced Bernoulli source such that if the source produces a zero, the up/down counter is decremented by one unless it is at zero in which case it remains at zero; if the source produces a one, the counter is incremented by one unless it is at n in which case it remains at n. What is the vector of steady-state probabilities? How would you pick the p_i and q_i in order that the steady-state probabilities would approximate the binomial probability density function with parameters $n - 1$ and $p = 0.5$? Can you do it exactly?

A Markov chain may have one or more accessible absorbing states. The behavior of these chains is very different from the ones we have just

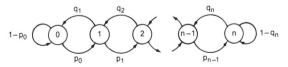

Figure 8.2. The birth/death Markov chain.

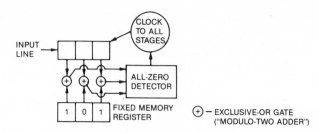

Figure 8.3. Electronic lock.

studied. To motivate the discussion, let us assume that we have an electronic "lock" that is composed of a three-stage shift register. The register accepts an input bit at each allowable transition interval and shifts its bits one stage to the right. If the register ever contains 101, the register locks up and no longer accepts new bits nor does it shift. It may also perform some external process such as locking or unlocking a vault, actuating a machine, or perhaps even triggering a detonator. The shift register lock is shown in Figure 8.3.

Figure 8.4 shows the Markov chain diagram exhibited by the machine of Figure 8.3. The three bits in the states correspond to the three bits in the register read in natural fashion, left to right. (We have assumed that the register was started in the all zero state.) The bits aside the transition arrows are the bits on the input line to the shift register at the allowed transition time. The absorbing state is shaded. Note that the minimum number of steps, or bits input to the lock, before the register can enter the locking state is 3. If we had started the chain in state 111, that is, had initially set the register to all ones, the minimum required number of steps would be 2. How long, on the average, will it take the register to enter the locking state if bits are entered randomly, from a balanced Bernoulli source? The mathematics needed to answer this question using Markov chain theory is quite straightforward and we follow an intuitive approach used by Kemeny et al. (1966) and also cite results from Kemeny and Snell (1960).

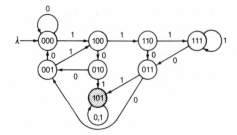

Figure 8.4. Markov chain diagram for the electronic lock of Figure 8.3.

For an absorbing chain, our transition probability matrix, T, can be partitioned as follows:

$$T = \begin{array}{c} \\ \{a\} \\ \{t\} \end{array} \begin{array}{c} \{a\} \quad \{t\} \\ \left[\begin{array}{c|c} I & 0 \\ \hline R & Q \end{array} \right] \end{array} \tag{15}$$

where $\{a\}$ represents the set of absorbing states and $\{t\}$ represents the set of transient states. A little reflection should convince the reader that the partition in (15) is both always possible and, indeed, natural.

Problem

Show that for a T of form (15),

$$T^k = \begin{bmatrix} I & 0 \\ (I + Q + Q^2 + \cdots + Q^{k-1})R & Q^k \end{bmatrix} \tag{16}$$

Now let us define a new matrix, N, whose elements, n_{ij}, are the mean number of times the chain will be in transient state j if it is started in transient state i. It is obvious that N must be of the same dimension as Q and it is not hard to see that

$$N = \sum_{k=0}^{\infty} Q^k \tag{17}$$

Here it is assumed that $\lim_{k \to \infty} Q^k$ exists, since Q is a probability transition matrix.

Expanding (17) we see that

$$N = I + Q + Q^2 + Q^3 + \cdots \tag{18}$$

Multiplying (18) by $I - Q$ we find that

$$N(I - Q) = (I + Q + Q^2 + Q^3 + \cdots)(I - Q) = I \tag{19}$$

and thus we have the important result that

$$N = (I - Q)^{-1} \tag{20}$$

Looking at our example in Figure 8.4, we see that we can immediately partition our matrix to comport to the form of (15) as follows:

$$
T = \begin{array}{c} \\ 101 \\ 000 \\ 001 \\ 010 \\ 011 \\ 100 \\ 110 \\ 111 \end{array}
\begin{array}{cccccccc}
101 & 000 & 001 & 010 & 011 & 100 & 110 & 111 \\
\end{array}
\left[
\begin{array}{c:ccccccc}
1 & 0 & 0 & 0 & 0 & 0 & 0 & 0 \\ \hdashline
0 & \frac{1}{2} & 0 & 0 & 0 & \frac{1}{2} & 0 & 0 \\
0 & \frac{1}{2} & 0 & 0 & 0 & \frac{1}{2} & 0 & 0 \\
\frac{1}{2} & 0 & \frac{1}{2} & 0 & 0 & 0 & 0 & 0 \\
\frac{1}{2} & 0 & \frac{1}{2} & 0 & 0 & 0 & 0 & 0 \\
0 & 0 & 0 & \frac{1}{2} & 0 & 0 & \frac{1}{2} & 0 \\
0 & 0 & 0 & 0 & \frac{1}{2} & 0 & 0 & \frac{1}{2} \\
0 & 0 & 0 & 0 & \frac{1}{2} & 0 & 0 & \frac{1}{2} \\
\end{array}
\right]
\tag{21}
$$

The matrix Q is the bottom right 7×7 component and we proceed to form $I - Q$:

$$
I - Q = \begin{bmatrix}
\frac{1}{2} & 0 & 0 & 0 & -\frac{1}{2} & 0 & 0 \\
-\frac{1}{2} & 1 & 0 & 0 & -\frac{1}{2} & 0 & 0 \\
0 & -\frac{1}{2} & 1 & 0 & 0 & 0 & 0 \\
0 & -\frac{1}{2} & 0 & 1 & 0 & 0 & 0 \\
0 & 0 & -\frac{1}{2} & 0 & 1 & -\frac{1}{2} & 0 \\
0 & 0 & 0 & -\frac{1}{2} & 0 & 1 & -\frac{1}{2} \\
0 & 0 & 0 & -\frac{1}{2} & 0 & 0 & \frac{1}{2} \\
\end{bmatrix}
\tag{22}
$$

Taking the inverse of $I - Q$ we obtain the following:

$$
N = \begin{bmatrix}
3 & 1 & 1 & 1 & 2 & 1 & 1 \\
2 & 2 & 1 & 1 & 2 & 1 & 1 \\
1 & 1 & \frac{3}{2} & \frac{1}{2} & 1 & \frac{1}{2} & \frac{1}{2} \\
1 & 1 & \frac{1}{2} & \frac{3}{2} & 1 & \frac{1}{2} & \frac{1}{2} \\
1 & 1 & 1 & 1 & 2 & 1 & 1 \\
1 & 1 & \frac{1}{2} & \frac{3}{2} & 1 & \frac{3}{2} & \frac{3}{2} \\
1 & 1 & \frac{1}{2} & \frac{3}{2} & 1 & \frac{1}{2} & \frac{5}{2} \\
\end{bmatrix}
\tag{23}
$$

We now form the column vector τ by the matrix multiplication

$$
\tau = N 1_R
\tag{24}
$$

where 1_R is a column vector of ones. The ith row of the vector τ is then simply the sum of the entries in the ith row of N. What does the ith row of τ mean? Recall that the elements of the ith row of the matrix N are the mean number of times that the chain will be in the different transient states if the chain is started in transient state i. Thus, the ith row of τ gives the mean time to absorption if the chain is started in transient state i. For our example

$$
\tau = \begin{array}{c} 000 \\ 001 \\ 010 \\ 011 \\ 100 \\ 110 \\ 111 \end{array} \begin{bmatrix} 10 \\ 10 \\ 6 \\ 6 \\ 8 \\ 8 \\ 8 \end{bmatrix} \tag{25}
$$

where we have again written the seven transient states to the left of the vector for clarity. From (25) we see that we expect the chain to reach absorption after ten steps, on the average, if started from transient state 000. But wait! Isn't this the same problem as observing a balanced Bernoulli source until the pattern 101 occurs? Indeed it is, and we recall from the problem following (3) that the mean waiting time is given by $d(2)$ where $d(x) = \sum_{r=1}^{k} e_r(1, 1)x^r$. For the pattern 101, $e_1(1, 1) = 1$, $e_2(1, 1) = 0$, $e_3(1, 1) = 1$, and $d(2) = 10$ which does agree with the approach via Markov chain theory. Note that starting from 000 is like starting from "scratch" for the Bernoulli source. Therefore, both methods should give the same result. What about the other entries in (25)? You can not directly use (25) for any three bits. What do you do? A column vector giving the variances, τ_{var}, is obtained by (Kemeny and Snell, 1960)

$$
\tau_{\text{var}} = (2N - I)\tau - \tau_{\text{sq}} \tag{26}
$$

where τ_{sq} is the τ vector but with its original elements squared. For our example,

$$
\tau_{\text{sq}} = \begin{bmatrix} 100 \\ 100 \\ 36 \\ 36 \\ 64 \\ 64 \\ 64 \end{bmatrix} \tag{27}
$$

and

$$(2N - I)\tau = \begin{bmatrix} 158 \\ 158 \\ 90 \\ 90 \\ 120 \\ 120 \\ 120 \end{bmatrix} \tag{28}$$

and thus

$$\tau_{\text{var}} = \begin{bmatrix} 58 \\ 58 \\ 54 \\ 54 \\ 56 \\ 56 \\ 56 \end{bmatrix} \tag{29}$$

Again, let us momentarily return to Section 8.1 and recall that the variance of the waiting time to a particular pattern from a balanced Bernoulli source is given by $d^2(2) + d(2) - 4d'(2)$. For our pattern of 101, $d'(x) = 1 + 3x^2$ and $d'(2) = 13$. Thus $d^2(2) + d(2) - 4d'(2) = 100 + 10 - 4 \cdot 13 = 58$ and again we certify agreement between the two approaches.

It is often useful to know the power spectral density exhibited by a Markov chain. Sittler (1956) authored an insightful and quite delightful paper on this subject and his results are easy to apply. The first step is to construct a "transfer function" matrix, P, from the transition probability matrix, T. This is done by the relation

$$P = (I - zT)^{-1} \tag{30}$$

where I is the identity matrix and z is a variable which we will set later. The second step is to determine the steady-state probabilities, S'. We are now ready to form what is sometimes called the right-half-plane power spectral density $\Phi^+(z)$ by evaluating

$$\Phi^+(z) = \sum_{i,j} a_i a_j s_i' P_{ij} \tag{31}$$

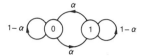

Figure 8.5. General two-state, symmetric Markov chain.

where a_k is the value assumed by the Markov chain in state k. The complete power spectral density, $\Phi(\omega)$, is formed by setting

$$\Phi(z) = \Phi^+(z) + \Phi^+(z^{-1}) \tag{32}$$

and evaluating (32) for $z = e^{j\omega}$.

As an example, let us evaluate the power spectral density of a balanced Bernoulli source. For this Markov chain

$$T = \begin{bmatrix} \frac{1}{2} & \frac{1}{2} \\ \frac{1}{2} & \frac{1}{2} \end{bmatrix} \quad \text{and} \quad P = \begin{bmatrix} 1 - z/2 & -z/2 \\ -z/2 & 1 - z/2 \end{bmatrix}^{-1}$$

$$= \frac{1}{1 - z} \begin{bmatrix} 1 - z/2 & z/2 \\ z/2 & 1 - z/2 \end{bmatrix}$$

The steady-state probabilities are, of course, $\mathbf{S}^{T'} = (\frac{1}{2}, \frac{1}{2})$ and $a_0 = 0$, $a_1 = 1$. Plugging in, we find that the only term that contributes is that for which $i = j = 1$ (Why?) and so

$$\Phi^+(z) = \frac{1}{2} \frac{1}{1 - z}\left(1 - \frac{z}{2}\right) \quad \text{and} \quad \Phi(z) = \Phi^+(z) + \Phi^+(z^{-1}) = \frac{3}{4} = \Phi(\omega).$$

Note that the power spectral density is flat across all the angular frequencies ω. Finally, the computation of (30) can be messy for large sizes of T. A simple method is to use $P = (1/z)(z^{-1}I - T)^{-1}$. This has the same general form as $[\lambda(\lambda I - A)^{-1}]$, where $z^{-1} = \lambda$ and $A = T$. See Chapter 6 for an efficient computation of $(\lambda I - A)^{-1}$.

Problem

Calculate $\Phi(\omega)$ for the general two-state, symmetric Markov chain shown in Figure 8.5 and plot $\Phi(\omega)$ for $\alpha = 0.4$ and $\alpha = 0.6$.

8.3. *m*-Sequences (Hershey, 1982)

The theory of maximum length (linearly generated) binary sequences, or *m*-sequences, is one of the most mathematically aesthetic disciplines of finite field theory. The theory offers far more than aesthetics, however, as m-sequences are extensively used by electrical engineers, particularly in the communications, radar, navigation, and computer disciplines. The theory behind m-sequences is sufficiently well developed and, for lack of a better word, "modular," so that even one who is not a mathematician can manipulate and apply powerful results to create new useful architectures and uncover new truths.

The sequences are an important subclass of recursively generated binary sequences which are defined by

$$s(t) = f(s(t-1), s(t-2), \ldots, s(t-n)), \qquad s(i) \in \{0, 1\} \qquad (33)$$

which states that the bit at time t is precisely dependent on the n bits preceding it. The sequence so produced is sometimes said to be of "span n" (Golomb, 1980). We see from (33) that the sequence is deterministically generated and we immediately deduce that the sequence will eventually give rise to a cycle or recurrent (in the Markov sense) set of states where a state is defined as the n-tuple

$$(s(\tau-1), s(\tau-2), \ldots, s(\tau-n)) \qquad (34)$$

We also deduce that the maximum possible cycle length must be bounded by the number of possible tuples of the form (34) which is 2^n. We also observe that because f in (33) is a function of n binary terms, f may be viewed as a boolean function of n variables. There are 2^{2^n} possible functions and thus 2^{2^n} possible recurrences. (See p. 12 of Golomb, 1967.)

If and only if f is expressible as a modulo-2 sum of terms, that is,

$$f = \sum_{i=1}^{n} \alpha_i s(t-i), \qquad \alpha_i \in \{0, 1\} \qquad (35)$$

then f is said to be a linear function and the recursively generated sequence is said to be linearly generated. We can, incidentally, view a linearly generated recursive sequence as a nonlinear difference equation using regular (nonmodular) mathematics. For example, the modulo-2 linear recursive sequence

$$s(t) = s(t-3) + s(t-5) \qquad (36)$$

where, according to our convention, the plus sign is modulo-2 addition, can be written

$$s(t) = s(t-3) + s(t-5) - 2s(t-3)s(t-5) \qquad (37)$$

where the plus and minus signs in (37) imply regular addition and subtraction. When, and only when, all the $s(i)$ are zeros and ones will (37) be the same as (36). In this one special case, a periodic solution obtains to the nonlinear difference equation (37).

One further general remark is in order. If and only if f can be written

$$f = g(s(t-1), s(t-2), \ldots, s(t-n+1)) + s(t-n) \qquad (38)$$

where g may be any boolean function of $n-1$ variables, will the sequence of tuples (34) be such that every tuple has a unique predecessor. This is an obvious, yet very powerful truth and is well presented by Golomb (1967, p. 116). Note that all span-n linear functions are of the form (38).

The study of sequences is in and of itself a tremendous undertaking. We do not pretend to even try. Why then do we select one particular class of sequences, namely, the m-sequences? The answer is twofold. First, the theory behind m-sequences is seasoned, tractable, and rich. Second, and most important to an engineer, m-sequences are useful, primarily because of their randomlike qualities. To motivate further we again return to Solomon Golomb, who, in his famous book (1967, pp. 25–26), sets three "randomness postulates" or three properties or characteristics one would expect or demand from sequences purported to be random. Before recounting these properties we must comment that there is a subtle "doublethink" involved. Because our sequences will be deterministically generated, they will exhibit a period. They are thus anything but random. What we are addressing is a study of their "short-term" behavior which is taken to be their statistical analysis over a single period only. Thus, we must (silently) preface our use of the word random with the prefix "pseudo." Golomb's postulates for sequence randomness are then:

1. $\left| p - 2 \sum_{i=1}^{p} s(i) \right| \leq 1$, where p is the sequence period.
2. To the extent that the period can be subdivided, the number of runs of zeros and ones exhibited must fall in inverse geometric proportion to their lengths, that is, half the runs should be 1-long, one quarter 2-long, and so on.
3. The autocorrelation, $R_{ss}(\tau) = \sum_{i=1}^{p} s(i)s(i+\tau)$, must be two valued, that is, $R_{ss}(0) = (p+1)/2$ and $R_{ss}(\tau) = \rho \neq (p+1)/2$, $\tau \neq 0$.

If a sequence comports to the above requirements, it is termed a pseudonoise or PN sequence. The following 31 bit period sequence (this sequence and all other sequences should be read from left to right)

$$1111100011011101010000100101100 \qquad (39)$$

meets all three postulates, that is,

1. There are 16 ones and 15 zeros.
2. There are 16 runs distributed as follows:

(a) four runs of zeros and ones each of length 1,
(b) two runs each of length 2,
(c) one run each of length 3,
(d) one run of zeros of length 4,
(e) one run of ones of length 5.
3. The autocorrelation is 16 for $\tau = 0$ and 8 otherwise.

The PN sequence (39) was generated by the recursion

$$s(t) = s(t-3) + s(t-5). \tag{40}$$

Because f is of the form (35), the recursion is linearly generated.

We thus have a hint that linearly generated sequences might be useful as PN sequence generators. Let us try another (arbitrarily chosen) linear generator, say, the recursion

$$s(t) = s(t-4) + s(t-5) \tag{41}$$

We easily find that (41) if started with the all ones tuple, gives rise to the sequence

$$111110000100011001010 \tag{42}$$

Sequence (42) is of period 21. Checking, we find that it satisfies randomness postulate (1). Sequence (42) exhibits a total of ten runs. Of the ten runs, half are of length one but it is required that the number of 1-long runs of ones must equal the number of 1-long runs of zeros which is impossible because 5 is an odd number. Hence, sequence (42) fails to meet the second randomness postulate. Sequence (42) also fails to meet the third postulate because it exhibits autocorrelation values of $\{10, 5, 4, \text{and } 3\}$.

Why has one linear generator of span equal to 5 produced a PN sequence and another linear span-5 generator failed? With further experimentation, we would come to the hypothesis that only and all linear generators of span n whose sequences exhibit periods equal to $2^n - 1$ produce PN sequences. Golomb (1967, pp. 43–45) establishes the sufficiency of the hypothesis, that is, a linear generator of span n that exhibits a period of length $2^n - 1$ must indeed produce a PN sequence. Not established, but *conjectured* by Golomb (1980) is the (even strengthened) necessity that the PN sequences are solely composed of maximum length, linearly generated sequences. Consider, for example, the nonlinearly generated sequence produced by the recursion

$$s(t) = s(t-1) + s(t-5) + \bar{s}(t-1)s(t-2)s(t-3) \tag{43}$$

(where the overbar on the s denotes complementation). Starting (43) with the all ones tuple we obtain the sequence

$$1111100101110101001101100010000 \qquad (44)$$

Sequence (44) meets randomness postulates (1) and (2) but exhibits autocorrelation values of $\{16, 9, 8, \text{ and } 7\}$ and hence fails postulate (3).

The longest, or maximum length, sequence that can be produced by a recursion of the form

$$s(t) = \sum_{i=1}^{n} \alpha_i s(t - i) \qquad (45)$$

is clearly $2^n - 1$ (as the all zero n-tuple will immediately perpetuate itself), hence the term m-sequence is given to such a sequence. [The authors believe that the term "m-sequence" was coined by Zierler (1959, p. 39).]

The period of a linearly generated recursive sequence can be straightforwardly analyzed by using generating functions (Golomb, 1967, pp. 30–33) or by Z-transform theory (Charney and Mengani, 1961). The essence of the theory is that the polynomial

$$1 + \sum_{i=1}^{n} \alpha_i x^i \qquad (46)$$

where the α_i in (46) are the same as in (45), is either reducible, divisible without remainder by a polynomial of degree d, $1 < d < n$, or irreducible. The recursion corresponding to a reducible polynomial, such as $x^4 + x^2 + 1 = (x^2 + x + 1)^2$, can not produce an m-sequence. The recursion corresponding to an irreducible polynomial will not necessarily produce an m-sequence, however. Irreducible polynomials are further dichotomized into those that are termed primitive, such as $x^4 + x + 1$, and those that are not, such as $x^4 + x^3 + x^2 + x + 1$. An irreducible degree n polynomial is primitive if and only if the smallest m for which it divides $x^m + 1$ is $m = 2^n - 1$. It is therefore redundant to say "irreducible and primitive" as primitivity implies irreducibility. Irreducibility implies primitivity only for Mersenne primes, that is, when $2^n - 1$ is prime.

The number of primitive polynomials of degree n is well known and equal to

$$\frac{\phi(2^n - 1)}{n} \qquad (47)$$

where ϕ is Euler's "totient" or phi function discussed in Chapter 3.

Finding primitive polynomials is then of great importance. Attempting to factor polynomials is, of course, the first step in proving primitivity as a primitive polynomial must be irreducible. Factoring has become an exciting field in and of itself. An early important method is due to Berlekamp (1967, 1970). A good review of the early methods along with some excellent exercises is given in Knuth (1969). For more recent work, the reader is invited to review Moenck's method (1977) which incorporates a time-saving refinement of Berlekamp's method. Another recent contribution is a new factoring algorithm by Cantor and Zassenhaus (1981) that is of the "probabilistic genre." The probabilistic algorithms are presently in vogue in all sorts of fields and portend, in the authors' opinion, to be a powerful and revolutionary approach to some classically difficult computations.

Tables of primitive polynomials are readily available. The earliest and most famous compendium is Marsh's set of tables (1957) which contains an exhaustive listing of all primitive polynomials through degree 19. Watson's list (1962) provides a primitive polynomial for each $n \leq 100$ and $n = 107$ and $n = 127$. Watson's list was followed by Stahnke's list (1973) which presents a primitive polynomial for all $n \leq 168$. Table 2A is an extract of Stahnke's compendium for the range $2 \leq n \leq 64$.

Table 2A. Extract of Primitive Polynomials from Stahnke's Compendium (1973). To interpret 19, 6, 5, 1, 0, for Example, Write $x^{19} + x^6 + x^5 + x + 1$

2, 1, 0	23, 5, 0	44, 27, 26, 1, 0
3, 1, 0	24, 4, 3, 1, 0	45, 4, 3, 1, 0
4, 1, 0	25, 3, 0	46, 21, 20, 1, 0
5, 2, 0	26, 8, 7, 1, 0	47, 5, 0
6, 1, 0	27, 8, 7, 1, 0	48, 28, 27, 1, 0
7, 1, 0	28, 3, 0	49, 9, 0
8, 6, 5, 1, 0	29, 2, 0	50, 27, 26, 1, 0
9, 4, 0	30, 16, 15, 1, 0	51, 16, 15, 1, 0
10, 3, 0	31, 3, 0	52, 3, 0
11, 2, 0	32, 28, 27, 1, 0	53, 16, 15, 1, 0
12, 7, 4, 3, 0	33, 13, 0	54, 37, 36, 1, 0
13, 4, 3, 1, 0	34, 15, 14, 1, 0	55, 24, 0
14, 12, 11, 1, 0	35, 2, 0	56, 22, 21, 1, 0
15, 1, 0	36, 11, 0	57, 7, 0
16, 5, 3, 2, 0	37, 12, 10, 2, 0	58, 19, 0
17, 3, 0	38, 6, 5, 1, 0	59, 22, 21, 1, 0
18, 7, 0	39, 4, 0	60, 1, 0
19, 6, 5, 1, 0	40, 21, 19, 2, 0	61, 16, 15, 1, 0
20, 3, 0	41, 3, 0	62, 57, 56, 1, 0
21, 2, 0	42, 23, 22, 1, 0	63, 1, 0
22, 1, 0	43, 6, 5, 1, 0	64, 4, 3, 1, 0

Unlike Watson's table, Stahnke's uses trinomials when a primitive trinomial exists for a particular n and a pentanomial otherwise. (Why are tetranomials excluded?) The trinomial listed is of the form $x^n + x^a + 1$ and the a listed is as small as possible. When Stahnke was forced to choose a pentanomial, he chose it to be of the form $x^n + x^{b+a} + x^b + x^a + 1$ with $0 < a < b < n - a$ and a as small as possible with b also as small as possible to comport with Scholefield's architecture (1960) which we shall examine later.

If we examine Stahnke's valuable list, we are struck by the lack of regularity or patterns in the polynomials. What, for instance, determines whether or not there exists a primitive trinomial for a given n? This and other similar questions broach the frontiers of knowledge of this corner of abstract algebra. There have been rents in the curtain of ignorance, however. Swan (1962) uncovered a number of criteria under which certain polynomial forms must be reducible and hence not primitive. Perhaps Swan's most general, at least most easily remembered, and, to the authors, most exciting, rule is that the trinomial

$$x^{8k} + x^m + 1, \qquad m < 8k \tag{48}$$

is always reducible. Thus, we at least understand the absence of primitive trinomials for $n = 8, 16, 24, \ldots$

Zierler and Brillhart (1968, 1969) have catalogued all the irreducible trinomials for $n \leq 1000$ and, where the factorization of $2^n - 1$ was known, have indicated those irreducible trinomials which were found to be primitive. Zierler (1969), adding mainly to work done by Rodemich and Rumsey (1968), has catalogued all primitive trinomials for the first 23 Mersenne primes (see Table 2B). Finally, Zierler (1970) has shown that the trinomial $x^n + x + 1$ is primitive for $n = 2, 3, 4, 6, 7, 15, 22, 60, 63, 127, 153,$ and 532.

Once we have a primitive polynomial we can construct a sequential machine out of memory elements and EXCLUSIVE-OR gates which will exhibit a cycle of $2^n - 1$ distinct states. Our first construction is a "natural" construction and to do it we use elementary matrix theory.

The characteristic polynomial of a matrix, M, where the entries in M are over GF(2), is, of course, obtained by evaluating the determinant

$$|M + \lambda I_n| \tag{49}$$

and is written

$$|M + \lambda I_n| = \lambda^n + \lambda^{n-1} c_{n-1} + \cdots + \lambda c_1 + c_0 = 0 \tag{50}$$

Table 2B. Zierler's Catalog of Primitive Trinomials for the First 23 Mersenne Primes. p is Given Where $2^p - 1$ is Prime along with the Primitive Trinomial

2	2, 1, 0	4253	None exist
3	3, 1, 0	4423	4423, 271, 0
5	5, 2, 0		4423, 369, 0
7	7, 1, 0		4423, 370, 0
	7, 3, 0		4423, 649, 0
13	None exist		4423, 1393, 0
17	17, 3, 0		4423, 1419, 0
	17, 5, 0		4423, 2098, 0
	17, 6, 0	9689	9689, 84, 0
19	None exist		9689, 471, 0
31	31, 3, 0		9689, 1836, 0
	31, 6, 0		9689, 2444, 0
	31, 7, 0		9689, 4187, 0
	31, 13, 0	9941	None exist
61	None exist	11213	None exist
89	89, 38, 0		
107	None exist		
127	127, 1, 0		
	127, 7, 0		
	127, 15, 0		
	127, 30, 0		
	127, 63, 0		
521	521, 32, 0		
	521, 48, 0		
	521, 158, 0		
	521, 168, 0		
607	607, 105, 0		
	607, 147, 0		
	607, 273, 0		
1279	1279, 216, 0		
	1279, 418, 0		
2203	None exist		
2281	2281, 715, 0		
	2281, 915, 0		
	2281, 1029, 0		
3217	3217, 67, 0		
	3217, 576, 0		

By the Cayley-Hamilton Theorem (see Chapter 6) then, we have

$$M^n + c_{n-1}M^{n-1} + \cdots + c_1 M + c_0 I_n = 0 \qquad (51)$$

We observe that (51) guarantees (constructively) that powers of M greater than or equal to n can be expressed by a linear combination of the set of matrices

$$\{I_n, M, M^2, \ldots, M^{n-1}, M^{n-1}\} \tag{52}$$

If the characteristic polynomial is primitive then each member of the set

$$\{I_n, M, M^2, \ldots, M^{2^n} - 2\} \tag{53}$$

will be distinct and $M^{2^n-1} = I_n$.

Given a primitive polynomial

$$f(\lambda) = \lambda^n + \lambda^{n-1}c_{n-1} + \cdots + \lambda c_1 + 1 \tag{54}$$

we can immediately write a companion matrix whose characteristic polynomial will be $f(\lambda)$:

$$M_c = \begin{bmatrix} 0 & \cdots & 0 & 1 \\ & & & c_1 \\ & & & c_2 \\ & I_{n-1} & & \vdots \\ & & & c_{n-1} \end{bmatrix} \tag{55}$$

Observe that M_c is merely the state transition matrix for the most common realization of an m-sequence generator shown in Figure 8.6. If we denote the content of the n-stage shift register's stage i at time t as $\mathbf{b}^T(t) = [b_{n-1}(t) \quad b_{n-2}(t) \quad \ldots \quad b_0(t)]$, then

$$\mathbf{b}^T(t+1) = \mathbf{b}^T(t)M_c \tag{56}$$

For the generator of an m-sequence, we can tap any one of the register stages. An example would be the sequence obtained from $\{b_0(t)\}$.

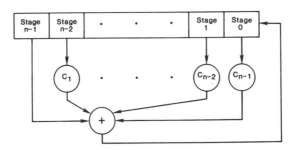

Figure 8.6. Companion matrix structure.

We must provide a note of caution. It has become traditional to list primitive polynomials as polynomials in x, that is, $f(x)$. To create a matrix of the form (55) which will implement the sequence corresponding to these polynomials we must make the transformation $f(x) \to \lambda^n f(1/\lambda)$ to convert $f(x)$ to the polynomial of form (54); see Golomb (1967, p. 35). Thus, if we wish to construct the companion matrix for the primitive polynomial $x^3 + x^2 + 1$ we would use $\lambda^3(1/\lambda^3 + 1/\lambda^2 + 1) = \lambda^3 + \lambda + 1$ as the characteristic polynomial.

As an example, let us set up a shift register network for the primitive polynomial $f(x) = x^3 + x^2 + 1$. The matrix M_c is given by

$$M_c = \begin{bmatrix} 0 & 0 & 1 \\ 1 & 0 & 1 \\ 0 & 1 & 0 \end{bmatrix}$$

and its characteristic polynomial is given by $|\lambda I + M_c| = \lambda^3 + \lambda + 1$. The shift register realization is given in Figure 8.11. Note that the diagram can be obtained by inspection from the primitive polynomial itself. There is a tap from $b_{k-1}(t)$ if the coefficient of x^k is nonzero.

It is at this juncture that the pure mathematicians lose interest for, as they correctly assert, there is only one finite field of 2^n elements and the set of elements

$$\{I_n, M_c, M_c^2, \ldots, M_c^{2^n-2}\} \tag{57}$$

is "as good as" any other choice since all finite fields of the same order are isomorphic. See, for example, MacDuffie (1940, p. 180). But the isomorphisms of the Galois field can be of significant practical importance and should not be dismissed as mere mathematical curiosities. The following from Berlekamp (1968, p. 104) captures the essence of this thought:

> From an engineering standpoint, it is misleading to overstress the uniqueness of GF(p^k), for this field may have many different representations .... The design and cost of circuitry to perform calculations in GF(p^k) depend critically on the representation. For this reason, some engineers prefer to think of different representations as different fields. This viewpoint is particularly justified in solutions where the cost of transforming from one representation to another is large.

Well, just how many field representations are there and how can they help us? Consider the general 3×3 matrix:

$$E = \begin{bmatrix} e_{11} & e_{12} & e_{13} \\ e_{21} & e_{22} & e_{23} \\ e_{31} & e_{32} & e_{33} \end{bmatrix} \tag{58}$$

If we take the determinant of $E + \lambda I$ we obtain

$$\lambda^3 + \lambda^2(e_{11} + e_{22} + e_{33}) + \lambda(e_{11}e_{22} + e_{11}e_{33} + e_{22}e_{33} + e_{23}e_{32} + e_{12}e_{21}$$
$$+ e_{13}e_{31}) + (e_{11}e_{22}e_{33} + e_{11}e_{23}e_{32} + e_{12}e_{21}e_{33} + e_{12}e_{23}e_{31}$$
$$+ e_{13}e_{21}e_{32} + e_{13}e_{22}e_{31}) \tag{59}$$

There are two primitive polynomials of degree 3, namely,

$$\lambda^3 + \lambda + 1 \quad \text{and} \quad \lambda^3 + \lambda^2 + 1 \tag{60}$$

Direct solution of the $\{e_{ij}\}$ for these cases yielding the polynomials in (60) uncovers no fewer than 48 distinct matrices [24 for each of the primitive polynomials in (60)] which are arrayed in eight fields each with 2^3 members (each field contains I_3 and 0_3, the multiplicative and additive identities, respectively, which do not, of course, exhibit a primitive characteristic polynomial). The eight fields are shown in Figure 8.7.

The counting problem has been solved for general n. Reiner (1961) has derived an expression for the number of matrices that exhibit a particular characteristic polynomial. Following a little manipulation of Reiner's result, we find that the number of $n \times n$ matrices that possess a specific primitive characteristic polynomial is

$$\frac{R(n)}{2^n - 1} \tag{61}$$

where $R(n)$ is the number of regular or nonsingular $n \times n$ matrices, that is, the number of matrices whose determinant is unity. Explicitly,

$$R(n) = 2^{n^2} \left(\frac{1}{2}\right)\left(\frac{3}{4}\right)\left(\frac{7}{8}\right) \cdots \frac{2^n - 1}{2^n} \tag{62}$$

Recalling that there are $\phi(2^n - 1)/n$ primitive polynomials of degree n, we find that the number of matrices that can serve as finite field generators, or, equivalently, as "wiring schematics" for m-sequence generators, is

$$c(n) = \frac{\phi(2^n - 1)}{n} \cdot \frac{R(n)}{2^n - 1} \tag{63}$$

Note that for Mersenne primes (p and $2^p - 1$ both prime), $c(p) \to R(p)/p$. Table 3 demonstrates just how very rich the potential architectural schematics are.

	Field A	Field B	Field C	Field D	Field E	Field F	Field G	Field H
	000	000	000	000	000	000	000	000
O_3	000	000	000	000	000	000	000	000
	000	000	000	000	000	000	000	000
	111	111	111	111	111	111	111	111
M	110	110	101	101	100	100	011	011
	100	011	100	011	110	101	110	101
	101	010	110	001	101	110	010	001
M^2	001	001	011	100	111	111	101	110
	111	101	111	110	011	010	100	010
	011	110	010	011	001	011	011	101
M^3	100	011	001	111	101	110	001	100
	101	100	110	010	010	100	111	011
	010	001	101	110	110	001	101	010
M^4	111	101	100	001	001	011	110	111
	011	111	010	101	100	111	010	110
	110	011	011	010	011	101	001	011
M^5	101	100	111	011	110	001	100	001
	010	010	101	100	111	110	011	100
	001	101	001	101	010	010	110	110
M^6	011	111	110	110	011	101	111	101
	110	110	011	111	101	011	101	111
	100	100	100	100	100	100	100	100
$M^7 = I_3$	010	010	010	010	010	010	010	010
	001	001	001	001	001	001	001	001

Figure 8.7. Eight field representations.

As an example of a specific architecture, different from the companion matrix structure of Figure 8.6, let us consider and analyze the following machine:

1. There are n flip-flop memory elements. Their states at time t are denoted by $b_{n-1}(t), \ldots, b_0(t)$.

Table 3. Architectural Richness

n	$\dfrac{\phi(2^n - 1)}{n}$	$\dfrac{R(n)}{2^n - 1}$	$c(n)$
2	1	2	2
3	2	24	48
4	2	1,344	2,688
5	6	322,560	1,935,360

2. Their states at time $t + 1$ are derived as follows. We add $b_0(t)$ to $b_{n-1}(t)$. The result is $b_{n-1}(t + 1)$. We then add $b_{n-1}(t + 1)$ to $b_{n-2}(t)$. The result is $b_{n-2}(t + 1)$. We add $b_{n-2}(t + 1)$ to $b_{n-3}(t)$. The result is $b_{n-3}(t + 1)$. We proceed in this fashion until we have attained $b_0(t + 1)$ by adding $b_1(t + 1)$ to $b_0(t)$.

A little thought will convince the reader that

$$\mathbf{b}^T(t + 1) = (b_{n-1}(t) \cdots b_0(t))M = \mathbf{b}^T(t)M \tag{64}$$

where

$$M = \begin{bmatrix} 1 & 1 & 1 & \cdots & 1 & 1 & 1 \\ 0 & 1 & 1 & \cdots & 1 & 1 & 1 \\ 0 & 0 & 1 & \cdots & 1 & 1 & 1 \\ & & & \vdots & & & \\ 0 & 0 & 0 & \cdots & 0 & 1 & 1 \\ 1 & 1 & 1 & \cdots & 1 & 1 & 0 \end{bmatrix} \tag{65}$$

To find that characteristic polynomial of M given in (65) we write the determinant

$$f_n(\lambda) = \begin{vmatrix} \lambda+1 & 1 & 1 & \cdots & 1 & 1 & 1 \\ 0 & \lambda+1 & 1 & \cdots & 1 & 1 & 1 \\ 0 & 0 & \lambda+1 & \cdots & 1 & 1 & 1 \\ & & & \vdots & & & \\ 0 & 0 & 0 & \cdots & 0 & \lambda+1 & 1 \\ 1 & 1 & 1 & \cdots & 1 & 1 & \lambda \end{vmatrix} \tag{66}$$

Problem

Expand (66) by evaluating the minors specified by the two nonzero entries of the first column and use simple matrix algebra to derive the following recursion:

$$f_n(\lambda) = (\lambda + 1)f_{n-1}(\lambda) + \lambda^{n-2} \tag{67}$$

By direct computation we find that

$$f_2(\lambda) = \lambda^2 + \lambda + 1 \tag{68}$$

and then by recursive computation we find

$$f_3(\lambda) = \lambda^3 + \lambda + 1$$
$$f_4(\lambda) = \lambda^4 + \lambda^3 + 1 \tag{69}$$

Thus, we know that the finite-state linearly sequential machine defined by (64) and (65) will exhibit a maximum length cycle of states if and only if (67) is primitive.

Let us examine the succession of states for one of the primitive polynomials, $f_4(\lambda) = \lambda^4 + \lambda^3 + 1$. We start the machine in the all ones state and observe the following state progression:

$$
\begin{array}{cccc}
1 & 1 & 1 & 1 \\
0 & 1 & 0 & 1 \\
1 & 0 & 0 & 1 \\
0 & 0 & 0 & 1^* \\
1 & 1 & 1 & 0 \\
1 & 0 & 1 & 1 \\
0 & 0 & 1 & 0^* \\
0 & 0 & 1 & 1 \\
1 & 1 & 0 & 1 \\
0 & 1 & 1 & 0 \\
0 & 1 & 0 & 0^* \\
0 & 1 & 1 & 1 \\
1 & 0 & 1 & 0 \\
1 & 1 & 0 & 0 \\
1 & 0 & 0 & 0^*
\end{array}
$$

The unity density states are starred and we note that their positions seem to be at approximately equidistant spacings throughout the cycle. Will this be true in general for those cases in which $f_n(\lambda)$ is primitive, or is it merely fortuitous for this case?

First, let us define the density one or unit weight vectors as

$$\mathbf{u}_i^T = (00 \ldots 010 \ldots 0) \tag{70}$$

where the single 1 is in the position i and $1 \leqslant i \leqslant n$. Second, consider the immediate successor states of the unit-weight vectors. On multiplying $\mathbf{u}_n^T$

by M we obtain

$$\mathbf{u}_n^T M = \mathbf{u}_1^T + \mathbf{u}_2^T + \cdots + \mathbf{u}_{n-1}^T \tag{71}$$

where the sums in (71) are vector (modulo 2 sums, component by component). Similarly, multiplying M by $\mathbf{u}_1^T$ yields

$$\mathbf{u}_1^T M = \mathbf{u}_1^T + \mathbf{u}_2^T + \cdots + \mathbf{u}_{n-1}^T + \mathbf{u}_n^T \tag{72}$$

Summing (71) and (72) we get

$$\mathbf{u}_n^T M + \mathbf{u}_1^T M = \mathbf{u}_n^T \tag{73}$$

In a manner similar to the above we also derive the equation

$$\mathbf{u}_1^T M + \mathbf{u}_2^T M = \mathbf{u}_1^T \tag{74}$$

Now suppose that $\mathbf{u}_n^T$ and $\mathbf{u}_1^T$ are on the same cycle, which, of course, they will be if $f_n(\lambda)$ is primitive. There then exists an integer d such that

$$\mathbf{u}_1^T = \mathbf{u}_n^T M^d \tag{75}$$

Now if we add (73) to (74) we obtain

$$\mathbf{u}_n^T M + \mathbf{u}_2^T M = \mathbf{u}_n^T + \mathbf{u}_1^T \tag{76}$$

Because the inverse of M exists we can convert (76) to

$$\mathbf{u}_2^T + \mathbf{u}_n^T + \mathbf{u}_n^T M^{-1} = \mathbf{u}_1^T M^{-1} \tag{77}$$

On substituting (75) into (77) we get

$$\mathbf{u}_2^T + \mathbf{u}_n^T + \mathbf{u}_n^T M^{-1} = \mathbf{u}_n^T M^{d-1} \tag{78}$$

Postmultiplying (73) by M^{-1} we find that

$$\mathbf{u}_n^T + \mathbf{u}_n^T M^{-1} = \mathbf{u}_1^T \tag{79}$$

Substituting (79) into (78) we get

$$\mathbf{u}_2^T + \mathbf{u}_1^T = \mathbf{u}_n^T M^{d-1} \tag{80}$$

Substituting (75) for u_1 in (80) we get

$$\mathbf{u}_2^T = \mathbf{u}_n^T M^d + \mathbf{u}_n^T M^{d-1} = (\mathbf{u}_n^T + \mathbf{u}_n^T M) M^{d-1} \tag{81}$$

Using (73) we can rewrite (81) as

$$\mathbf{u}_2^T = (\mathbf{u}_1^T M) M^{d-1} = \mathbf{u}_1^T M^d \tag{82}$$

Using (75) we immediately rewrite (82) as

$$\mathbf{u}_2^T = \mathbf{u}_n^T M^{2d} \tag{83}$$

Equation (83) demonstrates that $\mathbf{u}_2^T$ is the same distance from $\mathbf{u}_1^T$ as $\mathbf{u}_1^T$ is from $\mathbf{u}_n^T$. For our example, we note that this distance is 11 steps. We can easily extend the above argument for all $\mathbf{u}_i^T$ up to $i = n - 1$, that is, if $\mathbf{u}_1^T = \mathbf{u}_n^T M^d$, then $\mathbf{u}_2^T = \mathbf{u}_n^T M^d$, $\mathbf{u}_3^T = \mathbf{u}_2^T M^d, \ldots, \mathbf{u}_{n-1}^T = \mathbf{u}_{n-2}^T M^d$. The final unit-weight vector, $\mathbf{u}_n^T$, however, is found to satisfy

$$\mathbf{u}_n^T = \mathbf{u}_{n-1}^T M^{d+1} \tag{84}$$

Thus, we have found that for the particular sequential machine defined by (64) and (65), the unit-weight vectors are distributed at approximately equal distances around the cycle. This fact can be deduced without the matrix-oriented argument presented but is intended as an example of how matrix arguments can be simply and efficaciously used.

8.3.1. Special Purpose Architectures

As we have seen in Figure 8.6, the companion matrix shift register realization if the simplest realization of an m-sequence generator. It may not be the "best," however; that all depends on what consitutes value to

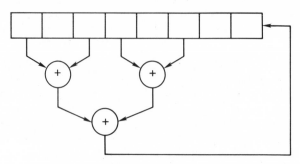

Figure 8.8. Implementation of $x^8 + x^6 + x^5 + x^3 + 1$.

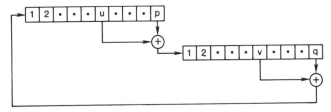

Figure 8.9. Parallel clocked structure.

the designer or implementer. As an example, let us assume that we wish to realize an m-sequence of period 255. As we have noted previously, Swan's criterion states that there are no primitive trinomials of degree n where n is divisible by 8. Thus, an implementation of the form shown in Figure 8.6 will require more than one two-input EXCLUSIVE-OR logic gate. As an example, let us choose the primitive pentanomial $x^8 + x^6 + x^5 + x^3 + 1$. It may be implemented as shown in Figure 8.8. Note that this implementation requires three modulo-2 adders. More important, the adders are layered to a depth of 2. The EXCLUSIVE-OR boolean function is not threshold realizable and, consequently, is often the time-limiting basic element in a logic family. How can we then obviate this annoying layering of relatively slow logic?

The answer can often be found through special architectures. Scholefield (1960) considers a variety of interconnected, parallel-clocked structures. For example, he shows that each stage of the structure shown in Figure 8.9 exhibits the recursion specified by the polynomial $x^{p+q} + x^{p+v} + x^{q+u} + x^{u+v} + 1$. By properly selecting the tetrad (p, q, u, v) it is possible to synthesize some polynomials in many ways with this particular structure. For our example, we can effect the realization of the pentanomial $x^8 + x^6 + x^5 + x^3 + 1 = (x^2 + x^5)(x + x^3) + 1$ with $(p, q, u, v) = (5, 3, 2, 1)$ as shown in Figure 8.10. The realization presented in Figure 8.10 has two

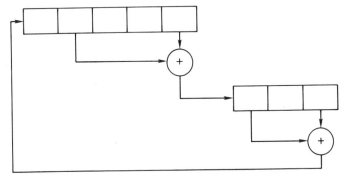

Figure 8.10. Realization of $x^8 + x^6 + x^5 + x^3 + 1$ by parallel clocked structure.

advantages over the realization presented in Figure 8.8. First, there is one less EXCLUSIVE-OR gate required. Second, and perhaps more important, there is no layering of EXCLUSIVE-OR gates.

Problem

(A Curious Architectural Property) Recall that the Cayley–Hamilton Theorem (51) requires that each element of M, the finite field generator matrix, exhibit the same recursion as the matrix as a whole as its successive powers are computed. In later sections it will become clear that the n sequences that describe any row or column of successive powers of M will be independent of each other, that is, no term-by-term sums of up to $n - 1$ of any of the row or column sequences will yield the nth remaining row or column sequence. These facts immediately yield the following theorem: Every power of M is expressible as the following matrix, that is, the following form is invariant over exponentiation:

$$
\begin{bmatrix}
m_1 & m_2 & \cdots & m_n \\
L_{2,1}(\mathbf{m}) & L_{2,2}(\mathbf{m}) & \cdots & L_{2,n}(\mathbf{m}) \\
\vdots & \vdots & & \vdots \\
L_{n-1,1}(\mathbf{m}) & L_{n-1,2}(\mathbf{m}) & \cdots & L_{n-1,n}(\mathbf{m})
\end{bmatrix}
\tag{85}
$$

The $m_i \in \{0, 1\}$. The $L_{i,j}(\mathbf{m})$ are linear combinations of $m_1, m_2, \ldots, m_n$.

Consider Field B of Figure 8.7. By solving some elementary linear equations, show that the general form (85) of the matrices in the field is

$$
\begin{bmatrix}
m_1 & m_2 & m_3 \\
m_3 & m_1 & m_2 + m_3 \\
m_2 + m_3 & m_3 & m_1 + m_2 + m_3
\end{bmatrix}
\tag{86}
$$

Note that all the Field B matrices can be derived by letting (m_1, m_2, m_3) assume all possible $2^3 = 8$ binary triples.

8.3.2. The Shift and Add Property

One of the most celebrated properties of an m-sequence is the so-called "shift and add property." It deserves study not only because it is of theoretical interest but also because it lies at the heart of special architectural techniques. Essentially, the shift and add property states that if an m-sequence is added, term-by-term, to a shift or phase of itself, the resulting sequence will be the same m-sequence but at yet another shift of itself. For example, consider the sequence generated by the primitive polynomial

$x^3 + x^2 + 1$: $\{1001011\}$. Let us delay the sequence by one clocktime: $\{1100101\}$. Adding these two sequences, we obtain $\{0101110\}$, which is the first sequence delayed by five clocktimes.

The proof of the shift and add property is simple and it is instructive to show it by two different methods, via algebra and via matrices. Consider first that the linear recurrence generating the m-sequence is the same as used in (45):

$$s(t) = \sum_{i=1}^{n} \alpha_i s(t - i) \tag{87}$$

Let us create a set of 2^n elements whose members are the set of sequences

$$
\begin{aligned}
&s(1), s(2), \ldots, s(2^n - 1) \\
&s(2), s(3), \ldots, s(1) \\
&s(3), s(4), \ldots, s(2) \\
&\quad\vdots \\
&s(2^n - 1), s(1), \ldots, s(2^n - 2) \\
&0, 0, \ldots, 0
\end{aligned}
\tag{88}
$$

Each of the 2^n sequences in (88) satisfies (87), and because (87) generates an m-sequence, we know that all n-tuples excepting the all zero n-tuple exist somewhere in the sequence $s(1), s(2), \ldots, s(2^n - 1)$. We realize then that all possible n-tuples, including the all zero n tuple, exist as the first n bits of one of the sequences in (88). Thus, we claim that all possible sequences or solutions to (87) are present in the set (88). These sequences or solutions are rotations or phases of each other. Now consider the term-by-term sum of any two of the sequences in (88). Let these sequences be denoted by $\{a_1, a_2, \ldots, a_{2^n-1}\}$ and $\{b_1, b_2, \ldots, b_{2^n-1}\}$. Because $a(t) = \sum_{i=1}^{n} \alpha_i a(t - i)$ and $b(t) = \sum_{i=1}^{n} \alpha_i b(t - i)$ linearity assures that the term-by-term sum sequence $\{a_1 + b_1, a_2 + b_2, \ldots, a_{2^n-1} + b_{2^n-1}\}$ also satisfies the recursion and is therefore contained in (88)—thus proving the shift and add property. Furthermore, it is clear that the set of sequences in (88) forms an abelian group under the operation of term-by-term addition. This proof and observation is given by Golomb (1967, pp. 44–45).

Weathers (1972, pp. 13, 15) has given a matrix proof. Assume that two m-sequence generators have the same state transition matrix, M. Assume that one machine is started with the initial state $\mathbf{b}^T(0)$ and that the other machine started at d clocktimes away from $\mathbf{b}^T(0)$, that is, at state $\mathbf{b}^T(0)M^d$. If an observer were to sum the contents of identical stages at each clocktime, he would observe the sequence

$$\mathbf{b}^T(0) + \mathbf{b}^T(0)M^d, \mathbf{b}^T(0)M + \mathbf{b}^T(0)M^{d+1}, \mathbf{b}^T(0)M^2 + \mathbf{b}^T(0)M^{d+2}, \ldots \tag{89}$$

which can be can rewritten

$$\mathbf{b}^T(0)(I + M^d), \mathbf{b}^T(0)(I + M^d)M, \mathbf{b}^T(0)(I + M^d)M^2, \ldots \qquad (90)$$

But, as Weathers points out, (90) is equivalent to starting the machine at the state $\boldsymbol{\beta}^T(0) = \mathbf{b}^T(0)(I + M^d)$ and observing the sequence $\boldsymbol{\beta}^T(0), \boldsymbol{\beta}^T(0)M, \boldsymbol{\beta}^T(0)M^2, \ldots$, which is either the all zero sequence or a rotation or phase shift of the m-sequence.

8.3.3. Phase Shifts and the Delay Operator Calculus

Consider again an m-sequence realized by the companion matrix of Figure 8.6. For specifics, let us consider the shift register arrangement of Figure 8.11 corresponding to the primitive trinomial $x^3 + x^2 + 1$. We have labeled the stages 0, 1, and 2. Consider that the observed m-sequence is taken from stage 0. Let D be the unit delay operator applied to a sequence. If the register of Figure 8.11 is started with all ones, the following m-sequence is observed:

$$S_0 = \{1001011\} \qquad (91)$$

At stage 1, the following sequence is observed:

$$S_1 = \{1100101\} \qquad (92)$$

which is D operating on (91). At stage 2 we observe

$$S_2 = \{1110010\} \qquad (93)$$

which is D^2 operating on (91). Consider now that we have a set of three switches $\{s_0, s_1, s_2\}$ as shown in Figure 8.12. Table 4 shows the sequences which are produced at point Σ according to the eight possible switch

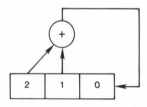

Figure 8.11. Realization of $x^3 + x^2 + 1$.

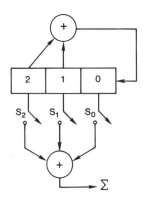

Figure 8.12. Phase switchable m-sequence generator.

configurations. Note that the set of eight sequences of Table 4 contain all seven phase shifts of the m-sequence produced by the primitive polynomial $x^3 + x^2 + 1$ and the all zero sequence. This set is of the form (88). The noteworthy point here is that all phase shifts are present and can be generated by summing, modulo 2, various combinations of the three stages of the machine shown in Figure 8.11. That this is true in general, that any phase of a $2^n - 1$ bit m-sequence can be formed by a linear combination of stages of the n-stage companion matrix shift register realization, is an amazing and useful property. [Tsao (1964) gives an excellent, elementary argument proving this.] Also implied is uniqueness, that is, no two linear combinations of stages will produce the same phase. This follows immediately from the "pigeon hole principle" (Schubeinfachprinzip) because there are $\sum_{i=0}^{n} \binom{n}{i} = 2^n$ possible linear combinations and 2^n possible phases including, of course, the all zero sequence.

We can now move naturally into the delay operator calculus. Observe that the machine in Figure 8.13 obeys the primitive polynomial $x^4 + x^3 + 1$.

Table 4. All Phase Shifts of $x^3 + x^2 + 1$

S_0	S_1	S_2	Σ
0	0	0	0 0 0 0 0 0 0
0	0	1	1 1 1 0 0 1 0
0	1	0	1 1 0 0 1 0 1
0	1	1	0 0 1 0 1 1 1
1	0	0	1 0 0 1 0 1 1
1	0	1	0 1 1 1 0 0 1
1	1	0	0 1 0 1 1 1 0
1	1	1	1 0 1 1 1 0 0

Observe that it can be viewed in terms of the delay operator notation by observing that

$$D^3\sigma + D^2\sigma = D^{-1}\sigma \qquad (94)$$

where σ is the m-sequence observed at stage 0. Rewriting (94) as

$$(D^4 + D^3 + 1)\sigma = 0 \qquad (95)$$

we see that the polynomial in x converts directly to a polynomial in D and this is true in general.

Davies (1965) presents the following algorithm to derive the numbers of the stages which must be added together to achieve a phase delay of d steps[†]:

1. Divide the degree n primitive polynomial, written left-to-right with terms of decreasing powers of D, into D^d.
2. Continue division until the remainder consists of powers of D all *of which* are less than n.
3. The powers of D in the remainder correspond to those stages that are to be summed modulo 2.

For example, let us say that we wish to generate the m-sequence according to the primitive polynomial $x^4 + x^3 + 1$ and that we simultaneously wish to generate the sequence delayed by six steps. Following Davies' algorithm

[†] We assume that the observed sequence is taken from the 0 stage.

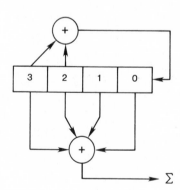

Σ Figure 8.13. Phase shift of $x^4 + x^3 + 1$.

we proceed as follows:

$$
\begin{array}{r}
D^2 + D + 1 \\
D^4 + D^3 + 1\ \overline{\smash)D^6} \\
\underline{D^6 + D^5 \qquad\qquad + D^2} \\
D^5 \qquad\qquad + D^2 \\
\underline{D^5 + D^4 \qquad\qquad + D} \\
D^4 \qquad + D^2 + D \\
\underline{D^4 + D^3 \qquad\qquad + 1} \\
D^3 + D^2 + D + 1
\end{array}
$$

The machine depicted in Figure 8.13 will produce an m-sequence at point Σ that is delayed by six steps from the sequence observed at stage 0.

Douce (1968) and Davies (1968) worked the inverse problem, that is, given the m-sequence generator polynomial and the stages of the companion matrix that are summed, they determined the phase delay of the sum. This algorithm is also very simple:

1. Divide the degree n primitive polynomial, written left-to-right with terms of *increasing* powers of D, into the stage specifying powers of D, also written as a polynomial in ascending powers (this polynomial will be of degree $<n$).
2. Continue division until a one term remainder is obtained.
3. The exponent of D of the remainder above is the phase delay.

Using our previous example, we perform the following division:

$$
\begin{array}{r}
1 + D + D^2 \\
1 + D^3 + D^4\ \overline{\smash)1 + D + D^2 + D^3} \\
\underline{1 \qquad\qquad + D^3 + D^4} \\
D + D^2 \qquad + D^4 \\
\underline{D \qquad\qquad + D^4 + D^5} \\
D^2 \qquad\qquad + D^5 \\
\underline{D^2 \qquad\qquad + D^5 + D^6} \\
D^6
\end{array}
$$

The first single-term remainder encountered is D^6 and our delay is therefore 6.

The above algorithms depend on the simple relation as expressed by Davies (1965):

$$D^d = f(D)q(D) + r(D) \tag{96}$$

where $f(D)$ is the primitive polynomial generating the m-sequence and $q(D)$ and $r(D)$ are the quotient and remainder polynomials, respectively. The 2^n possible residues, or equivalence classes that obtain upon the division, correspond to either the all zero sequence or the $2^n - 1$ possible phase delays.

Gardiner (1965) has devised, and Davis (1966) has further generalized, a sequential circuit that derives the numbers of the stages which must be added together to achieve a given phase shift delay.

8.3.4. Large Phase Shifts and the Art of Exponentiation

The problem we face here is that of determining the m-sequence after d steps. For this we need to determine the D^d modulo, the primitive polynomial generating the m-sequence. Given d, how many multiplications are required to obtain D^d? The answer in general, as far as the authors are aware, is unknown.

There are three important aspects we can consider. First, if $f(D) = \sum_{i=0}^{n} a_i D^i$, then $f^2(D) = \sum_{i=0}^{n} a_i D^{2i}$ as the cross products vanish due to modulo-2 arithmetic. Second, we can use the primitive polynomial to simplify the situation. Third, in simplifying $D^d = f(D)q(D) + r(D)$ we can use the Binary Exponentiation Algorithm by expressing d in normal binary form. This is discussed next.

Knuth (1969, pp. 401+) considers the problem at length and presents the Binary Exponentiation Algorithm which we have already seen in the chapter on number theory. The Binary Exponentiation Algorithm requires $\lfloor \log_2 d \rfloor + s(d) - 1$ multiplications where $s(d)$ is the number of ones in d's binary representation. The algorithm is not necessarily the "cheapest" in terms of multiplications required for general d. Knuth cites $d = 15$ as the smallest d for which there is a less costly procedure. The Binary Exponentiation Algorithm forms D^{15} from D with six multiplications. Let us, however, calculate D^{15} with only five multiplications as follows:

$$
\begin{aligned}
\text{START:} \quad & \delta \leftarrow D \\
& \delta \leftarrow \delta^2 & \text{(1 multiplication)} \\
& \delta \leftarrow \delta \cdot D & \text{(1 multiplication)} \\
& \gamma \leftarrow \delta & \text{(save } D^3) \\
& \delta \leftarrow \delta^2 & \text{(1 multiplication)} \\
& \delta \leftarrow \delta^2 & \text{(1 multiplication)} \\
& \delta \leftarrow \delta \cdot \gamma & \text{(1 multiplication)} \\
\text{END:} \quad & \delta = D^{15}
\end{aligned}
$$

Although the Binary Exponentiation Algorithm may not always be the cheapest in terms of multiplications required, it is easily programmed and its performance, in general, is quite good. Also, one must bear in mind that algorithms should not always be evaluated and selected by counting just one cost item, multiplications in this case. Total algorithmic complexity and "convenience" depend on many ancillary considerations such as storage requirements, indexing, sorting, and other housekeeping tasks.

The Binary Exponentiation Algorithm is quite useful for our task and is at the heart of a paper by VanLuyn (1978). VanLuyn's algorithm determines the stages to be summed to produce a phase shift of d steps. For consistency, we modify VanLuyn's algorithm slightly to comport with the definitions and architectures implied by the previous figures. The algorithm is motivated by the following: Assume that d is expressed in binary form as $d = a_0 2^0 + a_1 2^1 + a_2 2^2 + \cdots + a_s 2^s$; then we have

$$D^d = D^{a_0 + a_1 2^1 + a_2 2^2 + \cdots + a_s 2^s} = D^{a_0}(D^{a_1}(\cdots (D^{a_s})^2 \cdots)^2)^2 \quad (97)$$

But (97) is the recursive form of the Binary Exponentiation Algorithm. The embellishment needed is to reduce powers of D that equal or exceed n following each squaring and this can be done using Davies' (1965) long division method. VanLuyn's algorithm is then as shown in Figure 8.14.

As an example of VanLuyn's algorithm, consider that we wish to determine the setting of the switches s_0, s_1, s_2, s_3, s_4 of the machine depicted in Figure 8.15 so that the sequence at point Σ is delayed by 21 steps with respect to the sequence observed at stage 0. The machine of Figure 8.15 obeys the recursion $D^5 + D^2 + 1$ and the delay $s = 2^0 + 2^2 + 2^4$, hence

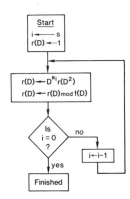

Figure 8.14. VanLuyn's algorithm (modified).

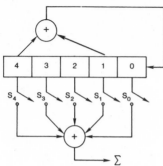

Figure 8.15. Phase switchable generator for $x^5 + x^2 + 1$.

$a_0 = 1$, $a_1 = 0$, $a_2 = 1$, $a_3 = 0$, and $a_4 = 1$. The algorithm proceeds as follows:

START: $r(D) = 1$

$a_4 = 1$: $r(D) = D^{a_4} \cdot (1)^2 = D$

$a_3 = 0$: $r(D) = D^{a_3} \cdot (D)^2 = D^2$

$a_2 = 0$: $r(D) = D^{a_2} \cdot (D^2)^2 = D^5$

$$
\begin{array}{r}
1 \\
\hline
D^5 + D^2 + 1\,\big|\,D^5 \\
\underline{D^5 + \quad\quad + D^2 \quad\quad + 1} \\
D^2 \quad\quad + 1 = r(D)
\end{array}
$$

$a_1 = 0$: $r(D) = D^{a_1} \cdot (D^2 + 1)^2 = D^4 + 1$

$a_0 = 1$; $r(D) = D^{a_0} \cdot (D^4 + 1)^2 = D^9 + D$

$$
\begin{array}{r}
D^4 \quad\quad + D \\
\hline
D^5 + D^2 + 1\,\big|\,D^9 \quad\quad\quad\quad\quad\quad + D \\
\underline{D^9 \quad\quad + D^6 \quad\quad + D^4} \\
D^6 \quad + D^4 \quad\quad + D \\
\underline{D^6 \quad\quad\quad + D^3 \quad + D} \\
\end{array}
$$

FINISHED: $\underline{D^4 + D^3 \quad\quad\quad = r(D)}$

Thus, we see that if switches 3 and 4 are closed, the sequence at point Σ will be delayed by 21 steps with respect to the sequence observed at stage 0. The reader is encouraged to try Davies' method (1965) on the above example for two reasons: first, to become convinced that the same results are achieved, and second, to gain an appreciation of the computational advantage that is provided by VanLuyn's algorithm, log d versus d.

For the sake of completeness and history it should be noted that Roberts et al. (1965) developed a better than linear d method but did not put it into an easy to manipulate form. Other authors have developed algorithms for special cases. Miller et al. (1977) have devised what is probably best captioned as a "coalescing pyramid" algorithmic structure that is efficacious when dealing with primitive trinomials. Hershey (1980) presented some time-saving shortcuts for those concerned with primitive trinomials of the form $x^n + x + 1$ where n is a Mersenne prime. Finally, Yiu (1980) has also developed a fast algorithm that is similar to VanLuyn's.

8.3.5. Decimation

Decimation, the creation of a new sequence by the periodic sampling of an old one, is a process of both great theoretical and practical import. Consider the m-sequence generated by the trinomial $x^3 + x + 1$:

$$10001110\ldots \tag{98}$$

Taking every second bit from (98) and starting from the zeroth, we get the sequence

$$1010011\ldots \tag{99}$$

But (99) is merely a phase shift of (98) and therefore displays the same primitive trinomial. If, however, we take every third bit from (98), again starting from the zeroth, we obtain

$$1100101\ldots \tag{100}$$

which is not a phase shift of (98) but an entirely new m-sequence. Indeed, it is the m-sequence generated by the other primitive trinomial of degree 3, $x^3 + x^2 + 1$.

8.3.6. Decimation by a Power of 2

Other experimentation would soon lead us to the hypothesis that decimating an m-sequence by a power of 2 always gives rise to the same m-sequence at some phase shift from the old. That this hypothesis is indeed true is a remarkable property of m-sequences. The validity of this theorem is quickly demonstrated by using abstract algebra. The following, however, is an excellent matrix-oriented proof by Weathers (1972, pp. 12–13). Consider that an m-sequence generator is started at state $\mathbf{b}^T(0)$ and we consider every other state, that is, decimation by 2. We observe

$$\mathbf{b}^T(0), \mathbf{b}^T(0)M^2, \mathbf{b}^T(0)M^4, \mathbf{b}^T(0)M^6, \ldots \tag{101}$$

But (101) is equivalent to the undecimated stepping from $\mathbf{b}^T(0)$ of a machine whose transition matrix is M^2 versus M. Following this observation, consider now that the characteristic equation is

$$M^n + \alpha_1 M^{n-1} + \cdots + I = 0 \tag{102}$$

We square (102), note that all the cross-terms disappear under modulo-2 arithmetic, and we have

$$M^{2n} + \alpha_1 M^{2(n-1)} + \cdots + I = 0 \tag{103}$$

Rewriting (103) as

$$(M^2)^n + \alpha_1 (M^2)^{n-1} + \cdots + I = 0 \tag{104}$$

we see that M^2 satisfies the characteristic equation (102) and thus the decimated sequence (102) is the original, undecimated m-sequence merely shifted in phase.

An m-sequence is a bit like the proverbial and material ring. It has no beginning, no end. A natural benchmark does exist, however.† It is called the "characteristic sequence." Recall that an m-sequence is changed in phase only, when decimated by 2. It turns out that there is a phase of the m-sequence that is left invariant upon decimation by 2. This phase is the characteristic sequence and was discovered by Gold (1966). Gold's straight-forward rules that construct the characteristic sequence for the nth degree primitive polynomial $f = f(x)$ are

$$\frac{d(xf)/dx}{f}, \qquad \text{if } n \text{ is odd} \tag{105}$$

$$\frac{d(xf)/dx}{f} + 1, \qquad \text{if } n \text{ is even} \tag{106}$$

For example, consider the trinomial $x^3 + x^2 + 1$ which gives the sequence presented in (100). Since n is odd, we apply formula (105) and see that

$$\frac{d(xf)/dx}{f} = \frac{d(x + x^3 + x^4)/dx}{1 + x^2 + x^3} + \frac{1 + x^2}{1 + x^2 + x^3} \tag{107}$$

We now perform the division specified in (107)

$$
\begin{array}{r}
1 \qquad\quad + x^3 \quad\; + x^5 + \quad x^6 + \cdots \\
\hline
1 + x^2 + x^3 \,\big|\, 1 \quad\;\; + x^2 \\
\underline{1 \quad\;\; + x^2 + x^3} \\
x^3 \\
\underline{x^3 \quad\;\; + x^5 + \quad x^6} \\
x^5 + \quad x^6 \\
\underline{x^5 \qquad\qquad + x^7 + x^8} \\
x^6 + x^7 + x^8 \\
\underline{x^6 \qquad\;\; + x^8 + x^9} \\
x^7 \qquad\; + x^9
\end{array}
\tag{108}
$$

† This is important for synchronization, which is discussed in a later chapter.

We obtain the characteristic sequence by sequentially reading the coefficients of 1, x, x^2, x^3, ... of the quotient. Thus, the characteristic sequence, or phase, of the primitive polynomial $x^3 + x^2 + 1$ is 1001011 .... Note that the characteristic sequence starts with zero for n even and one for n odd.

So enticing is this beautiful benchmark it has been independently discovered by Weinrichter and Surböck (1976) in an interesting and instructive way and will probably be rediscovered by other researchers in the future.

Finally, Arazi (1977) has developed the mathematical machinery to compute the initial setting or tuple of an m-sequence for its power of 2 decimation to achieve a desired phase shift. Arazi cites the characteristic sequence as the special case of zero phase shift.

8.3.7. General Decimation

If an m-sequence produced by an nth degree primitive polynomial is properly decimated, that is, if the period of decimation is relatively prime to $2^n - 1$, then another m-sequence will be produced. This new m-sequence will be described by the same primitive polynomial if and only if the decimation is a power of 2. For decimation other than a power of 2, there is no easy "paper-and-pencil" method of determining the polynomial with a few exceptions. The most famous of these exceptions is the "cubic transformation" introduced by Marsh (1957). This transformation is that produced by decimating by 3, or, as Golomb (1967, p. 79) calls it, "tertiation." Golomb (1969, pp. 363-366) has prepared a lucid algorithm for implementing Marsh's transformation. Golomb's algorithm proceeds as follows:

1. Create three "bins" A, B, and C. We will place, but not reduce modulo 3, all numbers that we encounter that are 0 modulo-3 into bin A. Into bin B we will place all numbers that are 1 modulo 3. Finally, we will place all numbers that are 2 modulo 3 into bin C.
2. The first set of numbers to be sorted and placed into the bins is the exponents of x in $f(x)$, the primitive polynomial.
3. For all distinct pairs (α_1, α_2) of exponents in $f(x)$ that are in the same bin (which is in reality a residue class) we form $(2\alpha_1 + \alpha_2)/3$ and $(2\alpha_2 + \alpha_1)/3$ and place the results into the appropriate bin.
4. For all distinct tuples $(\alpha_A, \alpha_B, \alpha_C)$ where α_A is an exponent of x of $f(x)$ which has been placed into bin A, α_B is an exponent of x of $f(x)$ which has been placed into bin B, and so on, we compute $(\alpha_A + \alpha_B + \alpha_C)/3$ and place it into the appropriate bin.
5. The occupants of the bins are now examined. If a particular occupant is present an odd number of times it is copied onto a list, L, otherwise it is discarded.

6. The new polynomial, the polynomial that describes the decimated-
by-3 recursion, is formed by summing together the terms consisting
of an x raised to each of the powers in the list, L.

An example is in order. Consider the primitive pentanomial

$$f(x) = x^5 + x^4 + x^2 + x + 1 \tag{109}$$

It produces the sequence

$$1110110011100001101010010001011 \tag{110}$$

Decimating the sequence (110) by 3, we obtain the sequence

$$1001011001111100011011101010000 \tag{111}$$

Let us use Golomb's method to discover the primitive polynomial which
(111) obeys:

	Bin A	Bin B	Bin C
Step 2	0	1, 4	2, 5
Step 3	3, 3	4	2
Step 4	3	1	2, 2
	$L = \{0, 3, 5\}$		

Thus, the polynomial that generates (111) is $g(x) = x^5 + x^3 + 1$. The reader
should note that in order for decimation-by-3 to yield a primitive polynomial,
n, the degree of the polynomial generating the original sequence must be
odd since $2^n - 1$, where n is even and greater than 0, is divisible by 3 and
the decimation would therefore be improper.

Golomb (1969, pp. 366–369) generalizes the cubic transformation to a
general kth power transformation but deriving the generator polynomial
for the decimated sequence quickly becomes overly involved with increasing
k. An easy way to derive the generating polynomial of a decimated sequence
is to solve a set of simultaneous equations derived as follows. We know
that if an m-sequence, that is generated by an nth degree primitive polynomial, is properly decimated, the resulting sequence will also be an m-
sequence generated by an nth degree polynomial. Thus, all we need to do
to uncover the polynomial is to produce a sequence of bits of the decimated

recursion

$$\{d(i)\}, \qquad i = 0, 1, 2, \ldots \tag{112}$$

and then solve for the α_i from $n - 1$ independent equations of the form

$$d(n + i) = \alpha_1 d(n + i - 1) + \cdots + \alpha_{n-1} d(i + 1) + d(i) \tag{113}$$

This method, for a different application, was suggested by Meyer and Tuchman (1972). As an example, let us apply the method to the sequence given in (111). The first four 6-tuples are

$$\begin{matrix} 100101 \\ 001011 \\ 010110 \\ 101100 \end{matrix} \tag{114}$$

From the four 6-tuples of (114) we immediately derive the four equations

$$\begin{aligned} 1 \quad + \alpha_2 \qquad &= 1 \\ \alpha_1 \quad + \alpha_3 \quad &= 1 \\ \alpha_1 + \alpha_2 \quad + \alpha_4 &= 0 \\ 1 \quad + \alpha_2 + \alpha_3 \quad &= 0 \end{aligned} \tag{115}$$

The solution of the equations in (115) yields $\alpha_1 = \alpha_2 = \alpha_4 = 0$ and $\alpha_3 = 1$. Thus, the primitive polynomial which generates the sequence in (111) is $x^5 + x^3 + 1$ as previously determined by the cubic transformation.

8.3.8. Inverse Decimation

The inverse problem of determining the undecimated sequence *given just a short segment* of the decimated sequence is considered by Arazi (1977). The following problem development and solution are different from Arazi's but more in keeping with our overall development. We assume that every kth bit is taken from the zeroth stage of a shift register which is implementing an nth degree primitive polynomial by the companion matrix architecture. The contents of the shift register, $\mathbf{b}(t)$, change according to $(t > 0)$

$$\mathbf{b}^T(t + 1) = \mathbf{b}^T(t)M \tag{116}$$

$$a(t) = \mathbf{b}^T(t)\mathbf{C} \tag{117}$$

where

$$\mathbf{C} = \text{col}\,[0, 0, 0, \ldots, 1] \tag{118}$$

and $a(t)$ corresponds to the output bit at time t. Note that (117) and (118) imply that the zeroth stage is tapped. From the hypothesis, we assume that we know

$$a(t), a(t + k), \ldots, a(t + (n - 1)k) \tag{119}$$

and we need to compute

$$\mathbf{b}^T(0) = [b_{n-1}(0) \quad b_{n-2}(0) \quad \cdots \quad b_0(0)] \tag{120}$$

from (119). In control theory this is usually referred to as the observability problem. [See Chapter 7; also see Luenberger (1979), pp. 285–287.]

From (116) and (117), we have

$$a(t) = \mathbf{b}^T M^t \mathbf{C} \tag{121}$$

and

$$[a(t), a(t + k), \ldots, a(t + (n - 1)k]$$

$$= \mathbf{b}^T[M^t \mathbf{C} \quad M^{t+k}\mathbf{C} \quad \cdots \quad M^{t+(n-1)k}\mathbf{C}] \tag{122}$$

Since it is known that the solution exists and is unique, it is clear that

$$\mathbf{b}^T(0) = [a^t, a^{t+k}, \ldots, a^{t+(n-1)k}]$$

$$\cdot [M^t \mathbf{C} \quad M^{t+k}\mathbf{C} \quad \cdots \quad M^{t+(n-1)k}\mathbf{C}]^{-1} \tag{123}$$

Problem

Consider the primitive polynomial $f(x) = x^3 + x^2 + 1$. If $a(3) = a(6) = 0$ and $a(9) = 1$, find the initial conditions on the vector $\mathbf{b}^T$. That is, find $b_2(0)$, $b_1(0)$, and $b_0(0)$.

8.3.9. Generation of High-Speed *m*-Sequences

The ability to generate high-speed *m*-sequences is important to a number of disciplines, especially direct sequence spread spectrum systems and precise ranging systems. The maximum rate at which the companion matrix realization of an *m*-sequence generator, Figure 8.6, can be run is determined by two parameters, the propagation delay of a shift register stage and the computation time required by the modulo 2 adder.

An "electronic" approach to speed increase was proffered by Harvey (1974) for certain trinomial feedback functions in which an analog delay

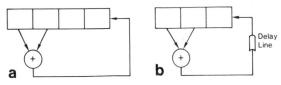

Figure 8.16. (a) Companion matrix realization of $x^4 + x^3 + 1$. (b) Delay line realization of
$$x^4 + x^3 + 1.$$

line was substituted for the zeroth stage, which must be untapped for
feedback to the modulo 2 adder. The delay of the analog line is chosen to
be the difference of the clocking period and the computation time of the
modulo 2 adder. The "before-and-after" diagrams illustrating Harvey's
method for implementing the m-sequence specified by $x^4 + x^3 + 1$ are shown
in Figures 8.16(a) and (b). Clearly Harvey's method also applies to the
substitution of more than one shift register stage by analog elements.

Ball et al. (1975) extended the above method and showed that for
trinomials of the form $x^n + x^{n-1} + 1$, no shift register stages are needed at
all as is shown in Figure 8.17. The delays indicated in the figure are
determined by

$$D_1 = T$$
$$D_2 = (n-1)T - D_g \tag{124}$$

where T is the bit period, n is the degree of the trinomial ($x^n + x^{n-1} + 1$),
and D_g is the computation time of the modulo 2 adder. Ball et al. noted
that the configuration shown in Figure 8.17 will run free and is, in their
words, "a special case of the delay-line oscillator."

Another approach to high-speed generation was offered by Lempel
and Eastman (1971). Unlike the methods just reviewed, the Lempel–
Eastman method depends on a mathematical, versus electronic, basis to
achieve high-speed generation. The method depends on a curious aspect
of decimated m-sequences. Essentially, what Lempel and Eastman proposed
was the creation of k shift registers each of which would be driven by the
feedback polynomial corresponding to the kth decimation of the sequence
$\{s(i)\}$, $i = 0, 1, 2, \ldots$ which is generated by the nth degree polynomial that
it is desired to implement at high speed. (The decimations need not be
proper, that is, k and $2^n - 1$ need not be relatively prime.) What is done is

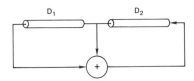

Figure 8.17. Analog (delay line) shift register.

to step, in succession, each of the k registers and modulo 2 add together the last stage from each. The registers are started with initial conditions so that register j exhibits the decimated sequence $\{s(ik + j)\}$, $i = 0, 1, 2, \ldots$. The sequence produced by this addition exhibits the desired polynomial and runs at a rate k times the rate of the individual shift registers. As an example, consider that we wish to synthesize the sequence produced by the primitive trinomial $x^3 + x + 1$ at a rate two times faster than we shift our registers. For this example, $k = 2$, the decimation is a proper one and, additionally, the polynomial driving our two registers must also be $x^3 + x + 1$ as the decimation is a power of 2. (Recall that this was shown in the previous section.) Thus, the diagram shown in Figure 8.18 will implement the recursion at twice the clock rate. If the top register is started with 010 and the bottom with 111, the machine will progress as follows:

Clock:	10101010101010
Top register stage 2:	01100001111110
Bottom register stage 2:	11111100110000
Mod 2 adder output:	10011101001110

The three implementations reviewed so far assume that the limiting parameter is the propagation delay of the shift register stage. Often this is not the case but it is rather the computation time required by the modulo 2 adder that caps the operating speed. One solution against this inherently different problem was given by Quan (1974) who proposed a modification of the Lempel–Eastman method. He suggested time division multiplexing of the decimated sequences instead of adding them modulo 2. Figure 8.19

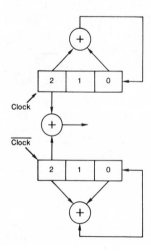

Figure 8.18. High-speed m-sequence generation by decimation and modulo 2 addition.

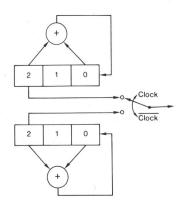

Figure 8.19. High-speed m-sequence generation by decimation and multiplexing.

depicts the scheme shown in Figure 8.18 recast under Quan's modification. The initial conditions of the registers are the same as for Lempel–Eastman version.

One final method for generating high-speed m-sequences is that proposed by Warlick and Hershey (1980). These authors first developed a special architecture for m-sequence generators based on primitive trinomials

$$x^n + x^a + 1 \qquad (125)$$

They found that the general architecture depicted in Figure 8.20, which they termed the "vanestream structure" (the repeated register blocks with feedback are termed the "vanes"), will progress through a maximum length cycle of $2^{vl+1} - 1$ states for a surprisingly rich set of triples (v, l, t) which they catalog, along with the primitive trinomials (125) upon which the structures are based, for $vl \leq 99$. The architecture of Figure 8.20 is then coupled with a parallel in (broadside load), serial out multiplexer as shown in Figure 8.21 to form what they term the "WINDMILL m-sequence generator." (The name WINDMILL derives from a predecessor sequential machine of theoretical interest only.) The slower logic vanestream generator is stepped at a rate R. After each step, the contents of the v stages are copied into the v high-speed shift register stages indicated and the v high-speed shift register is shifted v times in the direction indicated. The

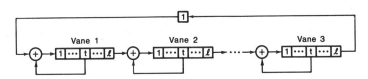

Figure 8.20. Vanestream structure.

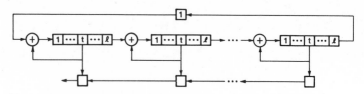

Figure 8.21. WINDMILL high-speed m-sequence generator.

high-speed stream exhibits the m-sequence specified by the trinomial

$$x^n + x^{n-a} + 1 \tag{126}$$

The WINDMILL thus achieves a speed ratio increase of v to attain a high-speed rate Rv. Warlick and Hershey briefly considered the electronic architectural implications of realizing a WINDMILL and concluded that the WINDMILL offers power advantages for certain hybrid MOS layouts and its periodic substructures are especially amenable to LSI blacksmithery.

8.3.10. On Approximating a Bernoulli Source with an m-Sequence

An m-sequence is, in general, an excellent approximation to a balanced Bernoulli source. However, if you attempt to use an m-sequence to approximate a binomial source by integrating the m-sequence, you may incur very poor results. Specifically, let us say you sum N consecutive bits from an m-sequence to approximate a binomial source. Ideally, the probability of observing a sum equal to k would be $2^{-N}\binom{N}{k}$. In general, you will be successful in $N \leq n$ where n is the degree of the primitive polynomial which generates the m-sequence. If, however, $N > n$, the distribution will skew and become distorted (see Tausworthe, 1965; White, 1967).

m-Sequence generators have been used in speech processing. To generate synthetic speech, we need periodic inputs for voiced speech and a noise input for unvoiced speech. m-Sequence generators have been found to be very effective as a noise source for unvoiced speech.

Problem

Use a computer to experiment on approximating a binomial source by integrating various m-sequences. Try cases for which $N \leq n$ and $N > n$.

References

Arazi, B. (1977), Decimation of m-Sequence Leading to Any Desired Phase Shift, *Electronics Letters*, Vol. 13, No. 7, pp. 213–215, March.

Ball, J., A. Spittle, and H. Liu (1975), High-Speed m-Sequence Generation: A Further Note, *Electronics Letters*, Vol. 11, No. 5, pp. 107–108, March.

Berlekamp, E. (1967), Factoring Polynomials Over Finite Fields, *BSTJ*, pp. 1853–1859, October.

Berlekamp, E. (1968), *Algebraic Coding Theory*, McGraw-Hill, New York.

Berlekamp, E. (1970), Factoring Polynomials Over Large Finite Fields, *Mathematics of Computation*, Vol. 24, No. 111, pp. 713–735, July.

Blom, G. and D. Thorburn (1982), How Many Random Digits are Required Until Given Sequences Are Obtained?, *Journal of Applied Probability*, Vol. 19, pp. 518–531.

Cantor, D and H. Zassenhaus (1981), A New Algorithm for Factoring Polynomials Over Finite Fields, *Mathematics of Computation*, Vol. 36, No. 154, pp. 587–592, April.

Charney, H. and C. Mengani (1961), Generation of Linear Binary Sequences, *RCA Review*, pp. 420–430, September.

Davies, A. (1965), Delayed Versions of Maximal-Length Linear Binary Sequences, *Electronics Letters*, Vol. 1, No. 3, pp. 61–62, May.

Davies, A. (1968), Calculations Relating to Delayed m-Sequences, *Electronics Letters*, Vol. 4, No. 14, pp. 291–292, July.

Davis, W. (1966), Automatic Delay Changing Facility for Delayed m-Sequences, *Proceedings of the IEEE*, Vol. 54, pp. 913–914, June.

Douce, J. (1968), Delayed Versions of m-Sequences, *Electronics Letters*, Vol. 4, No. 12, p. 254, June.

Gardiner, A. (1965), Logic P.R.B.S. Delay Calculator and Delayed-Version Generator with Automatic Delay-Changing Facility, *Electronics Letters*, Vol. 1, No. 5, pp. 123–125, July.

Gardner, M. (1979), Mathematical Games, *Scientific American*, pp. 120–125, October.

Gold, R. (1966), Characteristic Linear Sequences and Their Coset Functions, *Journal SIAM*, Vol. 14, No. 5, pp. 980–985, September.

Golomb, S. (1967), *Shift Register Sequences*, Holden-Day, Oakland, CA.

Golomb, S. (1969), Irreducible Polynomials, Synchronization Codes, Primitive Necklaces, and the Cyclotomic Algebra, in *Combinatorial Mathematics and Its Applications*, The University of North Carolina Press, Chapel Hill, Chap. 21, pp. 358–370. (Proceedings of the conference held at the University of North Carolina, Chapel Hill, NC, April.)

Golomb, S. (1980), On the Classification of Balanced Binary Sequences of Period $2^n - 1$, *IEEE Transactions on Information Theory*, Vol. 26, No. 6, pp. 730–732, November.

Harvey, J. (1974), High-Speed m-Sequence Generation, *Electronics Letters*, Vol. 10, No. 23, pp. 480–481, November.

Hershey, J. (1980), Implementation of MITRE Public Key Cryptographic System, *Electronics Letters*, Vol. 16, No. 24, pp. 930–931, November.

Hershey, J. (1982), Proposed Direct Sequence Spread Spectrum Voice Techniques for the Amateur Radio Service, NTIA Report 82-111.

Kemeny, J. and J. Snell (1960), *Finite Markov Chains*, Van Nostrand, New York.

Kemeny, J., J. Snell, and A. Knapp (1966), *Denumerable Markov Chains*, Van Nostrand, New York.

Knuth, D. (1969), The Art of Computer Programming, *Seminumerical Algorithms*, Addison-Wesley, Reading, MA, Vol. 2, pp. 381–396.

Lempel, A. and W. Eastman (1971), High Speed Generation of Maximal Length Sequences, *IEEE Transactions on Computers*, pp. 227–229, February.

Luenberger, D. (1979), *Introduction to Dynamic Theory—Theory, Models and Applications*, Wiley, New York, pp. 285–289.

MacDuffie, C. (1940), *An Introduction to Abstract Algebra*, Wiley, New York.

Marsh, R. (1957), *Table of Irreducible Polynomials Over GF(2) Through Degree 19*, U.S. Department of Commerce, Office of Technical Services, Washington, DC.

Meyer, C. and W. Tuchman (1972), Pseudorandom Codes Can Be Cracked, *Electronic Design*, Vol. 20, No. 23, pp. 74–76, November.

Miller, A., A. Brown, and P. Mars (1977), A Simple Technique for the Determination of Delayed Maximal Length Linear Binary Sequences, *IEEE Transactions on Computers* Vol. C-26, No. 8, pp. 808–811, August.

Moenck, R. (1977), On the Efficiency of Algorithms for Polynomial Factoring, *Mathematics of Computation*, Vol. 31, No. 137, pp. 235–250, January.

Quan, A. (1974), A Note on High-Speed Generation of Maximal Length Sequences, *IEEE Transactions on Computers*, Vol. 23, pp. 201–203, February.

Reiner, I. (1961), On the Number of Matrices with Given Characteristic Polynomial, *Illinois Journal of Mathematics*, Vol. 5, pp. 324–329.

Roberts, P., A. Davies, and S. Tsao (1965), Discussion on Generation of Delayed Replicas of Maximal-Length Binary Sequences, *Proceedings of the Institution of Electrical Engineers (London)*, Vol. 112, No. 4, pp. 702–704, April.

Rodemich, E. and H. Rumsey (1968), Primitive Trinomials of High Degree, *Mathematics of Computation*, Vol. 22, pp. 863–865.

Scholefield, P. (1960), Shift Registers Generating Maximum-Length Sequences, *Electronic Technology*, Vol. 37, pp. 389–394, October.

Sittler, R. (1956), Systems Analysis of Discrete Markov Processes, *IRE Transactions on Circuit Theory*, Vol. 3, pp. 257–266.

Stahnke, W. (1973), Primitive Binary Polynomials, *Mathematics of Computation*, Vol. 27, No. 124, pp. 977–980, October.

Swan, R. (1962), Factorization of Polynomials Over Finite Fields, *Pacific Journal of Mathematics*, Vol. 12, pp. 1099–1106.

Tausworthe, R. (1965), Random Numbers Generated by Linear Recurrence Modulo Two, *Mathematics of Computation*, Vol. 19, p. 201.

Tsao, S. (1964), Generation of Delayed Replicas of Maximal-Length Linear Binary Sequences, *Proceedings of the Institution of Electrical Engineers (London)*, Vol. 111, No. 11, pp. 1803–1806, November.

VanLuyn, A. (1978), Shift-Register Connections for Delayed Versions of m-Sequences, *Electronics Letters*, Vol. 14, No. 22, pp. 713–715, October.

Warlick, W. and J. Hershey (1980), High-Speed m-Sequence Generators, *IEEE Transactions on Computers*, Vol. C-29, No. 5, pp. 398–400, May.

Watson, E. (1962), Primitive Polynomials (Mod 2), *Mathematics of Computation*, Vol. 16, No. 79, pp. 368–369, July.

Weathers, G. (1972), Statistical Properties of Filtered Pseudo-Random Digital Sequences, Sperry Rand Corporation, Technical Report No. SP-275-0599, January. (Prepared for NASA George C. Marshall Space Flight Center, Huntsville, AL.)

Weinrichter, H. and F. Surböck (1976), Phase Normalized m-Sequences with the Inphase Decimation Property $\{m(k)\} = \{m(2k)\}$, *Electronics Letters*, Vol. 12, No. 22, pp. 590–591, October.

White, R. (1967), Experiments with Digital Computer Simulations of Pseudo-Random Noise Generators, *IEEE Transactions on Electronic Computers*, Vol. 16, pp. 355–357.

Yiu, K. (1980), A Simple Method for the Determination of Feedback, Shift Register Connections for Delayed Maximal-Length Sequences, *Proceedings of the IEEE*, Vol. 68, No. 4, pp. 537–538, April.

Zierler, N. (1959), Linear Recurring Sequences, *Journal SIAM*, Vol. 7, No. 1, pp. 31–48, March.

Zierler, N. and J. Brillhart (1968), On Primitive Trinomials (Mod 2), *Information and Control*, Vol. 13, pp. 541–554.

Zierler, N. and J. Brillhart (1969), On Primitive Trinomials (Mod 2): II, *Information and Control*, Vol. 14, pp. 566–569.

Zierler, N. (1969), Primitive Trinomials Whose Degree is a Mersenne Exponent, *Information and Control*, Vol. 15, pp. 67–69.

Zierler, N. (1970), On $x^n + x + 1$ over GF(2), *Information and Control*, Vol. 16, pp. 502–505.

Source Encoding

9.1. Introduction

Before architecture of a data communications network is started, there are a number of preparatory steps. One of these is to understand, characterize, and appropriately model the data source. Is it English text? A Markov source? Facsimile? Each source is unique and so are its output statistics. There is often great opportunity to remove redundancy from a source by using simple encoding methods and doing so may result in an appreciable savings in bits sent.

In this chapter we examine a variety of classical and useful methods for encoding the output of a very simple source. The methods presented can be used for other, less simple, sources and extensions of their use are natural. As with the other chapters, our exposition is not theoretical or formal but rather intended to allow the reader to rapidly gain an appreciation for the subject and master some basic tools that can be immediately applied to a host of "real world" situations.

9.2. Generalized Bernoulli Source

We define a generalized Bernoulli source as a source that produces the source symbols $\{s_i\}$ with probabilities $\{p_i\}$ at sample time t regardless of what has been produced prior to time t. The source, in other words, has no memory.

Let us assume that we have a generalized Bernoulli source that produces three symbols s_1, s_2, and s_3 with respective probabilities $p_1 = \frac{1}{2}$, $p_2 = p_3 = \frac{1}{4}$. How can we encode it? Well, there are all sorts of possibilities. We could simply choose to represent each source symbol with two bits, each dibit distinct as shown in Figure 9.1.

What do we observe as a result of this code? First, if we know when the encoding was started we can immediately and uniquely determine each source symbol after at most two bits. For example, if we observe the string $010001010001100101 \cdots$ we know that it represents the source symbol sequence $s_2 s_1 s_2 s_2 s_1 s_2 s_3 s_2 s_2 \cdots$. Suppose, however, we encountered the above string of bits but missed the first bit and thus observed $10001010001100101 \cdots$. We would have decoded the stream's first five dibits as representing the symbols $s_3 s_1 s_3 s_3 s_1$ but then we would have encountered the double unit dibit 11 and we would have to conclude that there was either an error in reception or that our dibit "framing" was improperly chosen.

Second, we can calculate the average number of transmitted bits per source symbol, $\bar{l}$. To do this we merely evaluate

$$\bar{l} = \sum_i (p_i) \cdot (\text{number of bits in code representing symbol } i) \qquad (1)$$

$$\tfrac{1}{2} \cdot \, 1 + \tfrac{1}{4} \cdot 2 = \tfrac{1}{2} + \tfrac{1}{2} = 1.$$

which for our example is $\frac{1}{2} \times 2 + \frac{1}{4} \times 2 + \frac{1}{4} \times 2 = 2$ coding bits per source symbol on the average.

The third observation we can make centers on our decoding procedure. If we encounter a dibit whose first bit is a zero we know that either symbol s_1 or s_2 is being sent and we must await the next bit to resolve the sent symbol. But, if the first bit of the dibit is a one, then we know *at that time* that symbol s_3 is being sent and that a zero must follow. Well, if a zero must follow, why bother sending it? In other words, why not code our source as shown in Figure 9.2.

What does this encoding do for us? Let us look at the average number of transmitted bits per source symbol, $\bar{l}$, as defined by (1). For our new code we have $\bar{l} = \frac{1}{2} \times 2 + \frac{1}{4} \times 2 + \frac{1}{4} \times 1 = 1.75$ coding bits per source symbol on the average. This certainly seems better than the previous case because we are using fewer coding bits on the average to encode the source symbols. Is this new code "costing" us anything? Consider, as we did before, that we miss a bit; we start decoding one bit late. For the previous code we

Source Symbol	Probability	Code
s_1	$\frac{1}{2}$	00
s_2	$\frac{1}{4}$	01
s_3	$\frac{1}{4}$	10

Figure 9.1. Code for the three-symbol source.

Source Symbol	Probability	Code
s_1	$\frac{1}{2}$	00
s_2	$\frac{1}{4}$	01
s_3	$\frac{1}{4}$	1

Figure 9.2. Another code for the three-symbol source.

stumbled upon our error when we encountered the 11 dibit. But for this new code we will never be able to determine that we did not start "in step" since all coding bit strings are possible. So it seems we have traded off synchronization slippage detection for greater coding "efficiency"—fewer coding symbols required per source symbol, on the average.

Let us suppose that our goal is strictly greater efficiency. Have we done the best we can? In the case of the second code we achieved a smaller $\bar{l}$ by assigning a single bit to symbol s_3. Symbol s_3 has a probability of occurrence of only $\frac{1}{4}$ as contrasted to symbol s_1 which is twice as likely. Would it not make for greater efficiency if we assigned the single bit code to source symbol s_1 instead? In this vein consider yet another code for the source as specified in Figure 9.3. Note that we now have $\bar{l} = \frac{1}{2} \times 1 + \frac{1}{4} \times 2 + \frac{1}{4} \times 2 = 1.5$ encoding bits per source symbol on the average.

We should now have formed a number of questions. For example, how should we assign code symbols or code words to source symbols to minimize $\bar{l}$? What is the smallest possible $\bar{l}$? Is there a way of encoding so that misframing or synchronization errors can be detected and corrected? How can we be sure that the code words we choose to represent source symbols will not lead to ambiguous decodings even if there are no synchronization errors? We explore some of these questions to some depth and in so doing cover only a small subset of coding theory, but as subset, nevertheless, that has proven to be of immense value and versatility in modern communications.

9.3. Unique Decodability

Let us assume that a generalized Bernoulli source has been coded by a set of code words of varying lengths. If we examine the code words we

Source Symbol	Probability	Code
s_1	$\frac{1}{2}$	0
s_2	$\frac{1}{4}$	10
s_3	$\frac{1}{4}$	11

Figure 9.3. Yet another code for the three-symbol source.

Source Symbol	Code
s_1	1
s_2	101
s_3	0101
s_4	0110

Figure 9.4. Code for a four-symbol source.

can determine if ambiguous messages are possible, that is, if unique decodability is impossible. As an introduction, consider that we have a generalized Bernoulli source with four possible symbols. Let us assume that the source is encoded as shown in Figure 9.4. Suppose we receive the following message: 101101. Are the source symbols $s_2 s_2$ or $s_1 s_4 s_1$? There is no way to tell and we conclude that the code is not uniquely decodable. Even (1963) has provided a useful test for unique decodability. We now present it by way of analyzing the code of Figure 9.4.

The first step is to relabel the code words with convenient letters, for example, A, B, C, and D versus s_1 through s_4. The next step is to rewrite the code words as follows: prefix and postfix the symbol S (for Start/Stop) to each code word. Then, for every code word, if and only if it consists of $n > 1$ bits, expand it by inserting L_i between bits i and $i + 1$ for all $i < n$ where L is the letter label of the code word. Thus, our code words take on the forms depicted in Figure 9.5.

We now proceed to construct a table (Table 1) that will develop an ambiguous or nonuniquely decodable sequence of bits if one exists. If one does not exist we declare the code uniquely decodable. Table 1 has three columns: the latter two are headed by the encoding symbols—a zero and a one for our case—and the first column is headed by the word "State". The table is developed row by row. The first entry under State is S. The table is filled in as follows. Starting from state S, with all code words, if we encounter a zero as the next symbol for one or more pairs of code words, we write all the pairs of symbols following the zero. If none exist, we enter a dash. We repeat this procedure for a one. Let us now derive the first entry.

Source Symbol	Code Word	Letter Label	Expanded Code Word
s_1	1	A	$S1S$
s_2	101	B	$S1B_10B_21S$
s_3	0101	C	$S0C_11C_20C_31S$
s_4	0110	D	$S0D_11D_21D_30S$

Figure 9.5. Expanded code words.

Table 1. Next State Table for Code of Figure 9.5

State	0	1
S	(C_1D_1)	(SB_1)
C_1D_1	—	(C_2D_2)
SB_1	$(B_2C_1), (B_2D_1)$	—
C_2D_2	—	—
B_2C_1	—	(SC_2)
B_2D_1	—	(SD_2)
SC_2	$(C_1C_3), (D_1C_3)$	—
SD_2	—	$(SD_3), (B_1D_3)$
C_1C_3	—	(SC_2)
D_1C_3	—	(SD_2)
SD_3	$(SC_1), (SD_1)$	—
B_1D_3	(SB_2)	—
SC_1	—	$(SC_2), (B_1C_2)$
SD_1	—	$(SD_2), (B_1D_2)$
SB_2	—	$(SS), (SB_1)$

All code words start with an S; but in only two instances is there a zero as the next symbol, that is, code words C and D. The symbols that follow the zero are C_1 for code word C, and D_1 for code word D. Thus, we enter the pair (C_1D_1) under the zero column. Now let us consider the other case; the case for which a one follows the S. The only code words for which this is true are code words A and B. The symbols that follow the one are S for code word A and B_1 for code word B. We thus enter the pair (SB_1) under the one column. We now form new rows under the State heading by writing, one per row, all state pairs produced that have not been written before. The next two rows are thus started with C_1D_1 and SB_1, respectively. Let us examine the pair C_1D_1. We see that C_1 and D_1 are both followed by ones and therefore there is no entry in the zero column and (C_2D_2) is entered in the one column. The state pair C_2d_2 is new and is therefore listed as the next row leader after SB_1. The pair SB_1 has two possible next pairs for a zero symbol: (B_2C_1) and (B_2D_1).

We continue to fill out the table until one of two things happens. If we encounter the pair (SS) we halt and declare the code nonuniquely decodable; if we do not encounter the pair (SS) but do not produce any new pairs we have shown the code to be uniquely decodable. We know that the code we are using as an example is nonuniquely decodable and we have produced codes in Table 1 up until the occurrence of the (SS) pair.

Table 1 can be used to produce an example of an ambiguous message. To find an ambiguous message all we do is write down the series of states that led to the SS pair. Next to these states we enter the column header, a zero or a one, according to the column in which the state appears. For our example, the state SS appeared in the one column, so we write SS-1. The SS state was produced by the state SB_2 which appeared under the zero column, so we write SB_2-0. The SB_2 state was produced by the state $B_1 D_3$ which appeared under the one column, so we write $B_1 D_3$-1 and so on until we get back to the start of the table at symbol S. Table 2 shows the list that is developed. The ambiguous message is recovered by reading the table from bottom (starting at SB_1) to top (ending at SS). The ambiguous message is thus 101101 which, as we commented before, can be interpreted as a message consisting of the string of source symbols $s_2 s_2$ or $s_1 s_4 s_1$.

Problems

1. Show that the coding

Source Symbol	Code Word
s_1	0
s_2	010
s_3	101

is not uniquely decodable and find an example of a nonuniquely decodable message.

2. Construct the table in the form of Table 1 for the following code and show that it is uniquely decodable:

Source Symbol	Code Word
s_1	0
s_2	010
s_3	1010

Table 2. Series of States That Produce an Ambiguous Message

SS-1
SB_2-0
$B_1 D_3$-1
SD_2-1
$B_2 D_1$-0
SB_1-1

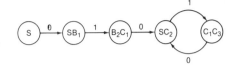

Figure 9.6. Graph for the code of
problem 2.

3. (Even, 1963) There is an interesting and useful directed graph associated
with the table of the form of Table 1. The nodes of the graph are the pairs
listed in the column headed by the word State. The arcs of the graph are
directed to the state(s) generated by the column symbol. The arcs are labeled
with a zero or a one according to the column in which the generated state
lies. For example, consider the code of Problem 2, above. Show that the
graph for this code is as shown in Figure 9.6.

If the graph for a uniquely decodable code is loop-free with the path
from Ⓢ to the farthest terminal node equal to l "steps", then a message
can be decoded as it arrives with a delay of at most l bits. If the graph for
a uniquely decodable code is not loop-free, as is the case with the present
example, then it may not be possible to decode even the first symbol until
the entire message has been received. For our example, the infinite message
$01010101010101010101010\cdots$ (repeated 10) $\cdots$ is such that we can not
recover the first source symbol. To see this more easily, consider that the
above message is terminated after k bits. The table below gives the only
possible decoded message as a function of odd k. (Why do we not need to
consider even k here?)

k	Unique Message
9	$s_1 s_3 s_3$
11	$s_2 s_3 s_3$
13	$s_1 s_3 s_3 s_3$
15	$s_2 s_3 s_3 s_3 s_3$

What are the unique messages for $k = 17$ and $k = 19$?

9.3.1. A Basic Inequality Required by Unique Decodability

If a set of code words is to compose a uniquely decodable code, then
there is a *necessary* (but *not sufficient*) condition that must be met by the
number and lengths of the code words. This inequality† is called the

---

† A particularly nice derivation of the inequality was later provided by Karush (1961). It is
recommended to the interested reader.

McMillan inequality after McMillan (1956). It is a good "first test" for unique decodability. For code words constructed from zeros and ones, the inequality requires that for the sum over all the code words

$$\sum_i 2^{-l_i} \leq 1 \tag{2}$$

where l_i is the length of the ith code word. Note that the inequality is sometimes exact, as it obviously is, for example, for the set of single bit, binary code words $\{0, 1\}$. Also remember that the inequality is a necessary but not sufficient condition. Observe, for example, that the inequality is satisfied for the code of Figure 9.4 which we saw was not a uniquely decodable code. (This caveat should not be surprising since the inequality addresses only the number and length of the code words and not their specific bits.)

9.3.2. Comma-Free Codes

As we have seen, there are codes which we can not decode unambiguously if we do not successfully decode each code word from the inception of the message. Such a circumstance could arise either from reception errors or from our starting our decoding "late". For those cases in which we must be able to segment code words, no matter where in the message we wish to start decoding, there is a class of codes called "synchronizable codes" which can be used. An important subset of synchronizable codes is found in a set of codes which have been classically designated "comma-free codes". These codes were introduced by Golomb et al. (1958). A comma-free code is defined to be a set of code words constructed from an alphabet of k encoding symbols (we use only zeros and ones in our discussion). The words are defined to be all the same length—for our purposes n-bit words. To be comma-free, the set of n-bit words had to have the following property: for any two (not necessarily distinct) code words, $c_1 c_2 \cdots c_n$ and $c'_1 c'_2 \cdots c'_n$, none of the n-bit overlaps

$$c_i c_{i+1} c_{i+2} \cdots c_n c'_1 c'_2 \cdots c'_{i-1} \qquad (1 < i \leq n) \tag{3}$$

may be a code word.

Problem

Show that the code consisting of the following six 5-bit code words is not a comma-free code: $\{00101, 01101, 01110, 10000, 10101, 11000\}$.

The first observation we make is that if $c_1 c_2 \cdots c_n$ is to be a code word of a comma-free code, then it can not be "factored" into the following form

$$\underbrace{c_1 c_2 \cdots c_d c_1 c_2 \cdots c_d \cdots c_1 c_2 \cdots c_d}_{n \text{ bits}} \tag{4}$$

for if it could, then the concatenation

$$c_1 c_2 \cdots c_n c_1 c_2 \cdots c_d \tag{5}$$

would not be free of overlaps.

We can set a nice upper bound on the number of possible comma-free code words in a set of n-bit words. We do this by partitioning all 2^n n-bit words into equivalence classes based on the operation of end-around rotation. We say that two n-bit words are equivalent if and only if one is a rotation of the other. Then the words

$$\begin{aligned}
&c_1 c_2 \cdots c_{n-1} c_n \\
&c_2 c_3 \cdots c_n \quad c_1 \\
&c_3 c_2 \cdots c_1 \quad c_2 \\
&\quad \vdots \\
&c_n c_1 \cdots c_{n-2} c_{n-1}
\end{aligned} \tag{6}$$

form an equivalence class of n/d members where d is as defined in (4). At most, only one of the words in (6) may be a code word of the comma-free code and then only if $d = n$. (Why?) Note also that d must be the same for all members in (6).

As we have said, there are 2^n possible distinct n-bit words. Let K_d be the number of members in all the equivalence classes exhibiting words with a particular d. Then we have

$$\sum_{d|n} K_d = 2^n \tag{7}$$

By using the Möbius inversion formula we can transform (7) to

$$K_n = \sum_{d|n} \mu(d) 2^{n/d} \tag{8}$$

Of each equivalence class of n members that contributes to (8), at most one member from each class may be chosen as a comma-free code word and thus the number of comma-free code words is upper bounded by

$$\frac{1}{n} \sum_{d|n} \mu(d) 2^{n/d} \tag{9}$$

This upper bound is, incidentally, the number of "necklaces" that can be created using n zeros and ones or, equivalently, the number of cycles realizable by an end-around n-stage shift register. The bound for many n is thus tabulated in the section on combinatorics.

Eastman (1965) showed that the upper bound could be realized if n is odd. Eastman also presented an algorithm for constructing a comma-free code out of n-bit words.

9.4. Synchronizable Codes

Comma-free codes are, as we have said, a subset of synchronizable codes. Synchronizable codes allow code words to be separated after a sufficient number of encoding bits have been observed. They thus do not require "commas" or knowledge of absolute symbol position with regard to the start of the message. In other words, if a listener has missed the message's beginning, or perhaps does not even know when the message began, the listener can still eventually begin to correctly decode the received message stream.

Synchronizable code words may be of different lengths as opposed to the comma-free code words of the previous section. A correct separation of two synchronizable code words can be determined through examination of at most s bits around any particular point in the message. The parameter s is a key parameter of a sychronizable code; it has the greatest impact on the complexity of the decoding algorithm and, of course, on the average amount of time required to determine code word boundaries, after a "break into" the middle of a message. As an example, the code consisting of only the two single-bit code words $\{0, 1\}$ is a synchronizable code for which $s = 0$; the code word boundaries are determined upon receipt of a single bit.

Scholtz (1966) showed how to generate a synchronizable code dictionary by a suffix construction method. What we do is start with a synchronizable code of parameter s. We remove a word from the initial set of code words, c. An augmented set of code words is created that consists of those code words remaining after the excision of word c and those new code words created by appending c, cc, ccc, ... to those code words from the original set sans c. This augmentation process can be repeated as often as desired.

We follow Scholtz's example. Consider the most elementary of the synchronizable codes; the code word set $\{0, 1\}$. Let us extract 1 from the set and use it to construct new code words up to, say, length 4. We thus have the set

$$0$$
$$01$$
$$011 \tag{10}$$
$$0111$$

Note that the parameter s for the code of (10) has gone from $s = 0$ to $s = 3$. In general, the difference between the new s following a suffix construction and the old s is the length of the suffix. In (10), the longest suffix, 111, is three bits long and thus the new s is 3.

Let us perform another suffix construction by extracting 01 from (10) and let us generate all code words up to length 5. We obtain the following synchronizable code

$$0$$
$$001$$
$$011 \tag{11}$$
$$0111$$
$$01101$$

Problems

1. Starting with the synchronizable code $\{0, 1\}$, show that it is possible to use Scholtz's suffix construction method to generate a synchronizable code having 11 code words all of length 5 or less.
2. Devise an algorithm for decoding the code of problem 1 above.

9.5. Information Content of a Bernoulli Source

Information can be defined in a strictly mathematical way or it can be approached a bit more viscerally. Raisbeck (1963) takes the latter tack and our presentation is guided by his exposition. We consider that we learn "something" every time we observe the outcome of a Bernoulli source. If the probability that a one will be produced is $\frac{1}{10}$ and, consequently, the probability that a zero will be produced is $\frac{9}{10}$, then we "learn" more when we observe a one than a zero; we are, in a manner of speaking, more "surprised" at encoutering a one than we are a zero. The question is, How do we quantify this surprise?

Let us, for the moment, assume we can quantify what we learn and let us call it information. It seems reasonable that if we observe two outcomes of a Bernoulli source, we learn twice as much on the average, and, if we observe n outcomes, we learn n times as much on the average. Thus, if we learn $B(1, p)$ units on the average of information from observing a single outcome of a Bernoulli source with probability p of producing a one, we learn $B(n, p)$ units on the average by observing n outcomes and

$$B(n, p) = nB(1, p) \qquad (12)$$

A solution to the functional equation (12) is provided by defining the information gained by observing the oucome of an event with a priori probability p as

$$\log_2 \frac{1}{p} \qquad (13)$$

For our Bernoulli source then, the information gained on observing a one is

$$\log_2 \frac{1}{p} \qquad (14)$$

and the information gained on observing a zero is

$$\log_2 \frac{1}{1 - p} \qquad (15)$$

The average amount of information gained by observing a single output is then the average of (14) and (15) or

$$B(1, p) = -p \log_2 p - (1 - p) \log_2 (1 - p) \qquad (16)$$

If we were to observe two outcomes of a Bernoulli source we would observe one of the outcomes tabulated in Table 3. The a priori probabilities of the outcomes along with the information that accrues with their individual occurrences are also listed.

Table 3. Table Describing Possible Outputs and Information Accruing from Two Outcomes of a Bernoulli Source

Event	A priori probability of occurrence	Information gained by observing event
00	$(1 - p)^2$	$-\log_2 (1 - p)^2$
01	$(1 - p)p$	$-\log_2 [(1 - p)p]$
10	$p(1 - p)$	$-\log_2 [p(1 - p)]$
11	p^2	$-\log_2 p^2$

The average information that accrues from the two observations is then

$$B(2, p) = -(1 - p)^2 \log_2 (1 - p)^2 - (1 - p)p \log_2 [(1 - p)p]$$
$$- p(1 - p) \log_2 [p(1 - p)] - p^2 \log_2 p^2$$
$$= -2(1 - p) \log_2 (1 - p) - 2p \log_2 p = 2B(1, p)$$

as expected per (12).

Problems

1. Prove that (12) holds for n observations of a Bernoulli source for $B(1, p)$ as defined in (16).

2. Graph $B(1, p)$ versus p. Note that $B(1, p)$ is not a "sharp" function around $p = 0.5$. For what approximate values of p does $B(1, p) = 0.5$?

We now extend our information measure to a generalized Bernoulli source which produces the source symbols $\{s_i\}$ with probabilities $\{p_i\}$. The average information or "entropy" produced by such a source, usually denoted by H rather than by $B(\ )$ as in our restricted, motivational example, is

$$H = -\sum_i p_i \log_2 p_i \qquad (17)$$

Thus, the three-symbol source of Figure 9.1 produces information at an average rate of $-\frac{1}{2}\log_2 \frac{1}{2} - \frac{1}{4}\log_2 \frac{1}{4} - \frac{1}{4}\log_2 \frac{1}{4} = 1.5$ bits per sample (or symbol).

It is fundamental to realize that no binary coding scheme can possibly do better than allocating 1.5 bits per source symbol on the average. We thus have a basic inequality that measures, to some extent, the efficiency of a binary encoding scheme. If $\bar{l}$ is the average encoding length or the average number of encoding bits per source symbol then

$$H \leq \bar{l} \qquad (18)$$

The *efficiency* of a code is sometimes defined as the ratio $H/\bar{l}$.

9.6. The Huffman Code

Perhaps the most celebrated code is one developed by Huffman (1952). Huffman's construction allows the encoding of a generalized Bernoulli source. Huffman's code produces an encoding which exhibits the minimum possible average length for symbols that are encoded as they are produced. It can be shown that $\bar{l}$ for a Huffman code for a generalized Bernoulli source of entropy H as defined in (17) satisfies the inequality $H \leq \bar{l} < H + 1$. Huffman's code is also a prefix code—a code in which no code word of length n is found in the *first* n bits of a longer code word.

Table 4. Source Symbol Probabilities

Source symbol	Probability
s_1	0.2
s_2	0.55
s_3	0.15
s_4	0.1

Huffman's coding method comprises two phases. The first phase is a recursion, each replication of which depends on two basic steps: (1) listing the symbols in order of nonincreasing probabilities and (2) combining the two sources and their probabilities at the bottom of the table (if the table has more than two sources). This recursive method is sometimes called "source reduction". It is also necessary to remember how the tables were relisted at each step.

Abramson (1963) uses a nice diagrammatic method which we adopt to study Huffman's method by example. Suppose we have a four-symbol source with the probabilities shown in Table 4.

The first step is to order the source symbols in order of nonincreasing probabilities. This is done in Table 5. Table 6 shows the progression of the first phase of Huffman's coding method applied to the source symbol probabilities table, Table 5.

Phase two of Huffman's coding method is the development of the code words. We start with the final reduced source on the right—the two final probability entries. We assign a zero to the top entry and a one to the bottom. We now backtrack. One of the two final probabilities came from its left (in the table of reduced source probabilities) without being combined with another probability—for this entry we copy or "push" the code word, a zero or a one, from the right to the left and do nothing further with it. For the final probability entry that came from combining two entries to its left, we push back the one or zero and then *suffix* a zero to the pushed back bit for the entry in the top of the combined pair and suffix a one to the pushed back bit for the entry in the bottom of the combined pair as shown

Table 5. Relisted Source Symbol Probabilities

Source symbol	Probability
s_2	0.55
s_1	0.2
s_3	0.15
s_4	0.1

Table 6

Source symbol	Probability	Probabilities of reduced sources
s_2	0.55 – – – – – → 0.55 – – – – – →0.55	
s_1	0.2	→ 0.25 → 0.45
s_3	0.15	→ 0.2
s_4	0.1	

in Table 7. The process is continued in just the same way until we are back to the original source symbols as shown in Table 8.

We have now developed the Huffman code for the four-symbol source. The Huffman code exhibits an average length of $0.55 \times 1 + 0.2 \times 2 + (0.15 + 0.1) \times 3 = 1.7$ bits per symbol.

Problems

1. What would be the impact on decoding a Huffman code if instead of suffixing, we prefixed, thus reversing the bit order of the code words?

2. Table 9 is a Huffman code, adapted from Reza (1961), for the English language based on probability (frequency) of individual letter and space occurrences. Verify that the table is the proper Huffman code for the English language as represented by a generalized Bernoulli source with the given symbol probabilities and calculate $\bar{l}$.

3. As we have said, one thing that must always be taken into consideration when using a variable length source encoding scheme (such as a Huffman code) is recovery from channel and timing errors. If you are considering a Huffman code, be sure that the increased efficiency sufficiently offsets the

Table 7

Source symbol	First reduced source	Second reduced source
s_2 ← – – – – – 0 – – – – – – – 0		
s_1 ←	10 ←	1
s_3 ←	11 ←	
s_4 · ←		

Table 8

Source symbol	First reduced source	Second reduced source

Table 9. Probability of Occurrence of the 27 Symbols of the English Language[a] and Their Huffman Code

Symbol	Probability of occurrence	Huffman code word
A	0.0642	0100
B	0.0127	011111
C	0.0218	11111
D	0.0317	01011
E	0.1031	101
F	0.0208	001100
G	0.0152	011101
H	0.0467	1110
I	0.0575	1000
J	0.0008	00111111000
K	0.0049	00111110
L	0.0321	01010
M	0.0198	001101
N	0.0574	1001
O	0.0632	0110
P	0.0152	011110
Q	0.0008	0011111101
R	0.0484	1101
S	0.0514	1100
T	0.0796	0010
U	0.0228	11110
V	0.0083	0011110
W	0.0175	001110
X	0.0013	001111111
Y	0.0164	011100
Z	0.0005	00111111001
Space	0.1859	000

[a] © F. Reza, *An Introduction to Information Theory*, 1961, McGraw-Hill Book Co. (Reproduced with permission).

extra inconveniences of extra complexity, such as table lookup schemes, and increased sensitivity to errors. Although a Huffman code will never result in using more coding bits per source bit than any other code for a generalized Bernoulli source, it does not always mean that a Huffman code will result in a "better situation" or even "as good a situation" overall. The purpose of this problem is to demonstrate this assertion. Consider that we have the following generalized Bernoulli source and source probabilities. Without using a Huffman code and simply assigning the tribits 000, 001, 010, ..., 111 to the eight symbols, we obviously use three coding bits per source symbol and all code words are the same length which makes framing very simple as well as recovery from errors. Code the same source with a Huffman code and compute the number of coding bits used per source symbol on the average. Comment on your result.

Source Symbol	Probability
s_0	$\frac{9}{45}$
s_1	$\frac{6}{45}$
s_2	$\frac{6}{45}$
s_3	$\frac{5}{45}$
s_4	$\frac{5}{45}$
s_5	$\frac{5}{45}$
s_6	$\frac{5}{45}$
s_7	$\frac{4}{45}$

9.6.1. Connell's Method of Coding

Let us order Table 9 according to decreasing symbol probability (see Table 10). Note that the Huffman code words increase in length as, of course, they must, but aside from this there is not much order in the code words. Connell (1973) devised a simple method of rewriting the Huffman code words so that, if interpreted as normal binary quantities, their values become strictly increasing, that is, the following two conditions obtain: if the probability of symbol x is greater than the probability of symbol y then (1) the length of the code word for symbol x will be less than or equal to the length of the code word for symbol y and (2) the normal binary value of the code word for symbol x will be less than the normal binary value of the code word for symbol y.

Connell's method requires the a priori knowledge of the lengths of the Huffman code words. Let n_i be the number of code words of length i. The concatenation of the values of n, that is, $n_1 n_2 n_3 \cdots n_M$, where M is the length of the longest code word, is sometimes called the "code index". Connell's method produces a new code with the same code index. Thus,

**Table 10. The 27 Symbols of the English Language
Ordered by Their Probability of Occurrence**

Symbol	Probability rank	Huffman code word
Space	0	000
E	1	101
T	2	0010
A	3	0100
O	4	0110
I	5	1000
N	6	1001
S	7	1100
R	8	1101
H	9	1110
L	10	01010
D	11	01011
U	12	11110
C	13	11111
F	14	001100
M	15	001101
W	16	001110
Y	17	001111
G	18	011101
P	19	011110
B	20	011111
V	21	0011110
K	22	00111110
X	23	001111111
Q	24	0011111101
J	25	00111111000
Z	26	00111111001

of Table 10 we have

i	n_i
1	0
2	0
3	2
4	8
5	4
6	7
7	1
8	1
9	1
10	1
11	2

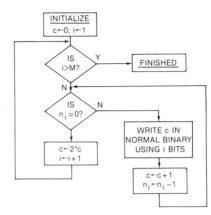

Figure 9.7. Algorithm for generating Connell code words.

for our code Connell's method is defined in the flow chart of Figure 9.7. Table 11 contrasts Connell's code with the Huffman code of Table 10.

Problem

Devise an algorithm that will allow you to encode and decode a Connell code without requiring the generation of all code words.

9.7. Source Extension and Its Coding

If we group the successive outputs of a generalized Bernoulli source into blocks of n symbols each, we form a new source which is called the nth source extension of the old source. By encoding the n-symbol blocks of the nth source extension it is possible to construct codes whose average code word length, divided by n, approaches the entropy, H, of the original source.

Let us again consider the Bernoulli source with probability of a one equal to $\frac{1}{10}$. The entropy of this Bernoulli source is 0.469 which means that the most efficient faithful code must use at least 0.469 encoding bits per source bit. Considering the second source extension, we find that there are four possible dibits with the following probabilities:

Extended Source Output	Probability
00	$0.9 \times 0.9 = 0.81$
01	$0.9 \times 0.1 = 0.09$
10	$0.1 \times 0.9 = 0.09$
11	$0.1 \times 0.1 = 0.01$

Table 11. Contrast of Connell's and Huffman's Codes

Code word length	Connell code word	Huffman code word
3	000	000
3	001	101
4	0100	0010
4	0101	0100
4	0110	0110
4	0111	1000
4	1000	1001
4	1001	1100
4	1010	1101
4	1011	1110
5	11000	01010
5	11001	01011
5	11010	11110
5	11011	11111
6	111000	001100
6	111001	001101
6	111010	001110
6	111011	001111
6	111100	011101
6	111101	011110
6	111110	011111
7	1111110	0011110
8	11111110	00111110
9	111111110	001111111
10	1111111110	0011111101
11	11111111110	00111111000
11	11111111111	00111111001

Building a Huffman code, we find that

00	can be represented by	0
01	can be represented by	11
10	can be represented by	100
11	can be represented by	101

The above coding results in an average of 0.645 coding bits per source bit. This simple encoding provides a great reduction toward the theoretical limit of 0.469 coding bits per source bit. If we extend our block size to

$n = 3$, we have the following tribits and their respective probabilities:

Extended Source Output	Probability
000	$0.9 \times 0.9 \times 0.9 = 0.729$
001	$0.9 \times 0.9 \times 0.1 = 0.081$
010	$0.9 \times 0.1 \times 0.9 = 0.081$
011	$0.9 \times 0.1 \times 0.1 = 0.009$
100	$0.1 \times 0.9 \times 0.9 = 0.081$
101	$0.1 \times 0.9 \times 0.1 = 0.009$
110	$0.1 \times 0.1 \times 0.9 = 0.009$
111	$0.1 \times 0.1 \times 0.1 = 0.001$

Again, building a Huffman code, we find that

000	can be represented by	0
001	can be represented by	100
010	can be represented by	101
011	can be represented by	11100
100	can be represented by	110
101	can be represented by	11101
110	can be represented by	11110
111	can be represented by	11111

The coding of the source's third extension results in an average of 0.533 coding bits per source bit. Another reduction has been achieved although it is not as dramatic a reduction as we observed in going from $n = 1$ to $n = 2$ *and* the Huffman code is quickly becoming very complicated.

Huffman codes, as we have noted, are optimal prefix codes in that they exhibit minimum average code word length. Huffman codes have two significant drawbacks, however, when the number of source messages is large as might arise when considering a large-order extension of a Bernoulli source. These drawbacks are (1) the complexity of deriving the Huffman code and (2) the large memory required to store the Huffman code words. Hankamer (1979) has suggested a procedure that overcomes these drawbacks by incurring what may be only a nominal penalty in average code word length. The procedure encodes the set of 2^L binary L-tuples by partitioning the messages into two groups A and B. The messages assigned to group A all have probability $> 2^{-L}$; the messages put into group B all have probabilities $\leq 2^{-L}$. Let $P_{\Sigma B}$ be the sum of all the probabilities of the messages in group B. We create a Huffman code for the elements in group A and an additional "escape character" whose probability is $P_{\Sigma B}$. The "escape character" serves as a flag or prefix. When a message from group

B is to be sent, the convention is that the "escape character" is sent followed by the message from group B. No compression coding is applied to messages within group B.

Recall that the average code length, $\bar{l}$, produced by the Huffman code for a source of entropy H is bounded $H \leq l < H + 1$. The average code word length produced by Hankamer's procedure is bounded

$$H \leq \bar{l} < H + 1 - P_{\sum B} \log_2 P_{\sum B} \tag{19}$$

As an example, let us consider tribit output blocks from the Bernoulli source with probability of a one equal to $\frac{1}{10}$. We find that group A has only one member, 000; the rest of the tribits all have probability ≤ 0.125. Building a Huffman code for the single member of group A and the "escape character" as prefix to the group B members we obtain the encoding table shown in Table 12.

The entropy of the source coded in Table 12 is 0.469 bits as previously determined. The average code word length of the Huffman code is 0.533 bits which does indeed satisfy the appropriate inequality $0.469 \leq 0.533 < 1.469$. The average code word length of the Hankamer code is 1.813 bits. Note that $P_{\sum B} = 0.271$ and that $-P_{\sum B} \log_2 P_{\sum B} = 0.510$ and that Hankamer's inequality is satisfied as $0.469 \leq 1.813 < 1.979$.

Problem

Derive the Huffman and Hankamer codes for the 4-tuples from the above source. Compare the average code word lengths. Comment on the memory requirements to code and decode using stored tables for each coding scheme. Demonstrate that the appropriate inequalities hold. Repeat the foregoing for 5-tuples.

Table 12. Huffman and Hankamer Codings of an Eight-Message Source

Source tribit	Probability	Huffman code word	Hankamer code word
000	0.729	0	0
001	0.081	100	1001
010	0.081	101	1010
011	0.009	11100	1011
100	0.081	110	1100
101	0.009	11101	1101
110	0.009	11110	1110
111	0.001	11111	1111

Another approach to encoding extended sources was advanced by Cover (1973). One of his ideas was to sequentially number all binary strings n bits long that are of weight w, that is, had exactly w ones.† The number of particular string of the $\binom{n}{w}$ possible strings is termed the "index" of the string. If we choose $w = np$ where p is the probability of a one, then the individually most likely strings encountered from the source will be in the set exhibiting this weight. A way in which we can use this property is to assign short code words to strings of weight $w = np$ and longer code words to strings of other weights as strings of other weights will be individually less likely. The possibilities for innovation based on this concept are manifold. Consider, for example, the following scheme for the Bernoulli source that we have been using, $p = 0.1$. We let $n = 6$, that is, the sixth source extension. If we were to construct a Huffman code we would have to create $2^6 = 64$ different code words. Let us instead code as follows: we number all 6-bit strings of weight zero and weight one and assign 3-bit code words which are merely sequential normal binary numbers as shown in Table 13. We use the remaining 3-bit code word, 111, as an "escape character". It is to be used when the 6-bit string to be encoded is of weight two or more. When 111 appears, it is followed by the 6-bit string of weight two or more. Thus, if the source were to output the following sequence

$$001000001100000000000000000000010000 \qquad (20)$$

the encoded stream would be

$$100111001100000000000101 \qquad (21)$$

which uses 24 encoding bits for 36 source bits or 0.667 encoding bits per source bit.

Table 13. Sequences of Weight Zero and One and Their Code Words

Weight	Sequence	Number	Code word
0	000000	0	000
1	000001	1	001
1	000010	2	010
1	000100	3	011
1	001000	4	100
1	010000	5	101
1	100000	6	110

† An algorithm for generating all n-bit words of weight w is given in Chapter 1.

The theoretical efficiency of the above coding scheme is easily analyzed. The probability that a particular 6-tuple from the source will be of weight zero or weight one is

$$\sum_{i=0}^{1} \binom{6}{i} (0.1)^i (0.9)^{6-i} = 0.8857$$

The average length of a code word is then

$$3 \times (0.8857) + (3 + 6) \times (0.1143) = 3.6858 \text{ bits}$$

(Where does the factor 0.1143 come from? Why is it multiplied by $3 + 6$?)

On the average then, the number of encoding bits per source bit is $3.6858/6 = 0.6143$.

We have chosen a very short density-based code for our example. Greater efficiencies can be obtained by choosing n large. [In the limit, as $n \to \infty$, it can be shown that codes based on enumeration of equiweight sequences approach the optimum number of $H(p)$ coding bits per source bit.] Finally, consider that a source block of length n bits would require 2^n Huffman code words to be computed; the corresponding effort required for an enumerative code may be much less.

9.8. Run Length Encoding

Still another approach to encoding a Bernoulli source for which p is significantly distant from 0.5 is to encode the length of the runs of the more likely bit. This approach belongs to the class of source coding techniques known as run length encoding. It is a very valuable method in that the encoding and decoding logics are simple and require only nominal memory. Run length encoding has been successfully used in many facsimile machines to provide compression and allow documents to be transmitted with much greater speed than they would if every sample point were independently encoded and sent.

A particularly interesting and potentially useful scheme was suggested by Ohnishi *et al.* (1977a) and investigated by Tanaka and Leon-Garcia (1982). Dubbed a truncated run length code, the scheme deals with runs of the most probable bits up to length R; that is, we truncate our runs to be less than or equal to R bits. The scheme is not universal, that is, it can not reach an efficiency of 100% for all sources; however, it does very well in general.

Consider that we have a Bernoulli source that produces a one with probability $p < \frac{1}{2}$. We first select a basic integer parameter $R \geq 1$. (A guide

**Table 14. Run Length Substitution
Code for R**

Source pattern	Code word
1	10
01	110
001	1110
$\vdots$	$\vdots$
000 $\cdots\cdots\cdots$ 1	1111 $\cdots$ 10
000 $\cdots\cdots\cdots$ 01	1111 $\cdots$ 11
000 $\cdots\cdots\cdots$ 00	0
|$\longleftarrow$ R bits $\longrightarrow$|	|$\blacktriangleleft$ R bits $\blacktriangleright$|

to selecting the appropriate R will be given shortly.) We then prepare
Table 14.

We encode by substitution as follows. We examine the first bit of the
source string. If it is a one, we represent it with 10. If it is a zero, we look
at the first two bits. If they are zero-one, we represent them with 110 and
so on. If, however, there are no ones in the first R bits, we represent the R
zeros by the single-bit code word 0. Note that this scheme encodes a variable
number of source bits with a variable length code word.

It has been shown by Ohnishi *et al.* (1977b) that all ranges of p ($0 < p \leqslant$
0.5) can be handled if R is taken to be an integral power of 2, $R = 2^r$. The
efficiency of this encoding scheme is plotted in Figure 9.8. The numbers
below the peaks are the appropriate values of R. The boundaries between
appropriate values of R fall at the points

$$p_r = 1 - \left(\frac{\sqrt{5}-1}{2}\right)2^{-r}, \qquad r = 0, 1, 2, \ldots \qquad (22)$$

Problems

1. Show that the average number of source output bits encoded per code
word is $(1/p)[1 - (1 - p)^R]$.

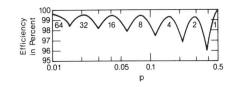

Figure 9.8. Efficiency of the encoding
scheme versus *p*.

Table 15. Substitution Table for $M = 8$

Source pattern	Code word
1	10
01	110
001	1110
0001	11110
00001	111110
000001	1111110
0000001	11111110
00000001	11111111
00000000	0

2. Show that the average code word length produced by coding the Bernoulli source with the code is

$$p\left(\frac{(1-p)^{2^r}}{1-(1-p)^{2^r}} + (r+1)\right)$$

3. Derive expression (22). [Hint: determine the points where the efficiency of the code is the same for r and $r + 1$.]

4. Show that this coding scheme gives rise to a Huffman code if $p = (\frac{1}{2})^{2^{-m}}$.

As an example, let us encode the previously considered 36-bit output string (20) from a Bernoulli source with $p = 0.1$. According to Figure 9.8 we choose $R = 8$ and prepare the substitution table, Table 15. The string to be encoded (20) is again 001000001100000000000000000000010000. Using Table 15 we see that the first 32 bits of the string encode to

$$1110111111010001111110 \qquad (23)$$

exhibiting a reduction of 22 encoding bits to 32 source bits.

9.9. Encoding to a Fidelity Criterion

If we can tolerate errors introduced by the source coding, we can use this latitude to effect a reduction in bits sent. The following simple example illuminates all the important concepts of this type of encoding. Suppose we have the third extension of a Bernoulli source with $p = 0.5$. Let us assume that we can afford only one coding bit for each 3-bit block from the source. What can we do and what do we incur by doing it.

Well, one thing we could do is ignore two of the three source bits and send the remaining bit. At the receiver we would know one of the source bits perfectly, without error, but we would have to guess as to the values

Table 16. Majority Encoding and Incurred Errors

Source tuple	Coding bit	Receiver assumption of source tuple	Number of errors
000	0	000	0
001	0	000	1
010	0	000	1
011	1	111	1
100	0	000	1
101	1	111	1
110	1	111	1
111	1	111	0

of the other two bits. For each bit we guessed, we would be incorrect half of the time and thus would experience an average error rate of 0.333.

Suppose, however, that we encode by sending the most common bit in the source 2-tuple. At the receiver we would assume that all source bits in the tuple were the same as the received bit. Table 16 shows all eight possible source tuples, the coding bit representing each tuple, the receiver assumptions, and the number of errors that result from this "majority" encoding. The average error rate incurred by the encoding scheme shown in Table 16 is $\frac{3}{4}$ errors per source tuple or an average error rate of 0.250.

The above example seems to say that by "spreading" the errors over all encoded bits we tend to incur less errors on the average. Additionally, the upper limit of incurred errors is lower for the majority encoding. With the majority encoding, the greatest error we can experience in a 3-tuple is 1 bit; using the former method of sending 1 bit perfectly and guessing at the other two, we can incur as many as two errors.

Coding the nth source extension of a Bernoulli source with $p = 0.5$ to a fidelity criterion is easily done by generating a "library" of n-bit random code words with each bit of each code word derived from a Bernoulli source with $p = 0.5$. We encode an n-bit source tuple by sending the number of the library code word which has the fewest disagreements with the n-bit source tuple. If the number of words in the library is R, then we require only $\log_2 R$ bits to encode n bits. This is a compression, C, of

$$C = \frac{n}{\log_2 R} \tag{24}$$

Problem

Assume we can tolerate an average bit error rate of ε. Show that as $n \to \infty$,

$$C \to \frac{\ln 2}{\ln \left[2(1 - \varepsilon)^{(1-\varepsilon)} \varepsilon^{\varepsilon} \right]}$$

References

Abramson, N. (1963), *Information Theory and Coding*, McGraw-Hill, New York.

Connell, J. (1973), A Huffman–Shannon–Fano Code, *Proceedings of the IEEE*, pp. 1046–1047.

Cover, T. (1973), Enumerative Source Coding, *IEEE Transactions on Information Theory*, Vol. 19, pp. 73–77.

Eastman, W. (1965), On the Construction of Comma-Free Codes, *IEEE Transactions on Information Theory*, Vol. 11, pp. 135–142.

Even, S. (1963), Tests for Unique Decipherability, *IEEE Transactions on Information Theory*, Vol. 9, pp. 109–112.

Golomb, S., B. Gordon, and L. Welch (1958), Comma-Free Codes, *Canadian Journal of Mathematics*, Vol. 10, pp. 202–209.

Hankamer, M. (1979), A Modified Huffman Procedure with Reduced Memory Requirement, *IEEE Transactions on Communications*, Vol. 27, pp. 930–932.

Huffman, D. (1952), A Method for the Construction of Minimum Redundancy Codes, *Proceedings of the IRE*, pp. 1098–1101.

Karush, J. (1961), A Simple Proof of an Inequality of McMillan, *IRE Transactions on Information Theory*, Vol. 7, p. 118.

McMillan, B. (1956), Two Inequalities Implied by Unique Decipherability, *IRE Transactions on Information Theory*, Vol. 2, pp. 115–116.

Ohnishi, R., Y. Ueno, and F. Ono (1977a), Optimization of Facsimile Data Compression, *National Telecommunications Conference Record*, pp. 49.1.1–49.1.6.

Ohnishi, R., Y. Ueno, and F. Ono (1977b), Efficient Coding for Binary Information Sources, *Transactions IECE, Japan*, Vol. 60-A, No. 12.

Raisbeck, G. (1963), *Information Theory*, The MIT Press, Cambridge, MA.

Reza, F. (1961), *An Introduction to Information Theory*, McGraw-Hill, New York.

Scholtz, R. (1966), Codes with Synchronization Capability, *IEEE Transactions on Information Theory*, Vol. 12, pp. 135–142.

Tanaka, H. and A. Leon-Garcia (1982), Efficient Run-Length Encodings, *IEEE Transactions on Information Theory*, Vol. 28, pp. 880–890.

Information Protection

10.1. Classical Cryptography

Modern (substitution) cryptography depends on a very simple concept—modular addition of unpredictable, noiselike quantities to plaintext elements. What do we mean by this and, understanding it, how do we implement it? The following example is intended to motivate the question and lead us to the answer.

Consider that we have two bits, b_p and b_k. We form a third bit, b_c, by summing b_p and b_k using normal arithmetic:

$$b_c = b_p + b_k \tag{1}$$

If we observe bit b_c what can we deduce about bits b_p and b_k? Table 1 depicts all the possible situations governed by (1).

On examining Table 1 we note that when $b_c = 0$ or $b_c = 2$ we can uniquely determine *both* b_p and b_k. Only when $b_c = 1$ do we encounter ambiguity in an attempt to resolve b_p and b_k. One further observation concerning (1) and Table 1 is that we can uniquely determine any single component of the triple (b_p, b_k, b_c) given the other two components.

Table 1. All Possible Cases Governed by Equation (1)

b_p	b_k	b_c
0	0	0
0	1	1
1	0	1
1	1	2

Now consider that we perform our addition modulo 2 and set

$$b_c \equiv b_p + b_k \quad (\text{mod } 2) \tag{2}$$

Table 2 depicts all the situations governed by Equation (2).

Looking now at Table 2 we find that we can no longer resolve either b_p or b_k from knowing b_c alone for any of the b_c values. But note also that we can still uniquely determine any component of the triple (b_p, b_k, b_c) given the other two components.

Now how do we tie all this into cryptography? The answer is quite simple. Assume that two physically separated parties wish to send messages to each other so that no third party who might observe the messages in transit would be able to understand or "read" the messages. All the parties need is a set of "key" bits $\{b_{k_i}\}$, which are known to both parties and held secret by both parties, which have been produced by a balanced Bernoulli source. Assume Party A wishes to send a message of "plaintext" bits $\{b_{p_i}\}$ to Party B. Party A performs the following operation:

$$
\begin{array}{llll}
b_{p_1} & b_{p_2} & \cdots & b_{p_n} \quad (\text{plaintext}) \\
b_{k_1} & b_{k_2} & \cdots & b_{k_n} \quad (\text{key}) \\
\hline
b_{c_1} & b_{c_2} & \cdots & b_{c_n} \quad (\text{cipher})
\end{array}
$$

where $b_{c_i} \equiv b_{p_i} + b_{k_i} \ (\text{mod } 2)$ and sends the "cipher" bits $\{b_{c_i}\}$ to Party B. Now, if an unauthorized party, Party C, were to observe the $\{b_{c_i}\}$ which is also called the "encrypted message," all Party C could determine would be that (1) Party A sent a message to Party B and (2) the message was no longer than n bits in length; but Party C could not determine a single bit of $\{b_{p_i}\}$. (Why?)

Party B recovers the plaintext by performing the following operation:

$$
\begin{array}{llll}
b_{c_1} & b_{c_2} & \cdots & b_{c_n} \quad (\text{cipher}) \\
b_{k_1} & b_{k_2} & \cdots & b_{k_n} \quad (\text{key}) \\
\hline
b_{p_1} & b_{p_2} & \cdots & b_{p_n} \quad (\text{plaintext})
\end{array}
$$

where $b_{p_i} \equiv b_{c_i} + b_{k_i} \equiv b_{p_i} + b_{k_i} + b_{k_i} \equiv b_{p_i} + 0 \equiv b_{p_i} \ (\text{mod } 2)$.

Problems

1. Suppose that the plaintext elements were not bits but rather discrete quantities that could assume any of m different values. Design a set of appropriate encryption/decryption operations to accommodate this new

**Table 2. All Possible
Cases Governed by
Equation (2)**

b_p	b_k	b_c
0	0	0
0	1	1
1	0	1
1	1	0

genre of plaintext elements. Specifically, comment on (a) the modulus to be used—must it be equal to m? What would happen if it were (i) greater than m (ii) less than m? (b) the key(stream) elements—how must they be related to the chosen modulus? From what type of source should they be generated?

2. If the set of key bits, or, as it it sometimes called, the "keystream," is not derived from a balanced Bernoulli source or a suitable approximation, then Party C may be able to deduce something further about the plaintext. This is due to the strong statistical coupling that is often present between the plaintext bits. It was shown by Captain Parker Hitt in 1917 that English text has so much redundancy that if one English text is used as a keystream for another English text, both texts can be recovered by analyzing the ciphertext. We can do this in a number of ways. A likely word such as "the" or "and" or a likely letter group such as "tion" or "ment" is presumed to be present. The presumed word or letter group is subtracted (modulo m) from the ciphertext. If the letter grouping resulting from subtraction is impossible in English, such as "XBR" or "QBL," then we reject the presumed hypothesis and try another likely group of letters. If the result is something that "looks like" English such as "ent" or "ome" perhaps, then we can continue by extending our hypothesized word or the word that is developing as a result of the subtraction. Assume the English letters are coded using normal binary as shown in Table 3.

**Table 3. Normal Binary
Encoding of the English
Alphabet**

A—00000
B—00001
C—00010
$\vdots$
Y—11000
Z—11001

(a) Two English texts were converted to binary using Table 3. One converted text was used as a keystream for enciphering the other. The following was the ciphertext produced:

10110   10011   10010   00000   10001   01001   01000

00110   01010   10100   00011   01010   00010   11100

Recover both texts. [Hint: The word "the" occurs in one of the messages. There are no spaces between words.]

(b) Part (a) above was not too difficult because of the great redundancy inherent in the English language. Suppose we "squeeze out" some of that redundancy by using the Huffman code of Table 10 in Chapter 9 to convert our English texts before enciphering. The following is the ciphertext produced:

00000100101001100000000000101000000

10000100110011000010011101110100100

Attempt to recover both texts and comment on any increase in complexity. [Hint: The same hint applies as applied in part (a).]

3. Suppose that the same keystream is used to encipher two different messages. What does this situation have in common with the situation considered in problem 2 above?

A classical, two-party cryptosystem thus requires that both parties generate or access a stored keystream in message "synchrony," that is, the keystream applied at the receiving party must be started precisely at the beginning of the received ciphertext. A classical binary cryptographic system also requires a keystream generated by a suitable approximation to a balanced Bernoulli source. It is also generally prudent to assume that the keystream generator logic will become "canonically" known, that is, its general contents or overall logic structure should be presumed known to any interested unauthorized party. The security of the keystream should reside in a small vector of secretly held bits called the "key." The key must be secretly transported to each authorized party *before* secure communications can be commenced. This consideration is the defining element of "classical" cryptography. The key is entered into the keystream generator to produce a keystream that resembles the output from a balanced Bernoulli source.

In addition to the foregoing, it should also be presumed that any interested unauthorized party may obtain significant access to the keystream. This overtly puzzling assumption follows in the spirit of problem 2 above;

that is, we may presume that the encrypted plaintext may often be accurately guessed and thus portions of the keystream laid bare.

Let us analyze the cryptographic system depicted in Figure 10.1.

The key for the system shown in Figure 10.1 is the 31-bit initial state of the shift register. The output from the 31-stage m-sequence generator is probably very similar, at first glance, to a balanced Bernoulli source. Recall that the defining characteristic of a Bernoulli source is that knowledge of previous outputs does not aid the prediction of future outputs. But suppose the plaintext in Figure 10.1 is English text coded according to Table 3. Knowledge of just 7 consecutive letters of the plaintext would allow us to know 31 consecutive bits of the keystream. These 31 consecutive bits would allow us to generate all future keystream bits. (Why? How?) Thus, the keystream generator of Figure 10.1 is not a good approximation to a balanced Bernoulli source and the system security is minimal.

How then do we approximate a balanced Bernoulli source for sufficient security? An answer has been provided through a remarkable effort undertaken by the National Bureau of Standards (NBS)—the nation's metrologists. The NBS, in response to national requirements to provide computer file and transmission security, solicited the public for a viable encryption architecture. The process set in motion was truly fantastic (see Hershey, 1983) and resulted in an encryption algorithm known as the Data Encryption Standard, or DES. The DES is specified in Federal Information Processing Standard Publication Number 46 (FIPS PUB 46). Although the internal workings of the algorithm are interesting, indeed fascinating, it is not necessary to recount and appreciate them; rather, we treat the algorithm in a systems context. This is, after all, how the algorithm is available to us— as an LSI chip, chipset, or firmware package with input and output system specifications.

10.1.1. The DES

The DES is an algorithm for converting 64 bits considered as an input into a 64-bit output under control of another 64 bits. The process of conversion is extremely complex and is considered a good cryptographic scheme. For enciphering, the input is called the plaintext, the output is called the ciphertext, and the 64 bits that control the conversion process is the keying variable or key. For deciphering, the reverse obtains. This process

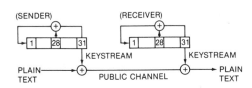

Figure 10.1. A cryptographic system.

is an example of what is known generically as a "codebook" cryptographic process.

The name "codebook" derives from the old (World War I vintage) cryptographic procedure of encrypting messages by using words, letter combinations, or other equivalents to stand for message text elements, usually words, letters, numbers, or phrases. The particular correspondence is defined by a specific codebook or system key. Calling the DES a codebook is an apt term for the following reason. Consider the 64-bit input. There are 2^{64} (approximately 18 quintillion) possible inputs. They can be easily and naturally listed lexicographically by counting in normal binary as shown in Figure 10.2. Each 64-bit input results in a 64-bit output. There are also 2^{64} possible outputs (the same form and number as the inputs). Note that the DES algorithm can be viewed as an array of 64 nonlinear 64-input boolean functions. The specific functions are specified by the keying variable.

The number of ways 2^{64} inputs could be mapped to 64-bit outputs is easily computed. There are 2^{64} choices for each input's assigned output and therefore $(2^{64})^{2^{64}}$ different ways. This is a truly astronomical number. However, most of these ways are unacceptable. Suppose two inputs result in (are mapped to) the same output. How can we go "backward?" That is, given the output, how can we uniquely determine the input? Clearly, we can not. Therefore, we must require that each input result in an output different from that resulting from any other input. Because there are the same number of outputs as inputs, the mapping becomes "one-to-one," that is, each output corresponds to one and only one input and vice versa. (See Figure 10.3.)

How many ways then can 2^{64} inputs be mapped to 2^{64} outputs under the above condition, one-to-one? Enumerating by sequential assignment, there are 2^{64} output possibilities for the first input, $2^{64} - 1$ possibilities for the second input, and so on. There are thus $2^{64}!$ ways. This is much smaller than $(2^{64})^{2^{64}}$ but is also "astronomical."

The DES is controlled by a 64-bit key. Not all 64 bits function as part of the codebook control, however. Eight of the bits serve as simple odd

Figure 10.2. The 2^{64} possible inputs to the DES.

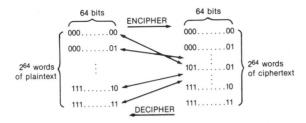

Figure 10.3. DES algorithm viewed as a codebook encipherment method.

parity checks over eight blocks of 7 bits each. This limits the number of different DES keys to 2^{56} or about 72 quadrillion. This is a large number but nowhere near 2^{64}! Thus, the DES can be considered an algorithm which implements a small fraction of the possible codebooks over 64 bits.

The DES as a cryptographic codebook must fulfill the following condition: there must be an easily implementable algorithm to reverse the codebook; that is, given the output determine the input if one has the key. If one does not have the key, then the process of reversing the codebook is termed cryptanalysis and must be "sufficiently" difficult and costly. What the cryptographic assessor has to do is to ensure three things:

1. The subset of 2^{56} possible codebooks achievable by a DES is a "good" subset. This entails showing that the codebooks corresponding to different keys are not related, that is, their structures are very different.
2. Recovery of a key, cryptanalysis, which allows reversal of the codebook without a priori knowledge of the key, will require, on the average, the expenditure of more valuable assets, money, time, personnel, and so on, than a particular minimum—the standard.
3. The DES is used in such a way (mode) that no severe shortcomings in cryptographic security will result.

10.1.2. Modes of the DES

A cryptographic algorithm does not, in itself, a cryptographic system make. Consequently, a number of "modes" have been developed around the DES algorithm. (See FIPS PUB 81, 1980.) The modes are different cryptosystems, each of which has advantages and disadvantages. The modes, their advantages and disadvantages, are lucidly treated by Pomper (1982). We make liberal use of his publication in studying a subset of three of the DES modes. The first mode we shall look at is the Electronic Codebook (ECB) mode. The ECB mode is the simplest of all the modes of the DES. A 64-bit input plaintext "block" of bits is encrypted to a 64-bit output ciphertext block under control of the DES keying variable, with the DES algorithm set to encipher. To decrypt, the receiver need only enter the 64-bit ciphertext block, insert the keying variable used by the sender, and set the DES algorithm to decipher.

Note that the ECB mode gives the same 64-bit ciphertext output for the same 64-bit plaintext input. Thus, if an unauthorized party is watching the enciphered traffic, the interloper can tell every time a 64-bit plaintext block is repeated. This may allow the interloper to learn something about the plaintext traffic which may be undesirable. For this reason the ECB mode is not recommended for high-volume traffic applications but rather only for encryption of sparse amounts of random or pseudorandom data.

Problems

1. If one does not have the encryption keying variable and attempts to search for it by exhaustion, how long will it take, on the average, to find a variable that "matches" a plaintext input and ciphertext output? Assume 10^6 trials per second. What is the probability that such a match arises from a variable different than that sought?

2. As we stated earlier, the DES algorithm can be viewed as an array of nonlinear 64-input boolean functions, that is, any cipher bit is a function of all the 64-input plaintext bits and also of each of the 56 keying variable bits. Hellman et al. (1976) have shown that if we consider each of the 64 nonlinear boolean functions as a function of the 64 plaintext bits *and* the 56 keying variable bits, that is, a function of 120 inputs, then the functions are all self-dual functions. What implications does this have?

It is useful to be able to encipher a continuous steam of plaintext traffic rather than the more constrained form of blocks of bits. The next two modes allow us to do just this. The first of these is the Cipher Feedback (CFB) mode. In this mode the DES algorithm is used to generate a pseudorandom binary sequence of bits to be used as a keystream to be added modulo 2 to a plaintext traffic "stream" creating a ciphertext stream. The intended receiver generates the same keystream, adds it modulo 2 to the received ciphertext stream, and recovers the plaintext stream. Figure 10.4, from Pomper (1982), depicts the operation of this mode. Note that both sender and receiver use the DES in the encipher setting. (Why are they set the same?) Note the parameter K in Figure 10.4. The CFB mode can be used for K in the range $1 \leq K \leq 64$. What we have done in Figure 10.4 is to use the block structure of the DES algorithm in such a manner that it produces a keystream; K bits at a time are "dumped" out of the output block in parallel and then serially shifted out. The DES performs one replication of the algorithm for each K bits produced. One other thing is needed to make

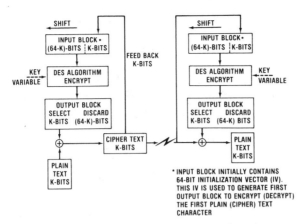

Figure 10.4. *K*-bit Cipher Feedback (CFB) mode.

the CFB mode work. This additional item is an Initialization Vector (IV). Each time the sender starts a new message (is initialized), a new IV must be used. The IV is a pseudorandom block of bits and is used as the first input to the DES. The IV used by the sender for the CFB mode must be different for each transmission but need not be kept secret. It may therefore be transmitted without protection to the receiver.

Problem

Study the CFB mode and discuss what would happen if a *single* cipher bit were inverted by the communications channel.

The second of the continuous keystream modes that we look at is the Output Feedback (OFB) mode. Figure 10.5, also from Pomper (1982), depicts the operation of this mode. The mode also requires an IV and the IV requirements are similar to those of the CFB requirements.

Problems

1. What are the differences between the CFB and OFB modes? What would be the effect if a single cipher bit were inverted by the communications channel?

2. Clearly, synchronization is crucial to the successful operation of OFB. Discuss the synchronization issue for CFB. Discuss specifically the case for which $K = 1$.

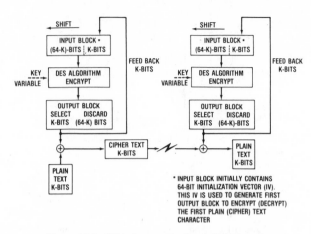

Figure 10.5. K-bit Output Feedback (OFB) mode.

3. If an interloper happens to know or correctly guess a stretch of plaintext traffic, can the interloper modify the ciphertext traffic to send a false, intelligible message to the receiver? Discuss for both the CFB and OFB modes. Also clearly outline how the interloper would go about the task.

4. Consider the OFB mode with $K = 64$. The keystream will be periodic. Why? What is the expected period?

10.2. Public Key Cryptography

Since the late 1970s, there has been a remarkable series of developments in mathematical information protection techniques that have come to be known as "public key" or "two key" cryptographic methods. Combining elements of abstract algebra, complexity theory, and communications protocols, public key cryptographic systems allow secure communications to take place without the a priori distribution of protected keying variables. This is what differentiates public key cryptosystems from classical cryptosystems. The concept of a public key cryptosystem was unveiled by Diffie and Hellman (1979). The field has been a boon for number theorists and computational mathematicians and has inspired much valuable research. This flurry of activity and the positive results produced have also rendered public key cryptography a "moving target." Many of the proposed public key cryptographic systems have been defeated or severely weakened within only a short period of time after their debut. What we do in this section is to look at three public key cryptosystems, Merkle's Puzzle System, the Diffie–Hellman (DH) System, and the system due to Rivest, Shamir, and Adleman (RSA). The first of these systems is of theoretical interest only and is included for motivational purposes. The second of these systems highlights the apparent nonstationarity of the security afforded by many of the public key cryptosystems. We then consider the RSA public key cryptosystem; we comment on its history and some points relating to its implementation. The RSA is perhaps the most "durable" of all the public key cryptosystems. Its security seems to be largely linked to the difficulty of factoring large composite numbers of a special form. Great advances have been made in factoring; however, it still appears that factoring remains a generally hard problem in spite of the centuries of attention it has received from the great mathematicians. Perhaps this is the reason that the RSA architecture has been so durable. Following these topics, we study an unusual weakness, not in the RSA itself, but in an implementation—a weakness brought about by a faulty protocol which is quite insidious in that it appears, at first blush, to be an eminently reasonable one. This particular example is only one of a number of interesting hazards that do not plague classical cryptography but often attend public key cryptosystems.

10.2.1. Merkle's Puzzle

One of the best ways to grasp the concept of public key cryptography is to study a concept propounded by Merkle (1978). The objective that Merkle set is the very objective of public key cryptography. It is to allow two parties, A and B, to "communicate securely" without the a priori establishment of a secret between them, that is, all communication between the parties is assumed to be available to any interested third party. To us, the phrase, "communicate securely" is taken in a very restricted sense for this introductory study. It shall mean only that an unauthorized or third party must expend sufficiently more work to "read" or decrypt the traffic between A and B than A or B need expend toward the encryption and decryption of their traffic.

Merkle's method is very straightforward. First A and B publicly choose and establish a cryptographic function. Such a function should be one-to-one and hard to invert; typically, a mapping from x to $f(x)$ such that given x, $f(x)$ is relatively easy to calculate, but given $f(x) = y$ there is no shortcut to finding x save exhaustively trying all y. The DES in the CFB mode, with a particular key and a published IV, is such a function.

We use the cryptographic function as a "puzzle." We do this by encrypting N frames of data of the form shown in Figure 10.6. Each of the N frames is encrypted by a different cryptographic function, for example,

Figure 10.6. Generic data frame.

| DECRYPTION CHECK | PUZZLE NUMBER | CRYPTO KEY |

if the DES is used, each data frame is encrypted using a different keying variable for the DES in the CFB mode.

The N encrypted frames are the system's "puzzles." The system operates as follows:

1. Either party, let us choose A, creates N puzzles and sends them to the other party, Party B. The N puzzles created are as depicted in Figure 10.7. The order in which the puzzles are sent is *random*.
2. Party B picks one of the N puzzles at random, solves it by trying all *applicable* DES keys. Party B knows when the puzzle is solved because the decryption yields the DECRYPTION CHECK, a publicly known k-bit vector.
3. Upon solution, Party B does two things:
   (a) Sends the PUZZLE NUMBER to Party A.
   (b) Uses the CRYPTO KEY as the keying variable for a classical cryptographic communications system.
4. Party A, which had created the puzzles, now knows which puzzle Party B selected and solved. Party A now uses the CRYPTO KEY, corresponding to the PUZZLE NUMBER returned by Party B, to communicate with Party B using a classical cryptographic communications system.

Let us analyze the above scheme from the point of view of the "legitimate parties," A and B, and from the point of view of a third party interceptor.

The first thing we might question would be the amount of work required to solve a puzzle. Certainly, if Party B were required to solve a DES encrypted message without any limiting restriction on the possible keying variables, Party B would have an essentially impossible task. For this reason, Parties A and B publicly agree to restrict the keying variables that may be used to "enpuzzle" or encrypt the data frames. This is easy to do. All that is needed is to agree to constrain a portion of the keying variable; for example, they might set the first k bits of the keying variable to zeros. Doing this will restrict Party B's search to 2^{56-k} different possible keys. If $k = 30$, for example, the search is tedious but eminently doable.

Now let us summarize in Table 4 what is known to the various interested parties.

DECRYPTION CHECK	1	CRYPTO KEY #1
DECRYPTION CHECK	2	CRYPTO KEY #2
⋮		
DECRYPTION CHECK	N	CRYPTO KEY #N

Figure 10.7. The N puzzles created by Party A.

Table 4

	Party A	Party B	Party C (Interceptor)
The N Puzzles	Knows	Knows	Knows
Restrictions on the keys used to encipher the N puzzles	Knows	Knows	Knows
The DECRYPTION CHECK	Knows	Knows	Knows
The number of the puzzle solved by Party B	Knows	Knows	Knows
The CRYPTO KEY corresponding to the puzzle solved by Party B	Knows	Knows	Does not know

Problems

1. What would be the impact on the security of Merkle's system if Party A sent Party B the puzzles in order and not randomly?

2. What guidelines would you set for the length of the DECRYPTION CHECK field? Why?

Now we ask what work is required of Parties A, B, and C.

Party A: Must create N puzzles. The work grows proportionately to N.

Party B: Must solve a puzzle. The work grows proportionately to the number of possible keys, 2^{56-k} in our example, allowed for puzzle enciphering.

Party C: Must solve half the puzzles, on the average, to find the puzzle that Party B solved and recover the CRYPTO KEY to be used later. The work grows in proportion to the *product* of N an the number of possible puzzle enciphering keys. For our example this is $N \cdot 2^{56-k}$. If $N \simeq 2^{56-k}$ then the work grows as the square of what is required of Parties A and B.

Although Merkle's system is not a practical or "real world" system, it is valuable in that it has shown us that two parties can, with a certain amount of work, establish a secret between them that would require an interloper to expend a significantly greater amount of work to recover.

10.2.2. Public Key Cryptography and Its Pitfalls—The Diffie–Hellman System

Public key cryptography rests on the asymmetric complexity of performing certain operations. Simply stated, we have a uniquely invertible operation, or mapping, f, such that we can compute

$$y = f(x) \tag{3}$$

with an amount of "forward" work, W_F. The computation of x given $f(x)$,

$$x = f^{-1}(y) \qquad (4)$$

requires a considerably greater amount of "backward" work, W_B, however. The ratio

$$\frac{W_B}{W_F} \qquad (5)$$

determines the degree of asymmetry in computational work. Not all asymmetrically complex functions, f, are useful for a public key cryptosystem. [Henze (1982), incidentally, derives general conditions that must be met by f in order for it to serve as the basis for a public key cryptosystem.]

In this brief section we look at a particular f and the system based on that f. We then describe the "nightmare" that plagues the public key cryptosystem theorists—the lack of a guarantee that there does not exist a "shortcut" method of performing (4) with its attendant and disastrous reduction of the ratio (5). At the risk of being canny, we might remark that "proof" in general rests on one of six general foundations:

1. Deduction
2. Induction
3. Convection ("handwaving")
4. Seduction ("It looked so appealing, I overlooked its faults.")
5. Intimidation ("An expert said it's so.")
6. Lack of counterexample

The last category of "proof" is often the one we find ourselves using. It is especially apropos of public key cryptography.

A function (3) that is suitable for a public key cryptosystem is that of exponentiation of a primitive root, a, of a prime p modulo p:

$$y \equiv a^x \pmod{p} \qquad (6)$$

Computing y given x according to (6) takes no more than $2\lceil \log_2 p \rceil$ multiplications if the Binary Exponentiation Algorithm is used. Computing x given y is the "discrete logarithm problem" and appears to be more difficult. Recall that we briefly considered this problem in the chapter on number theory. The method required $2\lceil \sqrt{p} \rceil$ modular multiplications along with a requirement to store some sizable tables. If this particular method of solution were the best that could be done, then

$$\frac{W_B}{W_F} = \frac{\lceil \sqrt{p} \rceil}{\log_2 p} \qquad (7)$$

	PARTY A		PARTY B
0	Picks X_A randomly from the interval $1 < X_A < p$. Party A keeps X_A secret.	0	Picks X_B randomly from the interval $1 < X_B < p$. Party B keeps X_B secret.
0	Computes a^{X_A} mod p and sends it to Party B.	0	Computes a^{X_B} mod p and sends it to Party A.
0	Receives a^{X_B} mod p from Party B.	0	Receives a^{X_A} mod p from Party A.
0	Computes $\alpha \equiv (a^{X_B})^{X_A}$ mod p.	0	Computes $\beta \equiv (a^{X_A})^{X_B}$ mod p.

Figure 10.8. The Diffie–Hellman public key cryptosystem.

and W_B/W_F could be made as large as desired with a reasonably sized prime modulus.

How could we use (6) for a public key cryptosystem? Consider the procedure from Diffie and Hellman (1979) sketched in Figure 10.8. Both parties have obtained the same quantity:

$$\alpha = (a^{X_B})^{X_A} = a^{X_B X_A} = a^{X_A X_B} = (a^{X_A})^{X_B} = \beta = Z \tag{8}$$

Now consider the situation as presented to Party C, a passive interceptor. To learn Z, Party C would have to determine either X_A or X_B; but to do this would require Party C to solve the inverse problem to (6) which is the discrete logarithm problem. Table 5 summarizes the situation that obtains with respect to what is known to all parties after the procedure in Figure 10.8 is executed. The quantity Z can now be used as a secret variable by Parties A and B to establish secure communications via a classical cryptographic system.

So it appears we have a nice public key cryptosystem. The ratio of W_B to W_F in (7) is impressive for reasonably large p. But is (7) "guaranteed"? Up until a short time ago no better way was known—there was no counterexample. In fairness, there was not much reason for many world class mathematicians to expend much time and effort on the problem. As it turns out, there has been a tremendous and fruitful assault on the discrete

Table 5

Quantity	Known by
a	All parties
X_A	Party A
X_B	Party B
a^{X_A}	All parties
a^{X_B}	All parties
$a^{X_A X_B}$	Parties A and B

logarithm problem over the last few years. The results have been very interesting. It seems, at this point anyway, that the ratio (5) is highly dependent on the character, rather than the size, of the prime p; a good survey is provided by Pohlig (1979). The *simplest* and most striking example of this character dependence is provided by Pohlig (1978) in which a beautifully simple method is shown for the (very limited) set of known primes of the form $p = 2^n + 1$. The first step is to represent x of (6) by

$$x = \sum_{i=0}^{n-1} b_i 2^i \tag{9}$$

We will see how to sequentially recover the $\{b_i\}$. We first raise y to the power $(p-1)/2$. If the result of this exponentiation modulo p is $+1$, then $b_0 = 0$; if -1, then $b_0 = 1$. This is a simple consequence of number theory. Recall that for any number α, $\alpha \neq 0$ or 1, $\alpha^{p-1} \equiv 1 \pmod{p}$. But if $\alpha = a$, where a is a primitive root of p, then

$$\alpha^{(p-1)/2} \equiv -1 \pmod{p} \tag{10}$$

and thus

$$y^{(p-1)/2} \equiv (a^x)^{(p-1)/2} \equiv (-1)^x \pmod{p} \tag{11}$$

To find b_1 we set

$$z \equiv y\alpha^{-b_0} \tag{12}$$

What (12) does is remove b_0 from x in (9). We can therefore write

$$z \equiv \alpha^{x_1} \pmod{p} \tag{13}$$

where

$$x_1 = \sum_{i=1}^{n-1} b_i 2^i \tag{14}$$

We can now divide by 2 and then proceed with the others.

Problems

1. Show that the following rule discerns b_1:

$$z^{(p-1)/4} \pmod{p} \equiv \begin{cases} +1, & \text{if } b_1 = 0 \\ -1, & \text{if } b_1 = 1 \end{cases}$$

2. Develop an algorithm for recovering all the $\{b_i\}$ in (9). Flow chart your algorithm.
3. Calculate the ratio (5) for a prime of the form $p = 2^n + 1$.

10.2.3. The RSA System

Rivest et al. (1978) announced a public key cryptosystem that now bears the initials RSA. The system depends on a set of public or published natural numbers and a set of private or secret natural numbers. The public, published set of natural numbers are, for each user, n and e. The number n is the product of two large, secretly held, distinct primes, p and q. The number e is the encryption key. The privately held, secret numbers are p, q, and d. The number d is the multiplicative inverse of e modulo $\phi(n) = (p-1)(q-1)$. To send a message, M, interpreted as a large natural number, *to* the party whose published numbers are n and e, one need only raise M to the eth power and reduce the result to a member of the principal residue set mod n. (The only restriction on M is that it lie in the range $2 \leq M \leq n - 1$.) When the receiving party receives $M^e \pmod{n}$, the party raises $M^e \bmod (n)$ to the secretly held natural number d and reduces the result to a member of the principal residue set modulo n, thus recovering the message M:

$$(M^e)^d \equiv M \pmod{n} \tag{15}$$

Problem

Why does (15) work?

If n could be factored, then $\phi(n)$ could be computed and then one could easily recover d given the publicly published e. Thus, we know that the RSA's security is tied to the difficulty of factoring (the special form) of a large composite number—the product of two large primes. But, we do not know if this is the only general way to recover d given e.

Some very general guidelines in implementing the RSA are the following (Williams, 1984):

1. Choose p and q to be of approximately the same size, but not equal.
2. The absolute difference of p and q, $|p - q|$, should be large.
3. There should be at least one large prime factor in (a) $p - 1$, (b) $p + 1$, (c) $q - 1$, and (d) $q + 1$. This is guaranteed if both of the primes p and q are "built" using the method developed by Williams and Schmid (see Chapter 3). In this method, $p = 2p' + 1$ and $q = 2q' + 1$ where p' and q' are both large primes.

4. The primes p and q should be sufficiently large to forestall the factoring of $n = pq$. The science of factoring is achieving ever more impressive feats. Consult Figure 10.9 for a note on the history of factoring.

5. (Floyd, 1982; Simmons and Holdridge, 1982.) For some cases it is conceivable that an interested party could simply try all possible M using the published e. If a posited M were raised to the published value e and resulted in the ciphertext, then the analyst would have recovered the message (but not the secret number d). To frustrate such an attempt, it has been suggested that the message M be augmented by prefixing some randomly chosen digits.

Problem

Digital Signatures (Diffie and Hellman, 1979). If Party A wishes to send a signed message, M, to Party B, she proceeds as follows:

1. Party A raises M to d_A (Party A's secret decryption key).
2. Party A now raises (M^{d_A}) to e_B (Party B's public encryption key).
3. Party B applies the secret encryption key upon receipt of the message and then applies Party A's public encryption key.

Show that Party B now has securely received a message from Party A that she knows *must have been authored* by Party A.

1903 Cole showed that $2^{67} - 1 = 193707721 \cdot 761838257287$

1975 (Morrison and Brillhart) $2^{128} + 1$, a 39 digit number, was factored into two primes; one of length 17 digits, the other of 22 digits

1981 (Brent and Pollard) $2^{256} + 1$, a 78 digit number, was factored into two primes; one of 16 digits, the other of 62 digits

1984 Simmons, Davis, and Holdridge factored $2^{251} - 1$, a 69 digit number, into three primes; one of 21 digits, one of 23 digits, and one of 26 digits

Further advanced success in factoring large composite numbers is dependent on:
  0 New factoring algorithms
  0 New computational hardware and architectures
  0 Programmer skill
  0 "Luck"—as is always true of algorithms that have probabilistic components

---

Figure 10.9. Concerning the factoring of very large composite numbers.

10.2.4. A Faulty Implementation Protocol for the RSA

"Network protocols and security are not independent issues. Rather, overall network architecture will be shaped by the interplay between them." This quote is from a paper by Stillman and Defiore (1980). Although the subject of the paper was not concerned with specific cryptographic architectures, it seems expressly appropriate as an introduction to this section.

Public key cryptosystems seemed to hold great promise to free network designers from many of the shackles that attend classical cryptographic tools. Public key cryptosystems may indeed eventually play an important role in network design, however, there are increasing indications that the true security afforded by a particular design may not always be quickly assessed. In the words of Tartakower, the great Russian chess master, "The mistakes are all out there, waiting to be made."

As an example of what we are alluding to, we present an intriguing result due to Simmons (1983). Suppose we have a particular broadcast message service available on a network. The network serves two classes of subscribers: those who have paid to receive the broadcast message service and those who have not. To protect their revenues, the broadcast message service decides to encrypt messages using the RSA public key cryptosystem. To ease key management and reduce overhead, the service decides to use the same modulus, $n = pq$, for all subscribers. Each subscriber is issued a secret decryption key—d_i is the decryption key held secret by subscriber i—and, as is common practice with the RSA, all the encryption keys, $\{e_i\}$, and the modulus n are available to any interested party.

All the above sounds eminently reasonable and efficient. Now for the problem. Suppose that the same broadcast message is sent to two different subscribers, say subscribers 1 and 2. Suppose further that e_1 and e_2 are relatively prime, that is, $(e_1, e_2) = 1$. (This latter assumption would not be an unusual circumstance.) We write

$$M^{e_1} \equiv C_1 \pmod{n}$$
$$M^{e_2} \equiv C_2 \pmod{n}$$

$$(16)$$

where M is the common message and C_1 and C_2 are the distinct ciphertexts. Because $(e_1, e_2) = 1$ there exist integer coefficients a and b such that

$$ae_1 + be_2 = 1 \qquad (17)$$

The coefficients a and b are easily determined by the algorithm shown in Figure 10.10. This algorithm is built on the Euclidean Algorithm and is adapted from Knuth (1973). The variables w_1, w_2, w_3, w_4, and w_5 are working variables.

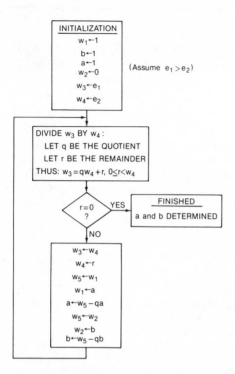

Figure 10.10 Algorithm for finding a and b in Equation (17).

Because e_1 and e_2 are both positive, exactly one of the two coefficients a and b must be positive and the other negative. If the negative coefficient is a, we attempt to compute C_1^{-1}; if b is negative, we attempt to compute C_2^{-1}. These computations are also performed using the Euclidean Algorithm. If $(C_1, n) = 1$, then C_1^{-1} exists and the Euclidean Algorithm will produce C_1^{-1}; if, however, C_1 and n are not relatively prime, application of the Euclidean Algorithm will produce one of the two prime factors of n, that is, either p or q. Such a result of course breaks the cryptosystem for all messages. Proceeding, let us assume that b is negative and we are not so fortunate to break the entire cryptosystem but that we recover C_2^{-1}. We are now able to write

$$(C_1)^a (C_2^{-1})^{|b|} \equiv (M^{e_1})^a ((M^{e_2})^{-1})^{|b|} \equiv M^{ae_1 + be_2} \equiv M \pmod{n} \quad (18)$$

and we recover the message M; a rather surprising result.

Problem

Devise an RSA system (with suitably small values of p and q) and demonstrate the procedures just covered. Work through one example for which $(C_1, n) = 1$ and one example for which $(C_1, n) \neq 1$.

We might suppose that Simmons' attack on the common modulus RSA system could be frustrated by padding each message with a field of random or address specific values. Unfortunately, there does not appear to be a simple fix. Subsequent to Simmons' work, DeLaurentis (1984) in essence showed that if one has knowledge of a single encryption/decryption pair (e_i, d_i), such as would be possessed by an authorized user, then one can, with some work, recover messages intended for other recipients or successfully forge the digital signatures of other net members. The message seems clear: an RSA architecture must not be designed around a common modulus even though such a practice would seem very attractive from a key management standpoint.

10.3. Secret Sharing Systems

The final cryptographic topic at which we look is a secret sharing system, sometimes called a "shadow" system. Such a system allows secret information to be split among a number of holders† so that a minimum number of them must conspire and pool their knowledge to recover the secret. Such a system can be very valuable in situations requiring distributed command, control, or access.

The specific system we use is one exhibited by Karnin et al. (1983). This system allows an m-bit secret to be shared among n "trustees" such that any $k(k < n)$ of the trustees may recover the secret by pooling their knowledge; but if fewer than k of the trustees pool their knowledge, they will know no more about the secret than any single trustee. The system is also operationally convenient because the data that must be retained and protected by a trustee is the same size, m bits, as the secret.

The system is very simple to exercise:

1. Pick n and k as desired.
2. Find a set of $n + 1$ zero-one matrices

$$\{A_0, A_1, A_2, \ldots, A_n\} \tag{19}$$

where $A_0 = (I_m \quad 0)$ where I_m is the identity matrix of order m and 0 is the matrix of dimension $(k - 1)m \times m$ of all zeros. The matrices in the set (19) are all of dimension $km \times m$. The set (19) must be devised so that each matrix of dimension $km \times km$, formed by concatenating any k of the matrices in (19), must have the full rank of km. The $\{A_i\}$ may be made public without compromising the system.

---

† Sometimes called "trustees."

3. Let the secret be denoted by the vector

$$\mathbf{s}^T = [s_1 \quad s_2 \quad \cdots \quad s_m] \tag{20}$$

We now construct a vector $\mathbf{u}$ which has the secret bits of $\mathbf{s}$ as its first m bits and randomly chosen bits as its last $(k-1)m$ entries. The vector $\mathbf{u}$ is thus km bits long. The randomly chosen bits in $\mathbf{u}$ are kept secret and later erased by the party who is dividing the secret among the trustees.

The vector $\mathbf{t}_i$ is given to the ith trustee, $1 \leq i \leq n$, where

$$\mathbf{t}_i^T = \mathbf{u}^T A_i \tag{21}$$

As an example let us build a system that will split a 2-bit secret among four trustees such that any two trustees may recover the secret upon pooling their knowledge. We thus have that $n = 4$, $m = 2$, and $k = 2$. Following step 2 we pick

$$A_0 = \begin{bmatrix} 1 & 0 \\ 0 & 1 \\ 0 & 0 \\ 0 & 0 \end{bmatrix}$$

$$A_1 = \begin{bmatrix} 0 & 0 \\ 0 & 0 \\ 1 & 0 \\ 0 & 1 \end{bmatrix}$$

$$A_2 = \begin{bmatrix} 1 & 0 \\ 1 & 1 \\ 1 & 0 \\ 0 & 1 \end{bmatrix} \tag{22}$$

$$A_3 = \begin{bmatrix} 0 & 1 \\ 1 & 0 \\ 1 & 0 \\ 0 & 1 \end{bmatrix}$$

$$A_4 = \begin{bmatrix} 1 & 0 \\ 0 & 1 \\ 1 & 1 \\ 0 & 1 \end{bmatrix}$$

Problem

Show that the five matrices in (22) satisfy the condition required in step 2.

Now let

$$\mathbf{s}^T = [s_1 \quad s_2] \tag{23}$$

and

$$\mathbf{u}^T = [s_1 \quad s_2 \quad r_1 \quad r_2] \tag{24}$$

where r_1 and r_2 are randomly generated bits. The secret is revealed by

$$\mathbf{s}^T = \mathbf{u}^T A_0 = [s_1 \quad s_2] \tag{25}$$

The four vectors given to the trustees are

$$\mathbf{t}_1^T = \mathbf{u}^T A_1 = [r_1 \quad r_2] \tag{26}$$

$$\mathbf{t}_2^T = \mathbf{u}^T A_2 = [s_1 + s_2 + r_1 \quad s_2 + r_2] \tag{27}$$

$$\mathbf{t}_3^T = \mathbf{u}^T A_3 = [s_2 + r_1 \quad s_1 + r_2] \tag{28}$$

$$\mathbf{t}_4^T = \mathbf{u}^T A_4 = [s_1 + r_1 \quad s_2 + r_1 + r_2] \tag{29}$$

Because the r_1 and r_2 are kept secret from the trustees, no single trustee can discern anything about s_1 or s_2. But any two trustees together can recover s_1 and s_2.

Problem

Verify that each pair of two trustees can, by working together, recover s_1 and s_2.

10.3.1. Matrix Construction for Secret Sharing Systems

In step 2 of the process described we need a set of $(n + 1)$ nonzero matrices. In our example we did not construct the matrices but rather chose them arbitrarily to fulfill the relevant constraints. Let us now consider a simple procedure for constructing such a set. We use a bit of graph theory (Seshu and Reed, 1961).

First, a set of interconnected line segments is called a linear graph. An example is given in Figure 10.11, where x_i are the line segments or "edges" and v_i are the nodes. Any subset of elements of a graph is called a subgraph. A circuit is a connected subgraph such that there are exactly two distinct paths between any pair of vertices. For example, the subgraph consisting of the edges x_1, x_4, x_5, and x_6 form a circuit. A tree T is a connected subgraph of a connected graph, say G, such that T contains all its vertices and has no circuits. It is clear that the tree of a graph is not unique. For example, the edges x_1, x_2, x_3, x_4 form a tree. Also, the edges x_1, x_2, x_5, x_6 also form a tree. However, the edges x_1, x_4, x_5, x_6 can not form a tree.

Second, we consider a vertex matrix description of a graph G with θ vertices and e edges. The vertex matrix $A_a = [a_{ij}]$ is a matrix with θ rows and e columns where

$$a_{ij} = 1 \quad \text{if edge } j \text{ is the incident at vertex } i$$

$$a_{ij} = 0 \quad \text{if edge } j \text{ is not incident at vertex } i$$

For the graph in Figure 10.11, the vertex matrix is given by

$$
A_a = \begin{array}{c} \\ V \\ e \\ r \\ t \\ e \\ x \\ \# \end{array}
\begin{array}{c} \\ 1 \\ 2 \\ 3 \\ 4 \\ 5 \\ \\ \end{array}
\begin{array}{c} \text{Element } \# \\ \begin{array}{cccccc} 1 & 2 & 3 & 4 & 5 & 6 \end{array} \\ \left[\begin{array}{cccccc} 1 & 0 & 0 & 0 & 1 & 0 \\ 1 & 1 & 0 & 0 & 0 & 1 \\ 0 & 1 & 1 & 0 & 0 & 0 \\ 0 & 0 & 1 & 1 & 1 & 0 \\ 0 & 0 & 0 & 1 & 0 & 1 \end{array} \right] \end{array}
\qquad (30)
$$

From this and from the definition of A_a, we see that each column has exactly two ones and $\theta - 2$ zeros. Using modulo 2 arithmetic, we can show that the rank of A_a in (30) is 4. In general, we state without proof that the

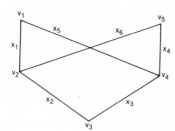

Figure 10.11. A five-node graph with six elements.

rank of a vertex matrix A_a of a connected graph G with θ vertices is $\theta - 1$. Considering this, we can simply delete one of the rows of A_a, say the last row, and write it as A. That is,

$$
A = \begin{array}{c} \\ \\ \\ \end{array}
\begin{array}{cccccc}
\text{Element} \ \# \\
1 & 2 & 3 & 4 & 5 & 6 \\
\left[\begin{array}{cccccc}
1 & 0 & 0 & 0 & 1 & 0 \\
1 & 1 & 0 & 0 & 0 & 1 \\
0 & 1 & 1 & 0 & 0 & 0 \\
0 & 0 & 1 & 1 & 1 & 0
\end{array}\right]
\end{array} \tag{31}
$$

which has the maximum rank which is 4.

Next let us ask ourselves which columns of A are independent. A few examples are given in Table 6. Interestingly, the subgraphs corresponding to the first three cases form trees, whereas the subgraph corresponding to the last case forms a circuit. We now state a theorem without proof (Seshu and Reed, 1961).

Theorem

A square submatrix of A of order $\theta - 1$ is nonsingular if and only if the edges corresponding to the columns in the submatrix constitute a tree of the graph.

Since the rank of $C = BA$ is equal to the rank of A when B is nonsingular, let us modify A by premultiplying by

$$
B = \begin{bmatrix}
1 & 0 & 0 & 0 \\
1 & 1 & 0 & 0 \\
1 & 1 & 1 & 0 \\
1 & 1 & 1 & 1
\end{bmatrix} \tag{32a}
$$

Table 6. Some Examples of Column Choices and Independence

Case	Columns	Independent?
1	1, 2, 3, 4	Yes
2	1, 2, 5, 6	Yes
3	3, 4, 5, 6	Yes
4	1, 4, 5, 6	No

and the result is

$$C = \begin{bmatrix} 1 & 0 & 0 & 0 & 1 & 0 \\ 0 & 1 & 0 & 0 & 1 & 1 \\ 0 & 0 & 1 & 0 & 1 & 1 \\ 0 & 0 & 0 & 1 & 0 & 1 \end{bmatrix} \tag{32b}$$

Note that the independence of the columns in (32b) indicated in Table 6 is unchanged. The matrix in (32b) is usually referred to as a cutset matrix and is associated with the tree containing the elements x_1, x_2, x_3, and x_4. The interpretation is that the edges corresponding to the nonzero entries in a particular row correspond to a cutset and the removal of these edges from the graph cuts the graph into two parts.

Let us now go back to our problem. From Table 6 and (32b), we see that

$$A_1 = \begin{bmatrix} 1 & 0 \\ 0 & 1 \\ 0 & 0 \\ 0 & 0 \end{bmatrix}, \qquad A_2 = \begin{bmatrix} 0 & 0 \\ 0 & 0 \\ 1 & 0 \\ 0 & 1 \end{bmatrix}, \qquad A_3 = \begin{bmatrix} 1 & 0 \\ 1 & 1 \\ 1 & 1 \\ 0 & 1 \end{bmatrix} \tag{33}$$

which fits our need. That is, the matrices formed by $[A_i \quad A_j]$, for $i, j = 1, 2, 3$ ($i \neq j$), are nonsingular. Can we extend this set? First, we can add only four more edges to the graph in Figure 10.11, as these edges will complete the graph. This means that we will have one edge between every pair of vertices. To add another matrix to the set in (33), we need to add two edges. Unfortunately, in adding two edges we do not achieve the indepenence we need. This is clear upon noting that the additional edges form additional circuits. To circumvent this problem, we use the A_i in (33) and add another matrix using a little matrix theory. We state that with the additional matrix

$$A_4 = \begin{bmatrix} 1 & 1 \\ 0 & 1 \\ \hline 1 & 1 \\ 0 & 1 \end{bmatrix} \tag{34}$$

we have the property that the matrices formed by $[A_i \quad A_j]$, $i, j = 1, 2, 3, 4$ ($i \neq j$), are nonsingular. We now prove this in general.

Consider the four ($2m \times m$) matrices

$$A_1 = \begin{bmatrix} I_m \\ 0_m \end{bmatrix}, \qquad A_2 = \begin{bmatrix} 0_m \\ I_m \end{bmatrix}, \qquad A_3 = \begin{bmatrix} T_1 \\ T_1^T \end{bmatrix}, \qquad A_4 = \begin{bmatrix} T_2 \\ T_1^T \end{bmatrix} \tag{35}$$

with I_m, an identity matrix of order m, 0_m, an $m \times m$ null matrix,

$$T_1 = \begin{bmatrix} 1 & 0 & 0 & \cdots & 0 & 0 \\ 1 & 1 & 0 & \cdots & 0 & 0 \\ 1 & 1 & 1 & \cdots & 0 & 0 \\ \vdots & \vdots & \vdots & & \vdots & \vdots \\ 1 & 1 & 1 & \cdots & 1 & 0 \\ 1 & 1 & 1 & \cdots & 1 & 1 \end{bmatrix} \quad \text{and} \quad T_2 = \begin{bmatrix} 1 & 0 & \cdots & 0 & 1 \\ 0 & 1 & \cdots & 0 & 0 \\ 0 & 0 & \cdots & 0 & 0 \\ \vdots & \vdots & & \vdots & \vdots \\ 0 & 0 & \cdots & 1 & 0 \\ 0 & 0 & \cdots & 0 & 1 \end{bmatrix}$$

$$(36)$$

where T_1 is a triangular matrix with ones on the diagonal and below the diagonal and zeros everywhere else and T_2 is a triangular matrix with ones on the diagonal and at the $(1, m)$ location and zeros everywhere else. By inspection, we see that T_1 and T_2 are nonsingular. Furthermore, $T_1 + T_2$ is also nonsingular, where the addition is modulo 2, element by element. This can be seen by expanding the determinant

$$|T_1 + T_2| = \begin{vmatrix} 0 & 0 & 0 & \cdots & 0 & 1 \\ 1 & 0 & 0 & \cdots & 0 & 0 \\ 1 & 1 & 0 & \cdots & 0 & 0 \\ \vdots & \vdots & \vdots & & \vdots & \vdots \\ 1 & 1 & 1 & \cdots & 0 & 0 \\ 1 & 1 & 1 & \cdots & 1 & 0 \end{vmatrix} \qquad (37)$$

By using the last column, we have $|T_1 + T_2| = 1$. By using the block matrix manipulations, we can easily see that the matrices $[A_1 \quad A_2]$, $[A_1 \quad A_3]$, $[A_1 \quad A_4]$, $[A_2 \quad A_3]$, and $[A_2 \quad A_4]$ are nonsingular. Now consider the matrix

$$[A_3 \quad A_4] = \begin{bmatrix} T_1 & T_2 \\ T_1^T & T_1^T \end{bmatrix}$$

which has the same rank as

$$\begin{bmatrix} T_1 & T_1 + T_2 \\ T_1^T & 0 \end{bmatrix}$$

where again the addition is modulo 2, element by element. Since T_1^T and $T_1 + T_2$ are nonsingular, it follows that the matrix formed by $[A_3 \quad A_4]$ is nonsingular, thus completing the proof.

A few comments are in order. First, we can modify the matrices in (35) by premultiplying by any nonsingùlar $(2m \times 2m)$ matrix and the independence will be unchanged. Secondly, suppose we divide the A_i in (35) into, say, eight matrices, where the new A_i are $2m \times (m/2)$ matrices. Will we still have the nonsingularity of the matrices generated from four arbitrary new A_i? Unfortunately not. For example, we can have eight matrices from (33) and (34) and cannot have the nonsingularity of any four columns because the first column from A_1, A_2, A_3, and A_4 together form a singular matrix.

References

Brent, R. and J. Pollard (1981), Factorization of the Eighth Fermat Number, *Mathematics of Computation*, Vol. 35, pp. 627–630.

DeLaurentis, J. (1984), A Further Weakness in the Common Modulus Protocol for the RSA Cryptosystem, *Cryptologia*, Vol. 8, pp. 253–259.

Diffie, W. and M. Hellman (1979), Privacy and Authentication: An Introduction to Cryptography, *Proceedings of the IEEE*, Vol. 67, pp. 397–427.

FIPS PUB 46 (1977), Data Encryption Standard, *Federal Information Processing Standards, Publication Number 46*, National Bureau of Standards.

FIPS PUB 81 (1980), DES Modes of Operation, *Federal Information Processing Standards, Publication Number 81*, National Bureau of Standards.

Floyd, D. (1982), A Survey of the Current State of the Art in Conventional and Public Key Cryptography, *Tech. Rpt. 81-10*, Dept. Computer Science, University of Pittsburgh.

Henze, E. (1982), The Solution of the General Equation for Public Key Distribution Systems, *IEEE Transactions on Information Theory*, Vol. 28, p. 933.

Hellman, M., R. Merkle, R. Schroeppel, L. Washington, W. Diffie, S. Pohlig, and P. Schweitzer (1976), Results of an Initial Attempt to Cryptanalyze the NBS Data Encryption Standard, SEL76-042, Stanford University.

Hershey, J. (1983), The Data Encryption Standard, *Telecommunications*, Vol. 17, Sept., pp. 77 ff.

Karnin, E., J. Greene, and M. Hellman (1983), On Secret Sharing Systems, *IEEE Transactions on Information Theory*, Vol. 29, pp. 35–41.

Knuth, D. (1973), *The Art of Computer Programming*, Vol. I, 2nd ed., Addison-Wesley, Reading, MA.

Merkle, R. (1978), Secure Communications Over Insecure Channels, *Communications of the ACM*, Vol. 21, pp. 294–299.

Morrison, M. and J. Brillhart (1975), A Method of Factoring and the Factorization of F_7, *Mathematics of Computation*, Vol. 29, pp. 183–205.

Pomper, W. (1982), The DES Modes of Operation and Their Synchronization, *Proceedings of the International Telemetering Conference*, pp. 837–851.

Pohlig, S. (1978), An Improved Algorithm for Computing Logarithms Over GF(p) and Its Cryptographic Significance, *IEEE Transactions on Information Theory*, Vol. 24, pp. 106–110.

Pohlig, S. (1979), An Overview of Secure Communications Using the Discrete Exponential, *EASCON Proceedings*, pp. 650–652.

Rivest, R., A. Shamir, and L. Adleman (1978), A Method for Obtaining Digital Signatures and Public-Key Cryptosystems, *Communications of the ACM*, Vol. 21, pp. 120–126.

Seshu, S. and M. Reed (1961), *Linear Graphs and Electrical Networks*, Addison-Wesley, Reading, MA.

Simmons, G. (1983), A "Weak" Privacy Protocol Using the RSA Cryptoalgorithm, *Cryptologia*, Vol. 7, pp. 180–182.

Simmons, G. and D. Holdridge (1982), Forward Search as a Cryptanalytic Tool Against a Public Key Privacy Channel, *Proceedings of the Symposium on Security and Privacy*, pp. 117–128.

Stillman, R. and C. DeFiore (1980), Computer Security and Networking Protocols: Technical Uses in Military Data Communications Networks, *Transactions on Communications*, Vol. 28, pp. 1472–1477.

Williams, H. (1984), An Overview of Factoring, in *Advances in Cryptology, Proceedings of Crypto 83*, Plenum Press, New York, pp. 71–80.

Synchronization†

1.1. Introduction

Synchronization is usually performed to establish either epoch or phase. Epoch synchronization refers to those techniques that seek to establish agreement on the epoch or occurrence of a particular instant of time. Phase synchronization comprises those processes that endeavor to determine the phase of a cyclic (steady-state deterministic) digital process. We consider a variety of epoch and phase synchronization methods, starting with epoch synchronization followed by the phase synchronization types.

11.2. Epoch Synchronization

Epoch synchronization is a process by which a transmitter communicates a reference time mark to a receiver. This is usually done in the time domain by sending a carefully constructed sequence. The sequence is such that it possesses an autocorrelation that has low sidelobes, thus allowing a large peak-to-sidelobe ratio when passed through a matched filter. There is some confusion in the literature that centers on sequence design according to the presence or absence of bit sense or "bit ambiguity." This difficulty is explored.

11.2.1. Autocorrelation

The autocorrelation function most commonly used for sequence design is as follows. Let the sequence be denoted by $s(1)$, $s(2), \ldots, s(n)$ where $s(i) \in \{0, 1\}$. We compute the autocorrelation, $r(k)$, by counting the agreements, A, between $s(i)$ and $s(i + k)$ over the range $0 \le i \le n - k$ and

---

† Much of this material is taken from Hershey (1982).

subtracting the disagreements, D, over the same range. From this definition it is clear that $r(k) = r(-k)$. As an example, let us determine the autocorrelation of the sequence

$$0000011010111001 \tag{1}$$

We calculate $r(1)$ by lining up the slipped sequence with the unslipped sequence as follows

0 0 0 0 0 1 1 0 1 0 1 1 1 0 0 1

<u> 0 0 0 0 0 1 1 0 1 0 1 1 1 0 0 1</u>

$A\,A\,A\,A\,D\,A\,D\,D\,D\,D\,A\,A\,D\,A\,D$ A indicates an
                                                 Agreement,
                                          D a Disagreement

From the above we see that $r(1) = 8 - 7 = 1$. Figure 11.1 depicts $r(k)$ for $k = 0, \pm1, \pm2, \ldots, \pm15$. Note that $\max\,(A - D)$ for $k \neq 0$ is 1. Note further that at $k = 8$, $A - D = -6$. This is a consequence of designing the sequence (1) to achieve

$$\min\left(\max_{k \neq 0}\,(A - D)\right) \tag{2}$$

What (2) implies is that bit "sense" is known, that is, the receiver knows the received bits exactly and not within the ambiguity of some differentially

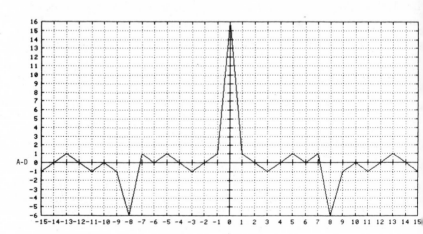

Figure 11.1. Autocorrelation.

Figure 11.2. Ambiguity of differentially coded systems.

coded systems (see Figure 11.2). This is a reasonable and practical assumption for many communications circuits and signaling architectures.

11.2.2. Bit Sense Known

Sequences or, as they are often called, "unique words," which are designed under (2), may be used in various telemetry schemes [see Maury and Styles (1964)] and for Time Division Multiple Access (TDMA) satellite communications. See, for example, Sekimoto and Puente (1968), Gabard (1968), Schrempp and Sekimoto (1968), and Nuspl et al. (1977). A common implementation that can be used for synchronization using sequences wherein bit sense is known, is a three step process (Schrempp and Sekimoto, 1968):

1. Allow carrier recovery by sending the receiver a stream of all zeros.
2. Provide bit timing recovery by transmitting a stream of zero/one alternations.
3. Mark epoch by sending a unique word.

As an example, let us assume we have carrier and clock and let us examine a few cases where we identify the epoch with the unique word specified in (1). We assume, for this case, that we are provided with a zero/one quantized bit stream and that we are passing this bit stream through a matched filter or replica of the unique word (1). (This is not the optimum method of detection, incidentally, but it is easy to realize in hardware.) Our experiment is depicted in Figure 11.3. The bitstream that is passed into the Bitstream Analysis Window is a stream of ones and zeros taken from a balanced Bernoulli source which is a source of zeros and ones such that the probability of a one at time t equals the probability of a zero at time t equals $\frac{1}{2}$; furthermore, the bit at time t is independent of all bits preceding it. At

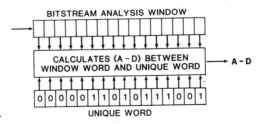

Figure 11.3. A matched filter.

$t = 30$, the unique word (1) is sent and thus at $t = 45$, the unique word i
lined up with its replica in the filter. At this point, if the unique word ha
not been corrupted by the channel, it will completely agree with its store
replica and cause a large pulse. Figure 11.4 depicts the action showin
$A - D$ as a function of time. Time is assumed to flow from left to right.

As Figure 11.4 depicts, the epoch is clearly defined by the $A - D$ puls
of height 16 at $t = 45$. (The correlation is not observed for the first 15 cloc
times as the window is being initially filled during this time.) To detect th
epoch then, we would set a threshold on the $A - D$ waveform and enforc
the rule that the epoch is declared when the $A - D$ waveform meets o
exceeds the threshold. This is obviously doable and effective if the uniqu
word is received, detected, and quantized without errors. Such is not th
case in general, however. Figure 11.5 depicts the action wherein the uniqu
word is passed through a binary symmetric channel (BSC) with $p = \frac{1}{16}$. (A
BSC with parameter p is a simple channel model that dictates that the bi
transmitted at time t is received correctly with probability $1 - p$ and i
independent of previous channel errors.†) In this case, the unique wor
suffered a single-bit error causing A to decrease by one epoch and D t
increase by one causing $A - D$ to drop from 16 to 14. For this case w
would have correctly detected the epoch only if we had set our threshold
in the range $8 < \text{threshold} \le 14$.

Figure 11.6 repeats the above experiment with a BSC with $p = \frac{1}{4}$. I
this example the unique word suffered four errors and the $A - D$ waveform
exhibited a peak value of only 8 at the correct epoch, $t = 45$. Note, however,
that a randomly occurring peak of height 8 also occurred at $t = 27$. For this

† We explore the BSC in greater depth in Chapter 12.

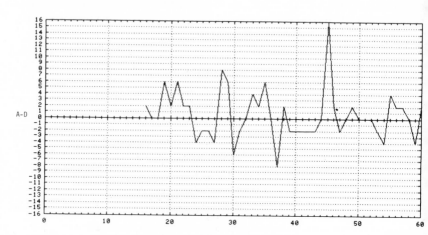

Figure 11.4. Matched filter action.

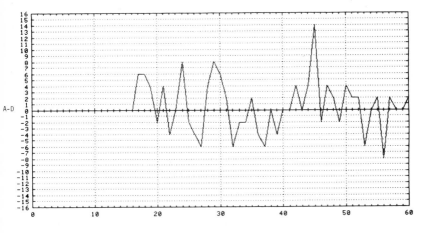

Figure 11.5. Matched filter action.

case it would have been impossible to select a threshold which would have correctly identified the epoch.

So far we have learned that we must be careful in choosing the structure of our unique word and we must be careful to correctly set the epoch detection threshold. There is another consideration that is worthwhile and this refers to "look time." The look time is defined as the amount of time a matched filter will be allowed to search for an epoch. The random local maxima that occur may cause a false epoch determination. The probability of a false epoch determination varies in a direct relationship to the look time. If a long look time is required, it may be necessary to select a high threshold. Doing so may require the user to pick a longer unique word

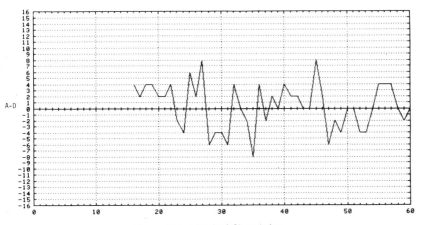

Figure 11.6. Matched filter action.

because a higher threshold may allow for fewer channel errors. As an example, Figure 11.7 depicts the action of the matched filter of Figure 11.3 for an input bitstream produced by a balanced Bernoulli source of 1000 bits. Note the many relatively large maxima that occur randomly.

Table 1 lists examples of unique words meeting condition (2) for lengths 3–29 bits.

11.2.3. Bit Sense Unknown

In many cases bit sense is unknown and unique word sequences are then designed by modifying (2) to the following

$$\min_{} (\max_{k \neq 0} (|A - D|)) \qquad (3)$$

By way of introduction, let us examine a 16-bit sequence designed according to (3).

$$0000011001101011 \qquad (4)$$

Figure 11.8 depicts $r(k)$ for this word for $k = 0, \pm 1, \pm 2, \ldots, \pm 15$. Note that for this word, the best that can be achieved under condition (3) is $|r(k)| \leq 2(k \neq 0)$ in contrast to $r(k) \leq 1(k \neq 0)$ for the word (1) which was designed under condition (2).

The great majority of research on unique words has addressed the case where bit sense in unknown. Probably this was a result of the great influence

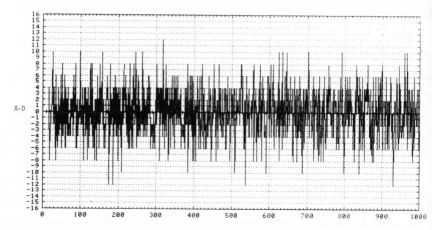

Figure 11.7. Matched filter action on random input bitstream.

Table 1. Examples of Unique Words When Bit Sense Is Known

Length	Maximum ACF ($k \neq 0$)	Sequence
3	0	001^a
4	1	0001^a
5	1	00010^a
6	1	000110
7	0	0001101
8	1	00001101
9	1	000011010
10	1	0001110010
11	0	00011101101^a
12	1	000010110011
13	1	0000011001010^a
14	1	00011011110010
15	2	000000110010100
16	1	0000011010111001
17	1	00000101100111010
18	1	000010101101100111
19	2	0000001011001110010
20	1	00000101110100111001
21	1	000000111001101101010
22	1	0001000111110011011010
23	2	00000011100101011011001
24	1	000001110011101010110110
25	1	0000001101101110001101010
26	2	00000001101010110011110010
27	2	000000011011001111001010100
28	2	0000000110110110001110101011
29	1	00000010110010011100111101010

^{a} Barker sequence.

that the radar field has had on special sequence development. A PSK modulated radar return is a good example of a case in which one would instinctively choose a pulse compression code or unique word designed under (3).

The most famous of the sequences designed under (3) are the Barker sequences. These sequences are those designed under (3) for which

$$\max_{k \neq 0} (|A - D|) = 1 \qquad (5)$$

Some of the sequences that meet (5) were introduced by R. H. Barker (1953) and bear his name. No Barker sequences beyond length 13 are known but

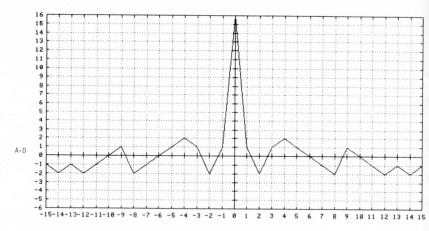

Figure 11.8. Autocorrelation.

it is known that if any do indeed exist, then they must have a length that is an even square. All possibilities up to $78^2 = 6084$ have been eliminated (see Petit, 1967).

Lindner (1975a) has compiled an exhaustive listing of all binary sequences or unique words that meet (3) for lengths 3–40 bits. Table 2 lists one of these sequences for each word length for $n = 3$ to $n = 40$ bits along with the $\max_{k \neq 0}(|A - D|)$ value achievable for n (Lindner, 1975b).

Many efforts have been made to create longer sequences from the Barker sequences that exhibit good values for the autocorrelation. Klyuyev and Silkov (1976), for example, proposed the construction

$$\alpha\alpha\alpha\bar{\alpha} \tag{6}$$

where α is a Barker sequence and the overbar indicates complementation. Constructions such as (6) can be viewed as examples of Kronecker constructions which are produced as follows (see Stiffler, 1971, or Turyn, 1968):

1. Let $s(1), s(2), \ldots, s(l)$ and $t(1), t(2), \ldots, t(m)$ be two sequences with $r(k)$ and $r'(k)$ the autocorrelations of $s(\ )$ and $t(\ )$, respectively.

2. Form the lm long sequence

$$s(1) + t(1), s(1) + t(2), \ldots, s(1) + t(m), s(2) + t(1), \ldots, s(l) + t(m) \tag{7}$$

Table 2. Examples of Unique Words When Bit Sense Is Unknown

Length	Maximum absolute ACF ($k \neq 0$)	Sequence
3	1	001[a]
4	1	0001[a]
5	1	00010[a]
6	2	000010
7	1	0001101[a]
8	2	00001011
9	2	000001101
10	2	0000011010
11	1	00011101101[a]
12	2	000001010011
13	1	0000011001010[a]
14	2	00000011001010
15	2	000000110010100
16	2	0000011001101011
17	2	00001100100101011
18	2	000001011010001100
19	2	0000111000100010010
20	2	00000100011101001011
21	2	000000101110100111001
22	3	0000000011100100110101
23	3	00000000111001010110010
24	3	000000000111001010110010
25	2	0001100011111101010110110
26	3	00000000011010011010101001110
27	3	000000000110011101000101 1010
28	2	0001100011111101010110110110
29	3	00000000011011001 0111000110101
30	3	000000000111000110101001 1011001
31	3	0000000001110001010100101 1011001
32	3	00000000010110101000110110011 1 10001
33	3	000000000111100101101010100110 0011
34	3	0000000001111001011010101001100 110
35	3	00000000011110001100110100100 101010
36	3	00000000011101000111001100110 10010100
37	3	00000000010110110010001101010 001110001
38	3	00000000011110000110100101010 1001100110
39	3	00000001001111000110110010100 111001010
40	3	00000010101011010010011110001 1011001100

[a] Barker sequence.

Unfortunately, the Kronecker constructed sequence (7) possesses an autocorrelation function $r''(k)$ such that

$$\max r''(k) \geq (\max r(k))(\max r'(k)) \tag{8}$$

and thus can never lead to a better normalized autocorrelation function than either of its component sequences.

A very interesting result due to Moser and Moon and cited on p. 198 of Turyn (1968) is that if a sequence is generated at random (from a balanced Bernoulli source) then one can expect $\max (|A - D|)$ to be on the order of the square root of the length of the sequence. Keeping this thought in mind, let us examine a technique that has often been suggested for finding a long sequence with good autocorrelation behavior. This technique is to generate an m-sequence and then pick the best phase to minimize the maximum value of $|A - D|$. We have done this with the 63-bit m-sequence generated according to $x^6 + x^5 + 1$. The best phase (one of several, actually) is

101101110110011010101011111110000010000011000101001111010001110010

$$\tag{9}$$

Figure 11.9 depicts the autocorrelation of this sequence. Note that the maximum absolute value is 6 which is a bit better than the square root of 63. Sarwate (1984) has shown, incidentally, that for an m-sequence of length $N = 2^n - 1$, $|A - D| < 1 + (2/\pi)(N + 1)^{1/2} \ln (4N/\pi)$. (Show that our example above satisfies Sarwate's inequality.)

We have not fully examined the behavior of epoch synchronization under channel noise nor have we even delved into the simple statistics

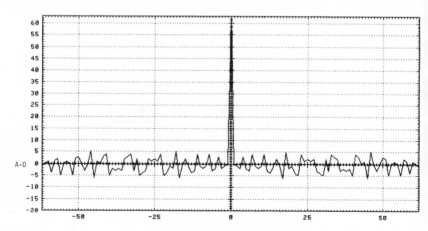

Figure 11.9. Autocorrelation.

relevant to choosing an appropriate threshold. The reasons for these omissions are twofold:

1. Epoch synchronization can be implemented by first hard quantizing the incoming data stream into zeros and ones. This hard quantization method may be practical but it is certainly not optimal.

2. The channel noise processes heavily influence the choice of parameters for the unique word once the matched filter or unique word detection system has been agreed upon. It is only then that the necessary statistics for describing the behavior of the unique word detection system should be derived.

One final observation is that unique word detection is a classical Type I/Type II decision process as diagramed in Figure 11.10. (See Kreyszig, 1967, p. 802.) In picking the unique word, the designer must assess appropriate costs for the Type I/Type II errors.

Problems

1. (A Faulty Implementation of the Barker Sequence). The following faux pas was actually committed and its system implementation used (for a time) in a communications system which suffered, not surprisingly, from synchronization problems. The design engineer needed to transmit a burst of binary data. The burst could be begun at any time. The channel, an unconditioned telephone line, was slightly noisy. The occasional noise caused the modem to convert the noise pulses into random ones and zeros. The engineer decided to preface the data transmission with a Barker sequence of length 13. The particular sequence chosen was 0000011001010. The data rate was sufficiently low that the synchronization was implemented in software. The

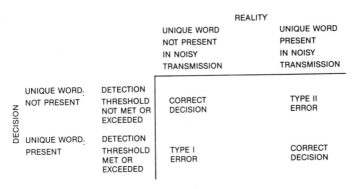

Figure 11.10. Type I/Type II errors.

engineer first created the following array which was nomenclatured as TABLE:

TABLE(1) = 0
TABLE(2) = 0
TABLE(3) = 0
TABLE(4) = 0
TABLE(5) = 0
TABLE(6) = 1
TABLE(7) = 1
TABLE(8) = 0
TABLE(9) = 0
TABLE(10) = 1
TABLE(11) = 0
TABLE(12) = 1
TABLE(13) = 0

Next the engineer programmed the algorithm shown in Figure 11.11.

(a) Verify that the algorithm works if there is no noise on the channel.
(b) Why does the synchronization scheme, as implemented, fail, indeed become extremely counterproductive, if the Barker sequence is preceded by noise?
(c) Derive a better algorithm for implementing this particular Barker sequence synchronization scheme for the same requirement. Construct the appropriate flow chart.

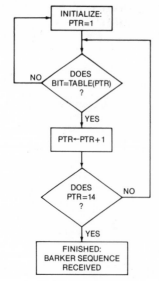

Figure 11.11. A poorly conceived synchronization algorithm.

2. Assume we have a BSC with parameter p. We wish to detect an epoch by using a unique word of n bits. We localize our search to a time window encompassing w bits ($w \gg n$). Find the length, n, of a suitably constructed unique word and the threshold T, $T \leq n$, so that the probability of a false detection (Type I error) within the w bits is $\leq \varepsilon_1$ and the probability of missing the unique word (Type II error) is $\leq \varepsilon_2$. Evaluate for $w = 10^6$, $p = 10^{-3}$, and $\varepsilon_1 = \varepsilon_2 = 10^{-6}$.

11.3. Phase Synchronization

Figure 11.12 depicts the generic diagram for the phase synchronization process. The top box represents a cyclic digital process, in effect a repetitive sequence of binary n-tuples. The middle box is a time-invariant (fixed) mapping from the binary n-tuples to a single binary unit. The sequence of bits thus produced constitutes the sequence. The period of the sequence, of course, is upperbounded by the cycle length of the cyclic digital process.

11.3.1. *m*-Sequence Synchronization

This phase synchronization process uses a shift register of length n with a primitive polynomial for feedback as the cyclic digital process of Figure 11.12, that is, an m-sequence generator. The combinatorial logic is simply a single tapped state of the shift register. The sequence is an m-sequence. This is depicted in Figure 11.13 to show comportment to the canonical structure of Figure 11.12.

As we have seen in Chapter 8, an m-sequence generated by a primitive polynomial of degree n has $2^n - 1$ distinct phases. To determine the phase and hereby establish synchronization it is necessary to know without error n consecutive bits and the polynomial that generates the m-sequence. If

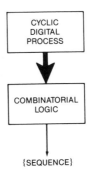

Figure 11.12. Phase synchronization: generic diagram.

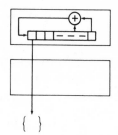

Figure 11.13. m-Sequence generator.

the signal-to-noise ratio is low, this may not be possible by direct demodulation. Consequently, we may have to rely on methods that work with more than n bits in making a decision.

The following sections deal with the three possibilities denoted by X's in Figure 11.14.

11.3.2. Polynomial Known: Errorless Reception

This is the simplest of the four possible cases and also the least likely because the result of a spread spectrum system is to spread the signal energy so that the individual chip possesses a low signal-to-noise investment, thus making errorless acquisition of n consecutive bits very unlikely.

Assuming, however, that n consecutive bits are obtained without error, then the future behavior of the m-sequence is determined and synchronization is established.

11.3.3. Polynomial Known: Errors in Reception

This is the most common case and actually consists of two subcases. First, if $2^n - 1$ is sufficiently small so that one can expect to see all phases of the m-sequence during synchronization, then one need only create a matched filter to look for m consecutive bits of the m-sequence where $m > n$ for moderate signal-to-noise ratios and $m \gg n$ for very low signal-to-noise

	DEMODULATED BITS ERRORLESS	SOME DEMODULATED BITS IN ERROR
PRIMITIVE POLYNOMIAL OF DEGREE n KNOWN	X	X
PRIMITIVE POLYNOMIAL OF DEGREE n UNKNOWN	X	

Figure 11.14. Three cases that will be considered for m-sequence synchronization.

ratios. When the m-sequence generates the m bits, the matched filter will detect this epoch and phase synchronization will be accomplished.

Second, if $2^n - 1$ is not sufficiently small so that one can not expect to see all the m-sequence phases during the synchronization period, then the above method is not guaranteed and we must have a better than random estimate of the phase of the m-sequence before we attempt synchronization. Our job then is to search through a limited number of phase candidates and winnow the correct one.

The way this is usually done is by means of a serial search procedure implemented via what has become known as a "sliding correlator." The sliding correlator integrates the product of the received sequence against what is presumed to be the correct phase of the m-sequence. If the phase is correct, the integral will show a large departure from the mean value. As an example, consider that we are trying to synchronize with the m-sequence generated by $x^6 + x^5 + 1$ and let us assume that our estimate of the phase is 4 clock times ahead of the true phase. For this experiment we correlate 6 bits at a time. Our rule for synchronization will be that we have achieved synchronization only if all 6 bits agree. Our procedure will be to test 6 bits. If all 6 agree, we declare that we are in synchrony; if all 6 do not agree, we retard our reference m-sequence by one clock time and compare (integrate) for another 6 bits. This process is depicted in Figure 11.15.

The number of bits to be cross-correlated against the reference m-sequence (6 in this example) will depend on the type of noise, the appropriate signal-to-noise ratio, the decision threshold, and the look time (in this case, the time allowed to achieve synchronization). Clearly, the sliding correlator can require an enormous amount of search time if the number of phase candidates is very large or if the signal-to-noise ratio is very low. Dixon (1976, pp. 181–183) discusses implementation of the sliding correlator by sliding the reference m-sequence in a continuous fashion (i.e., not discrete retardation per our example). Braun (1982) provides an excellent relevant theoretical analysis. He also points out that schemes such as that used in our example are "single dwell time" procedures and he considers a multiple dwell time approach that dynamically changes the width of the correlation window, that is, the number of bits correlated for each decision. This concept

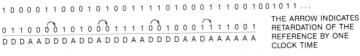

Figure 11.15. Sliding correlator.

of sequential hypothesis testing can bring far greater efficiency to our search. For background on sequential testing the reader is referred to Posner and Rumsey (1966).

Problem

Assume that the reception of the received m-sequence is errorless as regards Figure 11.15 and that we retard the reference every time the received bits and reference bits do not agree. Model the situation appropriately and derive the mean and variance of the time to synchronization if the sequences are initially displaced by a slip of S bits.

11.3.4. Polynomial Unknown: Errorless Reception

Yarlagadda and Hershey (1982) have proposed that Gold's characteristic sequence (see Chapter 8) be used as a benchmark for m-sequence synchronization. They found that there is a curious cross-correlation property between the truncated Rademacher sequences and an m-sequence at its characteristic sequence phase. The Rademacher sequences are those sequences that describe the bit sequences of a normal binary counter started from zero. For example, consider the 3-bit binary counter below:

$$
\begin{array}{ccc}
0 & 0 & 0 \\
0 & 0 & 1 \\
0 & 1 & 0 \\
0 & 1 & 1 \\
1 & 0 & 0 \\
1 & 0 & 1 \\
1 & 1 & 0 \\
1 & 1 & 1 \\
\end{array}
$$

The first Rademacher sequence is the sequence exhibited by the counter's first bit:

$$0 \quad 1 \quad 0 \quad 1 \quad 0 \quad 1 \quad 0 \quad 1 \tag{10}$$

the second Rademacher sequence is

$$0 \quad 0 \quad 1 \quad 1 \quad 0 \quad 0 \quad 1 \quad 1 \tag{11}$$

and the third is

$$0 \quad 0 \quad 0 \quad 0 \quad 1 \quad 1 \quad 1 \quad 1 \tag{12}$$

We drop the last bits of the Rademacher sequences to form sequences of length $2^n - 1$ where n is the number of bits or stages of the counter. These sequences are termed the truncated Rademacher sequences. We form a matrix R which is composed or partitioned of the n truncated Rademacher sequences of length $2^n - 1$. For our example,

$$R = \begin{bmatrix} 0 & 1 & 0 & 1 & 0 & 1 & 0 \\ 0 & 0 & 1 & 1 & 0 & 0 & 1 \\ 0 & 0 & 0 & 0 & 1 & 1 & 1 \end{bmatrix} \tag{13}$$

Now consider any m-sequence of period 7. Let us arbitrarily select the m-sequence generated by the primitive polynomial $x^3 + x + 1$:

$$1 \quad 0 \quad 0 \quad 1 \quad 1 \quad 1 \quad 0 \tag{14}$$

We now construct the $(2^n - 1) \times (2^n - 1)$ matrix, S, of all phases of (14)

$$S = \begin{bmatrix} 1 & 0 & 0 & 1 & 1 & 1 & 0 \\ 0 & 0 & 1 & 1 & 1 & 0 & 1 \\ 0 & 1 & 1 & 1 & 0 & 1 & 0 \\ 1 & 1 & 1 & 0 & 1 & 0 & 0 \\ 1 & 1 & 0 & 1 & 0 & 0 & 1 \\ 1 & 0 & 1 & 0 & 0 & 1 & 1 \\ 0 & 1 & 0 & 0 & 1 & 1 & 1 \end{bmatrix} \tag{15}$$

We now multiply R and S using conventional matrix multiplication, that is, we do not modularly reduce the row–column dot products:

$$RS = \begin{bmatrix} 2 & 1 & 3 & 1 & 2 & 1 & 2 \\ 1 & 3 & 2 & 1 & 2 & 2 & 1 \\ 2 & 2 & 1 & 1 & 1 & 2 & 3 \end{bmatrix} \tag{16}$$

Note that the fourth column's entries are all equal. This will be so only if the characteristic sequence of the m-sequence is

$$1 \quad 1 \quad 1 \quad 0 \quad 1 \quad 0 \quad 0 \tag{17}$$

A bank of n correlators will thus be able to determine the epoch at which an m-sequence, *regardless of the generating polynomial*, passes through its characteristic sequence phase. For a more involved example consider Figures 11.16 through 11.20 which show the output of the cross-correlation

$$\sum_{i=0}^{30} r_k(i)s(i+j) \tag{18}$$

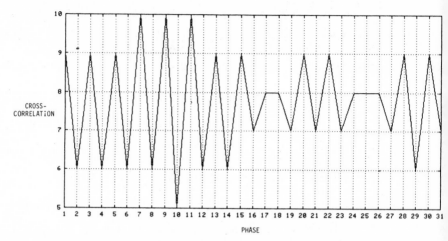

Figure 11.16. Cross-correlation of first Rademacher sequence.

where $r_k(i)$ is the ith bit of the kth Rademacher sequence and $\{s(\ )\}$ is the m-sequence generated by $x^5 + x^3 + 1$. The m-sequence is

$$
\underset{1\ 2\ 3\ 4\ 5\ 6\ 7\ 8\ 9\ 10\ 11\ 12\ 13\ 14\ 15\ 16\ 17\ 18\ 19\ 20\ 21\ 22\ 23\ 24\ 25\ 26\ 27\ 28\ 29\ 30\ 31}{1\ 1\ 1\ 1\ 1\ 0\ 0\ 0\ 1\ 1\ 0\ 1\ 1\ 1\ 0\ 1\ 0\ 1\ 0\ 0\ 0\ 0\ 1\ 0\ 0\ 1\ 0\ 1\ 1\ 0\ 0} \tag{19}
$$

The phases of the m-sequence are arbitrarily assigned by the smaller subscript numbers. The abscissa of 13 through 18 are the phase numbers.

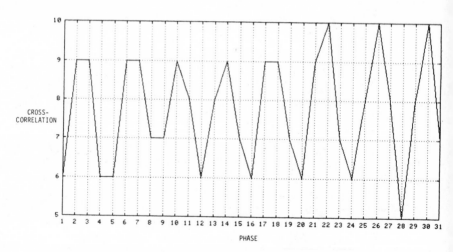

Figure 11.17. Cross-correlation of second Rademacher sequence.

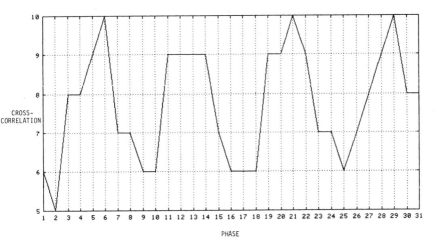

Figure 11.18. Cross-correlation of third Rademacher sequence.

Figure 11.21 is an overlay of Figures 11.16 through 11.20. These figures allow the reader to spot the point at which the cross-correlations are all equal and thereby to identify the phase (23) at which the characteristic sequence begins.

11.3.5. Rapid Acquisition Sequences

This synchronization process was introduced by Stiffler (1968). It uses a normal binary counter of n stages as the cyclic digital process. The counter

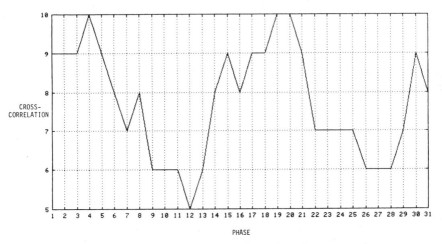

Figure 11.19. Cross-correlation of fourth Rademacher sequence.

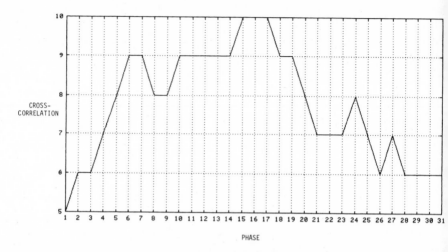

Figure 11.20. Cross-correlation of fifth Rademacher sequence.

is started at zero and incremented by one every clock time. When the count has reached $2^n - 1$, the next incrementation causes the counter to be reduced modulo 2^n and have all zeros in its stages. The contents of the counter stages $x_1, x_2, \ldots, x_n$ (x_1 is the least significant stage) are input to the following combinatorial logic:

$$f(x_1, x_2, \ldots, x_n) = \begin{cases} 0, & \text{if } \sum_{i=1}^{n} x_i \le \left[\dfrac{n}{2}\right] \\ 1, & \text{otherwise} \end{cases} \tag{20}$$

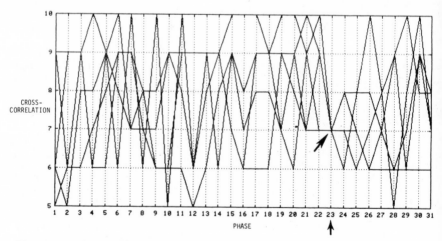

Figure 11.21. Coalescence of cross-correlations identifying the characteristic sequence.

where [ ] is the greatest integer function. The sequence produced, the output of the boolean function specified in (20), is termed the Rapid Acquisition Sequence (RAS) of length 2^n. Figure 11.22 depicts this configuration in terms of our canonical model (Figure 11.12).

Two items are worth mentioning. First, the function $f(\ )$ of (20) is a nonlinear boolean function that is threshold realizable. Because it is threshold realizable, it can be implemented in very fast hardware. Second, the sequence described by the successive contents of x_i is the ith Rademacher sequence.

It is easy to verify that the RAS of length 8 is

$$00010111 \tag{21}$$

Consider the full period cross-correlation of (21) against the first Rademacher sequence. Because the first Rademacher sequence has only two distinct phases, that is, the "normal" phase

$$01010101 \tag{22a}$$

and the other phase

$$10101010 \tag{22b}$$

it is clear that the cross-correlation will exhibit only two values. By direct computation

$$\begin{array}{l} 0\ 0\ 0\ 1\ 0\ 1\ 1\ 1 \\ \underline{0\ 1\ 0\ 1\ 0\ 1\ 0\ 1} \\ ADAAAADA \qquad A - D = 6 - 2 = 4 \end{array}$$

and

$$\begin{array}{l} 0\ 0\ 0\ 1\ 0\ 1\ 1\ 1 \\ \underline{1\ 0\ 1\ 0\ 1\ 0\ 1\ 0} \\ DADDDDAD \qquad A - D = -4 \end{array}$$

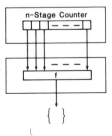

Figure 11.22. Rapid Acquisition Sequence (RAS) generator.

Thus, the cross-correlation is 4 for the Rademacher phase of (22a)—the "normal" phase—and −4 for the phase of (22b). The synchronization procedure begins to emerge. The first step is to cross-correlate the first Rademacher sequence of length 2^n over a full period of the 2^n long RAS. This first step gives us one of two equally probable answers and this "bit" of information resolves the phase of the RAS within modulo 2. Once the phase has been resolved modulo 2, the second Rademacher sequence is cross-correlated in both of its possible phases. (There are, of course, four distinct phases of the second Rademacher sequence, but we consider only the two phases that "survive" the first test, that is, only two of the four phases will be consonant with the determined phase of the first Rademacher sequence.) This computation tells us the phase of the RAS modulo 4. Thus, at each stage we gain a bit of information and sequentially recover the counter's stage sequences. Stiffler (1968) showed that the (normalized, i.e., $A - D$ divided by the sequence length) cross-correlation of the kth 2^n long Rademacher sequences with the 2^n long RAS is

$$\frac{1}{2^{n-1}} \begin{pmatrix} n-1 \\ \dfrac{n-1}{2} \end{pmatrix} \qquad n \text{ odd} \tag{23a}$$

and

$$\frac{1}{2^n} \begin{pmatrix} n \\ \dfrac{n}{2} \end{pmatrix} \qquad n \text{ even} \tag{23b}$$

if the $k - 1$ phases have been correctly resolved. Note that (23a) and (23b) are *independent* of k. Using Stirling's formula, Stiffler approximates (23a) and (23b) by

$$\left(\frac{2}{\pi}\right)^{1/2} n^{-1/2} \tag{24}$$

A graph of (23a) and (23b) and the approximation (24) is given in Figure 11.23.

As an example, consider that we wish to recover the phase of the 8-long RAS. We assume errorless reception for the example and that we are looking at the phase

1 1 1 0 0 0 1 0 1 1 1 0 0 0 1 0 1 1 1 0 0 0 1 0 1 1 1 0 0 0 1 0 $\cdots$

The first step is to compute the cross-correlation of the first Rademacher

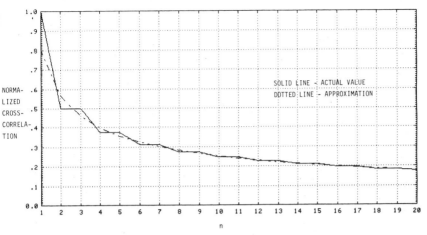

Figure 11.23. Normalized cross-correlation of the 2^n-long RAS.

sequence over a full period:

1 1 1 0 0 0 1 0 1 1 1 0 0 0 1 0 1 1 1 0 0 0 1 0 1 1 1 0 0 0 1 0 ···
0 1 0 1 0 1 0 1
DADDADDD $A - D = -4$

The cross-correlation is negative and therefore the first stage of the counter is out of phase with the normal phase of the first Rademacher sequence. The next step is to delay one bit and thus bring us into phase with the first Rademacher sequence component of the RAS and then cross-correlate the second Rademacher sequence with the RAS:

1 1 1 0 0 0 1 0 1 1 1 0 0 0 1 0 1 1 1 0 0 0 1 0 1 1 1 0 0 0 1 0 ···
0 1 0 1 0 1 0 1 0 0 1 1 0 0 1 1
      DDDDADDA $A - D = -4$

The cross-correlation is again negative so we bring our second Rademacher sequence into phase by delaying it $2^{2-1} = 2$, that is, two bits and then cross-correlate the third Rademacher sequence with the RAS:

1 1 1 0 0 0 1 0 1 1 1 0 0 0 1 0 1 1 1 0 0 0 1 0 1 1 1 0 0 0 1 0
0 1 0 1 0 1 0 1 0 0 1 1 0 0 1 1 0 0 0 0 1 1 1 1
          AAADDAAA $A - D = +4$

The cross-correlation is positive so here we find that our third Rademacher sequence is in phase and we are synchronized. Thus, a 2^n-long sequence can be synchronized with n decisions. A decision need not require a full period as per the example. We also need not "waste" time by delaying the cross-correlations as shown; we could instead have advanced the Rademacher sequences. We chose not to do these things in order to enhance the clarity of the presentation. Stiffler (1968) also examines the question of how many bits need be cross-correlated when operating in Additive White Gaussian Noise.

Now consider the RAS of length 32:

$$00000001000101110001011101111111 \tag{25}$$

Note that there are runs of bits (stretches of identical bits) up to length 7 in (25). Ipatov et al. (1975) noted that the RAS of length 2^n (n odd) will typically display runs of lengths up to $2^{(n+1)/2} - 1$. Ipatov et al. noted that these long run lengths could have deleterious effects on some systems that depend on transitions in the data for clock recovery. Ipatov et al. proposed a modification to the RAS (MRAS) that limits the maximum length of the runs to 2 bits without affecting the equal cross-correlation property (23a). The Ipatov et al. method is to change the combinatorial logic function from (20) to

$$f'(x_1, x_2, \ldots, x_n) = x_1 + g(x_2, x_3, \ldots, x_n) \tag{26}$$

where the addition is modulo 2, and

$$g(x_2, x_3, \ldots, x_n) = \begin{cases} 0, & \text{if } \sum_{i=2}^{n} x_i \leq \dfrac{n-1}{2} \\ 1, & \text{otherwise} \end{cases} \tag{27}$$

The MRAS of length 32 ($n = 5$) is then

$$01010101010101100101011001101010 \tag{28}$$

Problem

Consider that we wish to recover the phases of the n counter stages generating a RAS and that the RAS is received through a BSC with parameter p. We desire to recover the n phases with probability of error not exceeding P. Work out a formula for the number of bits that should be cross-correlated for each measurement. Work out the numbers for $n = 20$, $p = 10^{-3}$, and $P = 10^{-6}$.

11.3.6. The Thue–Morse Sequence

The Thue–Morse sequence (TMS) is that sequence produced by EXCLUSIVE-ORing or adding modulo 2 the contents of all the stages of an infinitely long binary counter started at zero and allowed to count indefinitely. The first few terms of the TMS are seen to be $0110100110\cdots$ from the following:

Counter	TMS
0	0
1	1
10	1
11	0
100	1
101	0
110	0
111	1
1000	1
1001	0
⋮	⋮

Figure 11:24 casts the TMS in terms of the canonical model (Figure 11.12).

Hershey (1979) reported that the TMS (1) never exhibits runs greater than length 2, (2) never repeats, and (3) is related to the coefficients of $\{x^i\}$ in expansion of the infinite product

$$\prod_{i=0}^{\infty} (1 - x^{2^i})$$

In his paper Hershey (1979) showed that the TMS could serve as a comma-free code to synchronize binary counters. Hershey and Lawrence (1981) suggested a follow-on method.

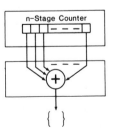

Figure 11.24. The Thue–Morse (TM) sequence generator.

11.3.7. A Statistical Property of the TMS

The TMS exhibits the following statistical property that is key to the method. Consider the two cases in which the counter's first m bits are either $011 \cdots 1$ or $111 \cdots 1$ (rightmost bit is least significant).

1. For the former case, the probability that the TMS will change at the next count is one if m is odd and zero if m is even. This is because the next count will result in the first m bits becoming $100 \cdots 0$ which represents a change in the number of ones in the counter from $m - 1$ to 1; clearly, the sum of ones modulo 2 is unchanged if and only if m is even.

2. For the latter case, the next count brings the first m bits to all zero and a carry propagates into the higher bits of the counter. The probability that the carry will stop at the $(m + 1)$st, $(m + 3)$rd, $(m + 5)$th, and so on position is $\frac{1}{2} + \frac{1}{8} + \frac{1}{32} + \cdots = \frac{2}{3}$. (The countersize is assumed large.) If the carry propagates in this manner, there will be a unit change, modulo 2, in the density of ones in the counter above the first m bits. In the lower part of the counter, the first m bits, the counter experiences a unit change in the density of ones modulo 2 if and only if m is odd. Combining these two effects, we note then that the probability that the TMS will change at the next count is $\frac{1}{3}$ if m is odd and $\frac{2}{3}$ if m is even.

11.3.8. Use of the TMS for Synchronization

The goal is to resolve the first s stages of the TMS transmitter's counter by using the above statistical property. The method is best and most easily presented as an example which can be extended in an obvious fashion. Consider the following 30-bit segment from somewhere in the TMS:

$$101101001100101101001011001101$$

(time order is left to right). We examine this segment in light of the above statistical property. We first make two counts, called A and B counts, of bit reversals or changes in the TMS. The B counts are one bit out of phase with the A counts. Starting at the beginning of our 30-bit segment, arbitrarily make only two counts each for A and B as shown in the top block of Figure 11.25.

Note that the TMS changed once during the A "windows" and twice during the B windows. From the TMS statistical property we conclude that the counter generating the TMS segment has a zero as its first bit every time a B window was begun. We have thus resolved, or phased, the first bit sequence from the transmitter's counter. Using this information, we can now proceed to resolve the second bit time sequence of the transmitter's counter.

To do this, we choose our sampling windows to begin at those times when the first bit in the transmitter's TMS counter is a one. What we are

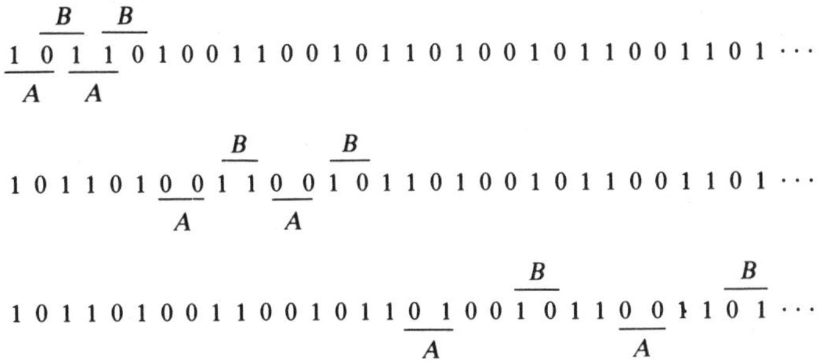

$$\begin{array}{cc} \underline{B} & \underline{B} \\ \end{array}$$

1 0 1 1 0 1 0 0 1 1 0 0 1 0 1 1 0 1 0 0 1 0 1 1 0 0 1 1 0 1 ···

$$\begin{array}{cc} \underline{A} & \underline{A} \end{array}$$

$$\begin{array}{cc} \underline{B} & \underline{B} \\ \end{array}$$

1 0 1 1 0 1 0 0 1 1 0 0 1 0 1 1 0 1 0 0 1 0 1 1 0 0 1 1 0 1 ···

$$\begin{array}{cc} \underline{A} & \underline{A} \end{array}$$

$$\begin{array}{cc} \underline{B} & \underline{B} \\ \end{array}$$

1 0 1 1 0 1 0 0 1 1 0 0 1 0 1 1 0 1 0 0 1 0 1 1 0 0 1 1 0 1 ···

$$\begin{array}{cc} \underline{A} & \underline{A} \end{array}$$

Figure 11.25. $A - B$ counts to determine first bit of TMS counter (top block), second bit of TMS counter (middle block), and third bit of TMS counter (bottom block).

trying to determine is which of the sampling windows, A or B, is "seeing" the first two bits in the transmitter's TMS counter change from 01 to 10. From the middle block of Figure 11.25, we note that the TMS changed once during the B windows and did not change during the A windows. Thus, we know that the counter generating the TMS segment had a zero as its second bit and a one as its first bit every time an A window was begun.

Going one step further, we can resolve the third bit. As we already know how the first two bits are progressing, we shall now sample the TMS whenever the first two bits in the TMS counter are both one as shown in the bottom block of Figure 11.25. This will allow us to find the 011 to 100 transitions and thus resolve the third bit.

From the bottom block of Figure 11.25, we note that the TMS changed once during the A windows and twice during the B windows. Thus, we know that the first three bits in the TMS counter were 011 every time a B window was begun. Other bits can be resolved in a similar manner.

The reader should note that in order to make n "$A - B$" counts to statistically resolve the sth bit of the TMS counter (n may have to be large under very noisy conditions) requires a TMS segment on the order of $n \cdot 2^s$ bits since each "$A - B$" count requires approximately 2^s bits. Thus, the number of bits required to sequentially resolve the first m bits using n hard decisions per bit is approximately $n(2 + 2^2 + 2^3 + \cdots + 2^m)$ plus a few "overhead" bits. The total is on the order of $n \cdot 2^{m+1}$ bits. As in the case of the RAS, we have recovered the same number of bits as decisions made.

11.3.9. Titsworth's Component Codes

Titsworth (1964) devised a method of "probabilistically" combining several sequences with "short," relatively prime periods to produce a sequence with a "long" period but a sequence which, if properly analyzed over only a "short" period, will yield the correct phases of the individual "short" sequences. Once the correct phases of the individual "short" sequences have been recovered, we can use the Chinese Remainder Theorem to solve for the phase of the "long" sequence. Sequence constructions based on Titsworth's procedure have become known as the "JPL codes." Figure 11.26 depicts Titsworth's scheme in terms of the canonical model, Figure 11.12.

We can construct a Titsworth Synchronization Sequence (TSS) by the following procedure. First, we pick an odd number of "short" sequences; they may themselves be generated by a recursion, however, they are depicted as stored in simple end-around shift registers in Figure 11.26. The "short" sequences should have periods that are pairwise relatively prime. The sequences should also be of approximately the same length. If the TSS is desired to be of length N, and if there are to be m (odd) component "short" sequences, then the length of a component sequence should be approximately $N^{1/m}$. The sequences should exhibit low cross-correlations. Second, we combine the m different sequences. Titsworth shows that the boolean function which has the largest and most uniform Walsh–Hadamard coefficients for the unit-weight vectors is the majority function. The impact of this discovery is that each "short" sequence will exhibit a high degree of cross-correlation to the TSS when the "short" sequence is at its proper phase.

Let us proceed with an example. We choose three sequences $A = 11000$, $B = 110101$, and $C = 1110010$ and arbitrarily designate the phase of a sequence with a subscript. All phases of the three sequences are given in

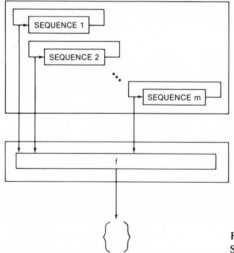

Figure 11.26. Titsworth's Synchronization Sequence (TSS) generator.

Table 3. All Phases of the Three "Short" Sequences

Phase	Sequence A	Sequence B	Sequence C
0	$A_0 = 11000$	$B_0 = 110101$	$C_0 = 1110010$
1	$A_1 = 10001$	$B_1 = 101011$	$C_1 = 1100101$
2	$A_2 = 00011$	$B_2 = 010111$	$C_2 = 1001011$
3	$A_3 = 00110$	$B_3 = 101110$	$C_3 = 0010111$
4	$A_4 = 01100$	$B_4 = 011101$	$C_4 = 0101110$
5		$B_5 = 111010$	$C_5 = 1011100$
6			$C_6 = 0111001$

Table 3. We combine the three "short" sequences by the majority function specified in Table 4.

Let us now calculate the Walsh–Hadamard transform of the majority function. This is shown in Table 5. Note that f' is obtained from f by substituting 1 for 0 and -1 for 1. The arrows in Table 5 identify the Walsh–Hadamard coefficients of the unit-weight vector. They are equal, as Titsworth proved they must be, and, for this case, equal to four. This means that if we cross-correlate either A or B or C against f we will find six agreements and two disagreements. The cross-correlation is then (Agreements-Disagreements)/8 = 0.5. Titsworth shows that for an m-input (m, odd) majority function, the Walsh–Hadamard coefficient of a unit-weight vector, the cross-correlation between the function's output and any of its inputs is

$$2^{1-m} \left(\begin{matrix} m-1 \\ \left[\dfrac{m}{2} \right] \end{matrix} \right). \tag{29}$$

Table 4. The Majority Function for Three Inputs.

A	B	C	$f(A, B, C)$
0	0	0	0
0	0	1	0
0	1	0	0
0	1	1	1
1	0	0	0
1	0	1	1
1	1	0	1
1	1	1	1

Table 5. The Walsh–Hadamard Transform of the Majority Function

A	B	C	f	f'	I	II	III	
0	0	0	0	1	2	2	0	
0	0	1	0	1	0	−2	4	←
0	1	0	0	1	0	2	4	←
0	1	1	1	−1	−2	2	0	
1	0	0	0	1	0	2	4	←
1	0	1	1	−1	2	2	0	
1	1	0	1	−1	2	−2	0	
1	1	1	1	−1	0	2	−4	

which for moderately large m can be well approximated by

$$\sqrt{(\pi/2)(m-1)} \tag{30}$$

For our case $m = 3$ and (29) is

$$2^{-2}\left(\begin{array}{c} 2 \\ \left[\dfrac{3}{2}\right] \end{array}\right) = 2^{-2}\binom{2}{1} = \frac{1}{2} \tag{31}$$

which checks.

In Figure 11.27 we show how the phases of the "short" sequences are resolved from a TSS. We start to sample the TSS with sequence A at its second phase, sequence B at its fourth phase, and sequence C at its first phase. The symbol (s, x_i) is the Agreements-Disagreements between the TSS and phase i of sequence x. We have measured the cross-correlation over a 21-bit stretch of the $5 \times 6 \times 7 = 210$-bit period TSS. Note that the highest values of (s, A_i), (s, B_i), and (s, C_i) correctly identify the "short" sequences. Note also that the cross-correlations are done, and can in general be done, in parallel. Thus, we can create a TSS of arbitrarily long period and recover its phase by analyzing only a relatively short stretch.

Problem

Assume that the TSS was started at the zero phases of the "short" component sequences. Use the recovered phases of the "short" sequences

A 0 0 0 0 1 1 0 0 0 0 0 1 1 0 0 0 1 0 ⋯

B 0 1 1 1 0 1 1 0 1 1 0 1 0 0 1 1 1 1 ⋯

C 1 0 0 0 1 1 0 1 0 1 0 1 1 1 0 0 1 1 ⋯

Maj (A, B, C) 0 1 1 1 1 1 1 0 1 1 0 1 1 0 1 1 1 1 ⋯

A–D

$(s, A_0) = 7 - 14 = -7$
$(s, A_1) = 11 - 10 = +1$
$(s, A_2) = 16 - 5 = +11\checkmark$
$(s, A_3) = 10 - 11 = -1$
$(s, A_4) = 6 - 15 = -9$

$(s, B_0) = 12 - 9 = +3$
$(s, B_1) = 6 - 15 = -9$
$(s, B_2) = 13 - 8 = +5$
$(s, B_3) = 10 - 11 = -1$
$(s, B_4) = 16 - 5 = +11\checkmark$
$(s, B_5) = 11 - 10 = +1$

$(s, C_0) = 10 - 11 = -1$
$(s, C_1) = 16 - 5 = +11\checkmark$
$(s, C_2) = 8 - 13 = -5$
$(s, C_3) = 10 - 11 = -1$
$(s, C_4) = 12 - 9 = +3$
$(s, C_5) = 10 - 11 = -1$
$(s, C_6) = 10 - 11 = -1$

Figure 11.27. Recovery of the phases of the "short" sequences by cross-correlation with the TSS.

and the Chinese Remainder Theorem to determine how long the TSS had been running at the time we started our sampling.

References

Barker, R. (1953), Group Synchronizing of Binary Digital Systems, in *Communication Theory*, Willis Jackson (Ed.), Academic Press, New York, Chap. 19, pp. 273–287.

Braun, W. (1982), Performance Analysis for the Expanding Search PN Acquisition Algorithm, *IEEE Transactions on Communications*, Vol. COM-30, No. 3, pp. 424–435, March.

Dixon, R. (1976), *Spread Spectrum Systems*, Wiley, New York.

Gabard, O. (1968), Design of a Satellite Time-Division Multiple-Access Burst Synchronizer, *IEEE Transactions on Communication Technology*, Vol. COM-16, No. 4, August.

Hershey, J. (1979), Comma-Free Synchronization of Binary Counters, *IEEE Transactions on Information Theory*, Vol. IT-25, No. 6, pp. 724–725, November.

Hershey, J. and W. Lawrence (1981), Counter Synchronization Using the Thue–Morse Sequence and PSK, *IEEE Transactions on Communications*, Vol. COM-29, No. 1, pp. 79–80, January.

Hershey, J. (1982), Proposed Direct Sequence Spread Spectrum Voice Techniques for the Amateur Radio Service, NTIA Report 82-111.

Ipatov, V., Yu. Kolomenskiy, and P. Sharanov (1975), On Modified Rapid Search Sequences, *Radio Engineering and Electronic Physics*, Vol. 20, pp. 135–136, September.

Klyuyev, L. and N. Silkov (1976), Periodic Sequences Synthesized from Barker Sequences, *Telecommunications and Radio Engineering*, Vol. 30, No. 4, pp. 128–129.

Kreyszig, E. (1967), *Advanced Engineering Mathematics*, Wiley, New York.

Lindner, J. (1975a), Binary Sequences up to Length 40 with Best Possible Autocorrelation Function, Part 1: Complete Tables, Institute für elektrische Nachrichten-technik der rheinisch-westfälischen Technischen Hochschule Aachen, Internal Report, September.

Lindner, J. (1975b), Binary Sequences up to Length 40 with Best Possible Autocorrelation Function, *Electronics Letters*, Vol. 11, No. 21, p. 50, October.

Maury, J. and F. Styles (1964), Development of Optimum Frame Synchronization Codes for Goddard Space Flight Center PCM Telemetry Standards, Proceedings of the 1964 National Telemetering Conference, Los Angeles, CA, June 2–4, 1964.

Nuspl, P., K. Brown, W. Steenaart, and B. Ghicopoulos (1977), Synchronization Methods for TDMA, *Proceedings of the IEEE*, Vol. 65, No. 3, March.

Petit, R. (1967), Pulse Sequences with Good Autocorrelation Properties, *Microwave Journal*, pp. 63–67, February.

Posner, E. and H. Rumsey, Jr. (1966), Continuous Sequential Decision in the Presence of a Finite Number of Hypotheses, *IEEE Transactions on Information Theory*, Vol. IT-12, No. 2, pp. 248–255, April.

Sarwate, D. (1984), An Upper Bound on the Aperiodic Autocorrelation Function for a Maximal-Length Sequence, *IEEE Transactions on Information Theory*, Vol. 30, pp. 685–687.

Schrempp, W. and T. Sekimoto (1968), Unique Word Detection in Digital Burst Communications, *IEEE Transactions on Communication Technology*, Vol. COM-16, No. 4, August.

Sekimoto, T. and J. Puente (1968), A Satellite Time-Division Multiple-Access Experiment, *IEEE Transactions on Communication Technology*, Vol. COM-16, No. 4, August.

Stiffler, J. (1968), Rapid Acquisition Sequences, *IEEE Transactions on Information Theory*, Vol. IT-14, No. 2, March.

Stiffler, J. (1971), *Theory of Synchronous Communications*, Prentice-Hall, Englewood Cliffs, NJ.

Titsworth, R. (1964), Optimal Ranging Codes, *IEEE Transactions on Space Electronics and Telemetry*, Vol. SET-10, pp. 19-30, March.

Turyn, R. (1968), Sequences with Small Correlation, in *Error Correcting Codes*, H. Mann (Ed.), Wiley, New York, pp. 195-228.

Yarlagadda, R. and J. Hershey (1982), Benchmark Synchronization of m-Sequences, *Electronics Letters*, Vol. 18, No. 2, pp. 68-69, January.

The Channel and Error Control

12.1. Introduction

Our information bits are transported from sender to receiver over a medium which we call a channel. The bits, or groupings of bits, are represented by waveforms and then either put directly onto a medium, as is the case with many local area networks, or sent through a modem (modulator/demodulator) which generates and resolves waveforms designed for use on a specific channel. A communications engineer characterizes the behavior, not of bits through the channel, but of waveforms and signal processing algorithms that are used to decide which particular symbol is being sent. The need for this arises, of course, because of noise on the channel. Here we are concerned only with the bit stream as it is presented to us after the modem, after all the decisions regarding received waveform processing have been made.

12.2. A Channel Model

We are thus dealing with channel models. One of the most persistent of these models is the Binary Symmetric Channel (BSC). The model is widely used primarily because of its simplicity although it is a reasonably good approximation to many important channels. But even for those channels which it does not approximate well, it still may be of use in a "first-cut" analysis of some error correction schemes that use a process called *interleaving*.

What then is the BSC? It is the simplest of models, characterized by only one parameter, p, the probability of an error in transmission. Figure 12.1 depicts the two paths that may be experienced by both a zero and a one. The error process is a simple Bernoulli process with parameter p. Thus, Figure 12.2 is immediately seen to be equivalent to a BSC.

Figure 12.1. The Binary Symmetric Channel (BSC).

Often we must send data through a series of channels as is the case for information passing through a network from the original sender node to the final receiver node. It behooves us to appreciate the effect of these additional links. The problem we now look at is the effect of cascading n-BSCs in *tandem*. We approach the problem in three different ways. The purpose of these redundant derivations is to consolidate different styles of proof and solution methods that have been previously introduced. Readers are also advised that, in their own work, solution of a problem by two or more methods serves not only as a check on one's work and the veracity of the solution but also often adds a degree of insight that can be extremely valuable.

Figure 12.3 depicts n-BSCs in tandem.

12.2.1. Combinatorial Solution

The channel depicted in Figure 12.3 is composed of symmetric channels and is therefore itself symmetric; the probability that a zero will be transmitted correctly is the same that a one will be transmitted correctly.

The probability that a bit will be transmitted incorrectly, p', is the probability that it will undergo an odd number of errors. This is

$$p' = \sum_{i \text{ odd}} \binom{n}{i} p^i (1-p)^{n-i} \tag{1}$$

Expanding (1) we see that

$$p' = \binom{n}{1} p(1-p)^{n-1} + \binom{n}{3} p^3 (1-p)^{n-3} + \binom{n}{5} p^5 (1-p)^{n-5} + \cdots \tag{2}$$

How do we perform this summation? The trick is to find some way to

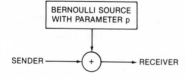

Figure 12.2. An equivalent representation of the BSC (+, Mod-2 addition).

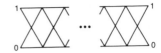

Figure 12.3. n Binary symmetric channels in tandem.

"select out" the even terms. This can be done quite easily by using the binomial expansion. Note that we are dealing with a sum that seemingly involves two terms, p and $1 - p$. Consider the following identities:

$$1 = ((1 - p) + p)^n = \binom{n}{0}(1 - p)^n + \binom{n}{1}(1 - p)^{n-1}p$$
$$+ \binom{n}{2}(1 - p)^{n-2}p^2 + \cdots \qquad (3)$$

$$(1 - 2p)^n = ((1 - p) - p)^n = \binom{n}{0}(1 - p)^n - \binom{n}{1}(1 - p)^{n-1}p$$
$$+ \binom{n}{2}(1 - p)^{n-2}p^2 + \cdots \qquad (4)$$

Consider the difference of (3) and (4):

$$1 - (1 - 2p)^n = 2\binom{n}{1}(1 - p)^{n-1} + 2\binom{n}{3}(1 - p)^{n-3}p^3$$
$$+ 2\binom{n}{5}(1 - p)^{n-5}p^5 + \cdots \qquad (5)$$

Therefore,

$$p' = \tfrac{1}{2}(1 - (1 - 2p)^n) \qquad (6)$$

and

$$1 - p' = \tfrac{1}{2}(1 + (1 - 2p)^n) \qquad (7)$$

Note that the resulting channel is also a BSC.

12.2.2. Solution by Difference Equations

Let $p'(j - 1)$ be the probability that a bit will be transmitted incorrectly over a $j - 1$ link tandem channel. We evaluate $p'(j)$ by noting that

$$p'(j) = p'(j - 1)(1 - p) + (1 - p'(j - 1))p \qquad (8)$$

Equation (8) says that a bit can be transmitted incorrectly in only one of two mutually exclusive ways. Either it is transmitted incorrectly over the first $j-1$ links and then transmitted correctly over the final link OR it is transmitted correctly over the first $j-1$ links and then corrupted on the final link. Regrouping (8) we see that

$$p'(j) + (2p-1)p'(j-1) = p \tag{9}$$

The homogeneous equation specified by (9) is

$$p'_h(j) + (2p-1)p'_h(j-1) = 0 \tag{10}$$

The solution to (10) is

$$p'_h(j) = k(-1)^j(2p-1)^j = k(1-2p)^j \tag{11}$$

Since the right-hand side of (9) is a constant, the general solution to (9) is easily determined to be

$$p'_g(j) = \tfrac{1}{2} \tag{12}$$

Thus,

$$p'(j) = \tfrac{1}{2} + k(1-2p)^j \tag{13}$$

For $j=1$, $p'(1) = p$, and thus $k = -\tfrac{1}{2}$. Therefore,

$$p' = p'(n) = \tfrac{1}{2}(1 - (1-2p)^n) \tag{14}$$

12.2.3. Solution By Markov Processes

A bit traversing the n-link tandem channel is in one of two states after passing through each BSC link. The bit is either CORRECT or INCORRECT. Since the links—BSC channels—are independent of each other, we can model the state of the transmitted bit by the two-state Markov process depicted in Figure 12.4. The matrix describing the process of Figure 12.4 is

$$M = \begin{bmatrix} 1-p & p \\ p & 1-p \end{bmatrix} \tag{15}$$

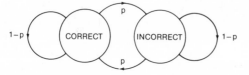

Figure 12.4. The two-state Markov process describing the state of a bit passing through the n-link tandem channels.

What we need to completely characterize the n-link channel is the nth power of M. This we do, as before, by setting up the characteristic equation

$$|M - \lambda I| = \begin{vmatrix} 1 - p - \lambda & p \\ p & 1 - p - \lambda \end{vmatrix} = 0 \tag{16}$$

Expanding (16) we have

$$\lambda^2 + 2\lambda(p - 1) + (1 - 2p) = 0 \tag{17}$$

Solving (17) we find that the eigenvalues are

$$\lambda_1 = 1 \quad \text{and} \quad \lambda_2 = 1 - 2p \tag{18}$$

The eigenvalues are distinct and therefore we are assured that the matrix M is diagonable. The eigenvectors are easily found to be

$$\begin{bmatrix} 1 \\ 1 \end{bmatrix} \quad \text{and} \quad \begin{bmatrix} 1 \\ -1 \end{bmatrix} \tag{19}$$

We now form the partitioned matrix of eigenvectors

$$A = \begin{bmatrix} 1 & 1 \\ 1 & -1 \end{bmatrix} \tag{20}$$

The inverse of A is easily found:

$$A^{-1} = \begin{bmatrix} \frac{1}{2} & \frac{1}{2} \\ \frac{1}{2} & -\frac{1}{2} \end{bmatrix} \tag{21}$$

Now let

$$B = A^{-1}MA = \begin{bmatrix} 1 & 0 \\ 0 & 1 - 2p \end{bmatrix} \tag{22}$$

We see that

$$B^n = A^{-1}M^nA = \begin{bmatrix} 1 & 0 \\ 0 & (1 - 2p)^n \end{bmatrix} \tag{23}$$

We solve (23) for M^n and find

$$M^n = AB^nA^{-1} = \begin{bmatrix} \frac{1}{2}(1 + (1 - 2p)^n) & \frac{1}{2}(1 - (1 - 2p)^n) \\ \frac{1}{2}(1 - (1 - 2p)^n) & \frac{1}{2}(1 + (1 - 2p)^n) \end{bmatrix} \tag{24}$$

which is all of the transition probabilities for the tandem channel in matrix form.

Problem

The asymmetric binary channel is defined in Figure 12.5. Consider that we concatenate n of these channels. Show, by any method, that the matrix of transition probabilities is

$$M^n = \frac{1}{p_1 + p_2} \begin{bmatrix} p_2 + p_1(1 - p_1 - p_2)^n & p_1(1 - (1 - p_1 - p_2)^n) \\ p_2(1 - (1 - p_1 - p_2)^n) & p_1 + p_2(1 - p_1 - p_2)^n \end{bmatrix}$$

One of the best known results from information theory states that the "capacity," C, of a BSC is

$$C = 1 + p \log_2 p + (1 - p) \log_2 (1 - p) \tag{25}$$

The capacity of a channel is the number of bits per second that can be transmitted without error. Shannon, the founder of information theory, using a nonconstructive proof, showed that coding schemes existed that would allow error-free transmission over a channel as long as the source's information rate does not exceed the channel's capacity. A heuristic way to look at (25) is to consider that what we are in effect doing in Figure 12.2 is combining two information sources, that of the sender and that of the noise process, and sending the two bit streams over a perfect channel with a capacity of unity. Recalling that the information rate of a Bernoulli source is $-p \log_2 p - (1 - p) \log_2 (1 - p)$, we subtract this rate from unity to find the remaining capacity for the sender which is given by (25).

Problem

The President is in need of a wise and laconic adviser. The adviser is to consider all sources of information and then respond with either a YES or a NO to questions. The Office of Personnel Management unleashes an extensive testing campaign of the three applicants. The results are shown below.

Applicant	Percentage of Correct Answers
Henry	80
Zbig	50
Al	10

Who is the best choice; who is the worst and why?

Figure 12.5. The asymmetric binary channel.

If the sender uses a BSC, the sender's bitstream will be corrupted by the Bernoulli noise process and will experience such statistics as an average of np errors for every n bits sent. How then can one use the BSC and still transmit information reliably?

Problem

One way to send information more reliably is simply to send the message an odd number of times and majority decode the received bits. For example, let us say that we wish to send the code word message 010 through a very noisy channel. We send the code word five times. At the receiver we observe, say, 110011010010011. We divide the received stream into triples and write, at the bottom, the bit which occurred most often in the column above it:

$$
\begin{array}{ccc}
1 & 1 & 0 \\
0 & 1 & 1 \\
0 & 1 & 0 \\
0 & 1 & 0 \\
0 & 1 & 1 \\
\hline
0 & 1 & 0 \\
\end{array}
$$

     0   1   0   Majority decoded message

Assume we want to transmit an m-bit code word over a BSC with error probability p by transmitting it $2k + 1$ times and majority decoding the bits. What is the minimum value of $2k + 1$ such that the probability of an error in the majority decoding will be less than or equal to E? Work out the numbers for

$$(p, m, E) = (0.001, 8, 0.01), (0.001, 8, 0.001), (0.001, 8, 0.0001),$$
$$(0.01, 8, 0.01), (0.01, 8, 0.001), (0.01, 8, 0.00001),$$
$$(0.1, 8, 0.01), (0.1, 8, 0.001), (0.1, 8, 0.0001)$$

The majority decoding method, while simple, is not very efficient. A better method is to use one of a host of techniques developed over the last

40 years known as forward error correction coding. Error correction coding is a mature and rich field. There is simply no way to do justice to it within a few pages; but we do try to give the reader a peek into the spirit of error correction coding and a sense of what it can do.

12.3. The Simplest Hamming Code

Error correction coding got seriously started with the publication of Hamming's milestone article in 1950. We can gain an excellent glimpse into error correction coding by studying Hamming's original work.

Hamming proposed to encode m bits of information using n bits where $n = m + k$. The additional k bits were to serve as check bits for error correction. It is useful to compute and consider m/n which is interpretable as the "rate" or proportion at which information bits are sent.

What can we do with k extra bits? If arrayed linearly and considered as a number, the bits can represent the integers from 0 through $2^k - 1$. What Hamming sought and found was a method whereby any single error within the n bits sent could be identified and corrected with the help of the k additional bits. This train of thought leads us to a very useful inequality. We must be able to "point to" the bit in error if there is a one bit error and, if there is no error, we must be able to also "point to" this condition. If we use the k extra bits to do this, the k bits must be capable of "pointing to" or indicating one of $n + 1$ different cases. Thus,

$$2^k \geqslant n + 1 \tag{26}$$

[The "1" in (26) covers the case in which there are no errors.]

Problem

Prepare a table of (n, m, k) triples giving the maximum m for $n = 3, 4, 5, \ldots, 16$ so that (26) is satisfied.

Hamming originally developed the code structure shown in Figure 12.6. Encoding is done by loading the m information bits into the word shown

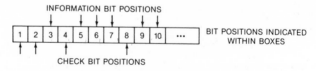

Figure 12.6. Original structure of the Hamming code word.

in Figure 12.6 and then computing and inserting the k check bits. (It does not matter, of course, where the check bits are actually placed as long as the sender and receiver have agreed. Contemporary theory imbeds the Hamming codes into a much larger class of codes and usually segregates the check bits instead of allowing them to be interspersed with data bits.) Hamming chose the placement shown in Figure 12.6 and it leads to a pretty result.

Hamming's scheme was to set the check bit in position one equal to the modulo 2 sum of all the bits (excluding itself) in positions that have a one in the least significant bit of the normal binary representation of their position. Thus, the check bit in position one is the modulo 2 sum of bits in positions $3, 5, 7, 9, 11, \ldots$. The check bit in position two is the modulo 2 sum of all the bits (again excluding itself) in positions that have a one in the next to the least significant bit of the normal binary representation of their position. Thus, the check bit in position two is the modulo 2 sum of all the bits in positions $3, 6, 7, 10, 11, 14, 15, 18, \ldots$. This process is logically extended and the third check bit, which is in position four, is the modulo 2 sum of all bits in positions $5, 6, 7, 12, 13, 14, 15, 20, \ldots$. (Note that by using the above scheme, none of the check bits involve any of the other check bits.)

Table 1 lists all 16 original form Hamming code words for $(n, m, k) = (7, 4, 3)$.

To determine the presence of and correct any single error in a received code word, we form what Hamming called the "checking number." The

Table 1. The 16 Code Words of Original Form for the Hamming Code with $(n, m, k) = (7, 4, 3)$

0	0	0	0	0	0	0
1	1	0	1	0	0	1
0	1	0	1	0	1	0
1	0	0	0	0	1	1
1	0	0	1	1	0	0
0	1	0	0	1	0	1
1	1	0	0	1	1	0
0	0	0	1	1	1	1
1	1	1	0	0	0	0
0	0	1	1	0	0	1
1	0	1	1	0	1	0
0	1	1	0	0	1	1
0	1	1	1	1	0	0
1	0	1	0	1	0	1
0	0	1	0	1	1	0
1	1	1	1	1	1	1

checking number is a k-bit number, $c_k c_{k-1} \cdots c_2 c_1$, and is formed by the following rule: bit c_i is the modulo 2 sum of check bit i (the bit in position 2^{i-1}) and the bits that are summed by check bit i. For example, c_3 is the modulo 2 sum of the bits in positions 4, 5, 6, 7, 12, 13, 14, 15, 20, ... Having formed the checking number we proceed to interpret and evaluate it as a normal binary number. If the value of the number is zero, then we declare that the word has been received error-free. If the value is nonzero, we declare a single error has occurred and the word is corrected by inverting the bit in the position whose number is equal to the word's value. For example,

$$1 \quad 1 \quad 0 \quad 0 \quad 1 \quad 1 \quad 0 \qquad\qquad (27)$$

is a valid Hamming code word for the $(n, m, k) = (7, 4, 3)$ code. Note that the checking number for (27) is 000 indicating no errors. Now assume that the bit in the sixth position has been inverted in transmission and we receive

$$1 \quad 1 \quad 0 \quad 0 \quad 1 \quad 0 \quad 0 \qquad\qquad (28)$$

We now derive 110 as the checking number. Interpreting the checking number as a normal binary number, we declare that the bit in position 6 is in error and we invert it, thus recovering (27).

Note that the code can not correct two errors and indeed may worsen the situation. For example, assume the code word of (27) suffered two bit inversions in, say, positions 3 and 5. We would receive

$$1 \quad 1 \quad 1 \quad 0 \quad 0 \quad 1 \quad 0 \qquad\qquad (29)$$

Forming the checking number for (29) we obtain 110 which says we should invert the bit in position 6 giving us

$$1 \quad 1 \quad 1 \quad 0 \quad 0 \quad 0 \quad 0 \qquad\qquad (30)$$

which now has three bits in error. We have thus hinted at a valuable lesson, namely, in picking an error correction code we have to exercise extreme caution. If we pick too powerful a code, one capable of correcting many errors, we will waste the throughput or capacity allowed by our channel. But if, on the other hand, our code is insufficient for much of the needed error correction required on our channel, we may do *more* harm than simply sending the data without error correction coding.

Problem

What would you suspect if a single error correcting code were, on the average, correcting half of the code words received? Now putting intuition aside and donning a theoretician's hat, determine an expression for the *values* of p for a BSC that could cause the above phenomenon. Solve for some of the values of p for a Hamming code with $(n, m, k) = (7, 4, 3)$. Comment on the region for which p is close to 0.5.

Quite often we encounter channels that can not be reasonably modeled as a BSC. Many channels exhibit errors that are highly dependent and occur in "bursts." A burst of length x means *any* pattern where the first bit in error is no further than $x - 1$ positions from the last bit in error. (Why, for x greater than 1, are there 2^{x-2} possible error patterns in a burst of length x?) One very simple coding method that can be used to great effect in countering bursts is that of "block interleaving." We space all bits of each Hamming code word so that they are separated by at least $x + 1$ bits. Thus, if the traffic is hit by an error burst of length no greater than x, the Hamming code words affected by the burst will be affected in at most one bit, which can, of course, be corrected. Diagrammatically, our block interleaving looks like

$$b_1^1 b_1^2 \cdots b_1^x b_2^1 b_2^2 \cdots b_2^x b_3^1 b_3^2 \cdots b_n^1 b_n^2 \cdots b_n^x \qquad (31)$$

The name "block interleaving" comes about by considering a block formed by x n-bit code words listed in a column as shown in Figure 12.7. The bits are extracted from this block by proceeding down the columns, left to right.

Problem

Determine the extra delay between sending and reception introduced by the block interleaving scheme of (31).

Interleaving need not be used for Hamming codes only. It is a powerful general technique. An article by Brayer and Cardinale (1967) reports an improvement of one to three orders of magnitude in performance through the use of interleaving on some HF channels.

$$
\begin{aligned}
b_1^1 b_2^1 b_3^1 &\cdots b_n^1 \\
b_1^2 b_2^2 b_3^2 &\cdots b_n^2 \\
&\vdots \\
b_1^x b_2^x b_3^x &\cdots b_n^x
\end{aligned}
$$

Figure 7. A block of xn-bit code words

12.4. The Hamming Code—Another Look

We now derive the Hamming single error correction code in a very different way. The purpose of this section is to show how something we have learned about m-sequences can lead to a very useful result.

Recall that the autocorrelation, agreements minus disagreements, of the m-sequence $\{s_i\}$ of length $2^n - 1$, has two distinct values. The autocorrelation is $2^n - 1$ only at one point, when the offset is zero. This remarkable property of m-sequences allows us to create an error correction code. To see why this is so consider that we form the set of words

$$
\begin{array}{ccccc}
s_1 & s_2 & \cdots & s_{2^n-2} & s_{2^n-1} \\
s_2 & s_3 & \cdots & s_{2^n-1} & s_1 \\
s_3 & s_4 & \cdots & s_1 & s_2 \\
& & \vdots & & \\
s_{2^n-1} & s_1 & \cdots & s_{2^n-3} & s_{2^n-2}
\end{array}
\tag{32}
$$

Each pair of words in (32) has 2^{n-1} bits that disagree or, equivalently, the Hamming distance between each pair of words in (32) is 2^{n-1}.

Now let us expand the list of words in (32) by also listing the bit-by-bit complements of the words in (32) and forming the set of words in (33):

$$
\begin{array}{ccccc}
s_1 & s_2 & \cdots & s_{2^n-2} & s_{2^n-1} \\
s_2 & s_3 & \cdots & s_{2^n-1} & s_1 \\
& & \vdots & & \\
s_{2^n-1} & s_1 & \cdots & s_{2^n-3} & s_{2^n-2} \\
\bar{s}_1 & \bar{s}_2 & \cdots & \bar{s}_{2^n-2} & \bar{s}_{2^n-1} \\
\bar{s}_2 & \bar{s}_3 & \cdots & \bar{s}_{2^n-1} & \bar{s}_1 \\
& & \vdots & & \\
\bar{s}_{2^n-1} & \bar{s}_1 & \cdots & \bar{s}_{2^n-3} & \bar{s}_{2^n-2}
\end{array}
\tag{33}
$$

Table 2. The 16 Code Words

Code word	Code word number	Code word	Code word number
0 0 0 0 0 0 0	0	1 1 0 0 0 1 0	8
0 0 1 1 1 0 1	1	1 0 0 0 1 0 1	9
0 1 1 1 0 1 0	2	0 0 0 1 0 1 1	10
1 1 1 0 1 0 0	3	0 0 1 0 1 1 0	11
1 1 0 1 0 0 1	4	0 1 0 1 1 0 0	12
1 0 1 0 0 1 1	5	1 0 1 1 0 0 0	13
0 1 0 0 1 1 1	6	0 1 1 0 0 0 1	14
1 0 0 1 1 1 0	7	1 1 1 1 1 1 1	15

Each pair of words in (33) all exhibit a *minimum* Hamming distance of $2^{n-1} - 1$. (Why?) Finally, we augment our list of words by two additional words, those consisting of all zeros and all ones, and obtain the code word list shown in (34).

$$
\begin{array}{ccccc}
0 & 0 & \cdots & 0 & 0 \\
s_1 & s_2 & \cdots & s_{2^n-2} & s_{2^n-1} \\
s_2 & s_3 & \cdots & s_{2^n-1} & s_1 \\
& & \vdots & & \\
s_{2^n-1} & s_1 & \cdots & s_{2^n-3} & s_{2^n-2} \\
\bar{s}_1 & \bar{s}_2 & \cdots & \bar{s}_{2^n-2} & \bar{s}_{2^n-1} \\
\bar{s}_2 & \bar{s}_3 & \cdots & \bar{s}_{2^n-1} & \bar{s}_1 \\
& & \vdots & & \\
\bar{s}_{2^n-1} & \bar{s}_1 & \cdots & \bar{s}_{2^n-3} & \bar{s}_{2^n-2} \\
1 & 1 & \cdots & 1 & 1
\end{array}
\tag{34}
$$

The minimum Hamming distance between any pair of words from (34) is still $2^{n-1} - 1$. Now for the observation that makes the set of words in (34) an error correction code. Because the Hamming distance between any pair of words in (34) is $2^{n-1} - 1$, any word may have up to $2^{n-2} - 1$ bits in error and still be recognized. The recognition is done by finding the word in (34) that has the shortest Hamming distance to the $2^n - 1$ bit received word. The "closest word" is declared to be the code word sent. Now how many code words are in the set (34)? Clearly, there are $2(2^n - 1) + 2 = 2^{n+1}$ and thus the code is capable of conveying $n + 1$ bits of information.

Let us look at an example. We pick $n = 3$ and use the 7-bit m-sequence created by $x^3 + x + 1$: 0 0 1 1 1 0 1. Forming the set of words specified in (34), we obtain the code words listed in Table 2. A pair of words taken from Table 2 have a minimum Hamming distance of $2^{3-1} - 1 = 3$ and thus the code is capable of correcting any single error. This code is then equivalent to the Hamming code (7, 4, 3). For example, let us pick code word number 5, 1 0 1 0 0 1 1, and suppose that bit 2 is in error, resulting in 1 1 1 0 0 1 1. We measure the Hamming distance of this word to all the words in Table 2. These distances are listed in Table 3. We see that 1 1 1 0 0 1 1 is closest to code word number 5, 1 0 1 0 0 1 1, and we thus conclude that code word number 5 was the code word sent and that bit 2 was garbled (inverted) by the channel.

Suppose we had created the set of code words (34) using $n = 4$ and a 15 bit m-sequence. The resulting code would have 32 code words, be able to convey 5 bits of information, and be able to correct up to $2^{4-2} - 1 = 3$ bits in error. (Try to develop a simple, automatic encoder.)

Table 3. The Hamming Distance from the Code Words to 1 1 1 0 0 1 1

Code word	Hamming distance to code word	Code word	Hamming distance to code word
0	5	8	2
1	5	9	4
2	3	10	4
3	3	11	4
4	3	12	6
5	1	13	4
6	3	14	2
7	5	15	2

12.5. The z-Channel and a Curious Result

Another simple channel model that has become popular is the "z-channel." Named for its appearance, it is an asymmetric BSC (see Figure 12.5) with either p_1 or p_2 set to zero; errors are thus unidirectional. For discussion purposes we will consider the z-channel shown in Figure 12.8. From the figure it is clear that a zero will always pass uncorrupted while a one will be corrupted with probability p. In the style of Figure 12.2, the z-channel of Figure 12.8 can be represented as in Figure 12.9.

The z-channel is a good model for many processes such as reading from some optical disks where the reading process may cause burn-through and change of a one to a zero. A set of "unidirectional" error correction codes has been developed for the z-channel but we shall not study them; instead we use the z-channel as a means of highlighting an important consideration of information theory. Our little expose is motivated, indeed guided, by a neat example developed by Mela (1961).

Consider that there are three "bins," A, B, and C. In one bin there is a one; in the other two are zeros. We are allowed three "looks" through a z-channel. The z-channel will allow us to see a zero, if present in a bin, with unity probability; it will allow us to see a one, if present, with probability $1 - p$. The symmetry of the problem allows three sampling strategies:

Strategy I: Look thrice at bin A. If the one is seen then it must be in bin A. If not detected, we assume (arbitrarily) that the one is in bin B.

Figure 12.8. A z-channel.

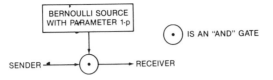

Figure 12.9. An equivalent representation of the z-channel of Figure 12.8.

Strategy II: Look twice at bin A and once at bin B. If the one is detected, then we know of course which bin it is in. If the one is not detected then we assume it is in bin C.

Strategy III: Look once at each bin. If the one is detected, we know where it is. If undetected, we assume (arbitrarily) that the one is in bin A.

We will now examine all three strategies as to their performance on three key issues: (1) probability of detecting the one, (2) the expected information gain, and (3) the probability that we will correctly identify the bin which contains the one.

12.5.1. Strategy I

The probability of detecting the one, P_d, is

$$P_d = P_A(1 - p^3) = \frac{1 - p^3}{3} \tag{35}$$

where P_A is the a priori probability that the one is in bin A $(=\frac{1}{3})$. The a posteriori probability that the one is in bin A given that it is not detected,

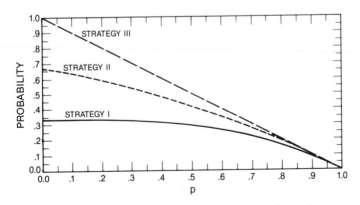

Figure 12.10. The probability that the one will be detected.

$P_{A|\bar{A}\bar{A}\bar{A}}$, is

$$P_{A|\bar{A}\bar{A}\bar{A}} = \frac{P_{\bar{A}\bar{A}\bar{A}|A} P_A}{P_{\bar{A}\bar{A}\bar{A}}} = \frac{P_{\bar{A}\bar{A}\bar{A}|A} P_A}{P_{\bar{A}\bar{A}\bar{A}|A} P_A + P_{\bar{A}\bar{A}\bar{A}|\bar{A}} P_{\bar{A}}}$$

$$= \frac{p^3}{2 + p^3} \tag{36}$$

The a posteriori probability that the one is in bin B given that it is not detected is $1/(2 + p^3)$. The a posteriori probability that the one is in bin C given that it is not detected is $1/(2 + p^3)$. The probability of correctly identifying the bin containing the one, P_k, is

$$P_k = \frac{1 - p^3}{3} + \left(1 - \frac{1 - p^3}{3}\right)\left(\frac{1}{2 + p^3}\right) = \frac{2 - p^3}{3} \tag{37}$$

The expected information gain from Strategy I is

$$H\left(\frac{1}{3}, \frac{1}{3}, \frac{1}{3}\right) - \left(1 - \frac{1 - p^3}{3}\right) H\left(\frac{p^3}{2 + p^3}, \frac{1}{2 + p^3}, \frac{1}{2 + p^3}\right) \tag{38}$$

12.5.2. Strategy II

The probability of detecting the one is

$$P_d = \frac{2 - p - p^2}{3} \tag{39}$$

The a posteriori probability that the one is in bin A given that it is not detected is

$$\frac{p^2}{p^2 + p + 1} \tag{40}$$

The a posteriori probability that the one is in bin B given that it is not detected is

$$\frac{p}{p^2 + p + 1} \tag{41}$$

The a posteriori probability that the one is in bin C given that it is not detected is

$$\frac{1}{p^2 + p + 1} \tag{42}$$

The expected information gain from Strategy II is

$$H\left(\frac{1}{3}, \frac{1}{3}, \frac{1}{3}\right) - \frac{p^2 + p + 1}{3} H\left(\frac{p^2}{p^2 + p + 1}, \frac{p}{p^2 + p + 1}, \frac{1}{p^2 + p + 1}\right) \tag{43}$$

The probability of correctly identifying the bin which contains the one is

$$\frac{3 - p - p^2}{3} \tag{44}$$

12.5.3. Strategy III

The probability of detecting the one is

$$1 - p \tag{45}$$

The a posteriori probability that the one is in bin A given that it is not detected is

$$\frac{1}{3} \tag{46}$$

The a posteriori probability that the one is in bin B given that it is not detected is

$$\frac{1}{3} \tag{47}$$

The a posteriori probability that the one is in bin C given that it is not detected is $(1/3)$. The expected information gain from Strategy III is

$$H\left(\frac{1}{3}, \frac{1}{3}, \frac{1}{3}\right) - pH\left(\frac{1}{3}, \frac{1}{3}, \frac{1}{3}\right) \tag{48}$$

where

$$H(p_1, p_2, p_3) = -\sum_i p_i \ln p_i. \tag{49}$$

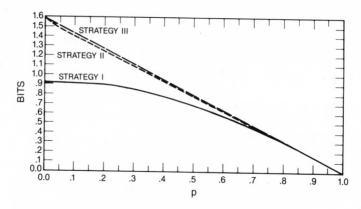

Figure 12.11. The expected information gain in bits.

The probability of correctly identifying the bin which contains the one is

$$\frac{3-2p}{3} \tag{50}$$

Problem

Prove that the equations and expressions from (39)–(50) are correct.

In Figure 12.10 we have plotted the probability that the one will be detected by Strategies I, II, and III. In Figure 12.11 we have plotted the expected information gain from Strategies I, II, and III. In Figure 12.12 we have plotted the probability of correctly identifying the bin which contains the one.

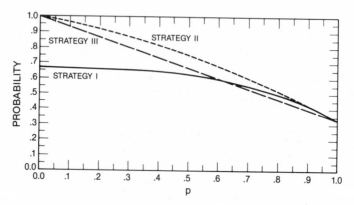

Figure 12.12. The probability of correctly identifying the bin containing the one.

Now note the apparent conundrum. The strategy that always provides the maximum expected information gain is Strategy III; but the strategy that always maximizes the probability of correctly identifying the bin which contains the one, the "bottom line," is Strategy II. The point here is an interesting one. Information means little unless properly referenced. It is easy, mathematically, to apply information theoretic procedures to many problems. The common forms encountered in information theory are friendly, convex functions. But beware! The results of such activity may be meaningless.

12.6. The Data Frame Concept

Let us assume that Party A wishes to transmit data to Party B. A qualitative characterization of the channel between them could be one of the following three descriptions: (1) errorless, (2) many errors, or (3) some errors. If the channel is errorless then we have an ideal "data pipe." We put a bit in at one end and *always* get it out uncorrupted at the other. For this felicitous situation no error control would be necessary. For a channel beset by many errors, such as encountered in very low bit rate, low power transmissions from deep space probes or operations in a jammed, electronic warfare environment, it may be necessary to employ error correction coding. But for the third case, which is typical of many extant data links, there is a very popular and practical method of sending a set of data bits. This method is the data frame technique which requests frames in error to be retransmitted. This technique of error control requires a channel in the reverse direction, that is, if Party A is sending data frames to Party B, then Party B must be able to send some data to Party A. The method is formally recognized by the International Standards Organization's (ISO's) Open Systems Interconnection (OSI) model as Layer 2; the data link layer.

The data frame concept is very simple. Data are "packaged" as a field, denoted here by D, into a frame which comprises other special fields. One of these other special fields might be an address field, denoted by A, so that the data frame will be properly distributed upon receipt. Still another of the special fields is the control field, denoted here by C. This field is used for a variety of functions including frame sequencing, designation of frame type, and so on. Still another of the special fields, and one we shall pay close attention to, is the error check field, denoted here by EC. The error check field is the cornerstone of the error control mechanism. The final special field is the repeated flag field, denoted by F. The flag field signifies the beginning and the end of a data frame. It too is a field that we shall study in depth. The generic form of a frame is diagramed in Figure 12.13.

| F | A | C | D | EC | F |

Figure 12.13. A generic data frame.

The flag field, F, alerts the receiver that a frame is either beginning or ending†. The specific flag "character" must therefore never occur as a substring of any other parts of the data frame. The usual way this is accomplished is through a process called "bit stuffing" or "bit escaping." Bit stuffing can be done in many ways. We study just one example but our chosen case is used in many contemporary data transmission architectures or protocols

1. We assume that we shall use the commonly used flag character 01111110.
2. Sender Stuffing Rule: If five sequential ones have been sent, then stuffing zero must be sent as the next bit (even if the traffic being sent has a zero as the next bit).
3. Receiver Destuffing Rule: If five sequential ones are received then the next bit is examined. If the bit is a one, the receiver expects it is receiving a flag. If the bit is a zero, the receiver concludes the zero was inserted as a stuffing zero and disregards it.

Figure 12.14 shows the results of stuffing various bitstreams. (Time moves from left to right.)

What does bit stuffing cost us and what analysis is appropriate? Clearly, bit stuffing usually increases the number of bits sent for a frame. The following Markov chain model allows us to calculate the average traffic extension or expansion factor for the case in which the frame bits can be modeled as coming from a balanced Bernoulli source. The parameter of interest, and the quantity that drives the formulation of our model, is the number of times we would see 5, 10, 15, 20, ... ones in a row. Consider then the Markov chain depicted in Figure 12.15: The chain models a balanced Bernoulli source in terms of runs of ones. The numbers within the circles are the number of ones in a row succeeding the most recent zero. The starred states, 5, 10, 15, and so on, are those states that lead to insertion of a zero stuffing bit. We can determine the fraction of time that the model

† Some communication protocols also allow the flag to be used during periods of no traffic as an idling or "interframe" fill character.

Bitstream Before Stuffing	Bitstream After Stuffing
0 1 0 1 1 1 1 1 0 0 0 1 1 1 1 0 0	0 1 0 1 1 1 1 1 0 0 0 0 1 1 1 1 0 0
0 1 1 1 1 1 1 0 0 0	0 1 1 1 1 1 0 1 0 0 0
0 1 1 1 1 1 1 1 1 1 1 0	0 1 1 1 1 1 0 1 1 1 1 1 0 0

Figure 12.14. Some examples of bit stuffing.

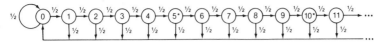

Figure 12.15. Markov chain modeling runs of ones.

will be in the starred states by computing the steady-state probabilities and summing those of the starred states.

Problem

Set up the transition matrix for the chain of Figure 12.15 and show that the steady-state probability of state i is $\frac{1}{2} \cdot 2^{-i}$.

Summing the steady-state probabilities of the starred states, P_s^*, we find

$$P_s^* = \frac{1}{2}(2^{-5} + 2^{-10} + 2^{-15} + \cdots) = \frac{1}{62} \tag{51}$$

Thus, on the average, for a balanced Bernoulli source, the sender will have to insert one zero stuffing bit for every 62 bits of the unstuffed traffic. The traffic will thus be expanded by the factor 63/62 and the ratio of zeros to ones will be 32/31.

Problems

1. The traffic after bit stuffing can be modeled by the Markov chain depicted in Figure 12.16. Again, the original traffic is taken from a balanced Bernoulli source and the numbers within the circles are the numbers of ones in a row succeeding the most recent zero. Derive the power spectral density of the bit-stuffed traffic stream.

2. Consider that the unstuffed traffic is not generated by a balanced Bernoulli source but rather by the Markov chain in Figure 12.17. States 0_1 and 0_2 both produce zeros; states I_1 and I_2 both produce ones. What we have is a chain that weakly couples three subchains. A source that might be approximated by such a chain is a linear delta modulator. Solve for the probability of bit stuffing for this source and the traffic expansion factor. Is it better or worse than for a balanced Bernoulli source? How could this source be precoded before bit stuffing to make the traffic expansion factor smaller?

Figure 12.16. Markov chain modeling runs of ones in the bit-stuffed traffic.

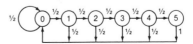

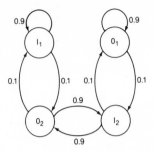

Figure 12.17. A Markov chain.

12.6.1. The Error Check Field and the CRC

Consider that we have a sequence of (message) bits $m_1 m_2 m_3 \cdots m_n$ and that we feed these bits to the machine in Figure 12.18 which consists of an EXCLUSIVE-OR logic function operating with a single shift register stage. Before the bits are passed to the machine, the shift register stage is preset to zero. What does the end state of the shift register tell us after the bits have been processed? It can convey only one of two messages, of course, and they are as follows:

1. The shift register's content is zero if there were an even number of ones in the bit sequence.
2. The shift register's content is one if there were an odd number of ones in the bit sequence.

The end content of the shift register is then the parity of the bit sequence. But it can be viewed in another way. Let us associate the following polynomial with the message sequence:

$$m_1 x^{n-1} + m_2 x^{n-2} + \cdots + m_{n-1} x + m_n \tag{52}$$

We now ask when the polynomial (52) has $x + 1$ as a factor, that is, leaves a remainder of zero when divided by $x + 1$. The answer, of course, from that we have learned in prior chapters is that (52) has $x + 1$ as a factor if and only if $\sum_{i=1}^{n} m_i \equiv 0$ modulo 2.

Problem

Prove the above assertion.

Figure 12.18. Running sum shift register stage.

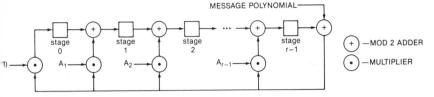

Figure 12.19. Sequential machine that divides the message polynomial by (53).

We can draw a thread between the machine in Figure 12.18 and the division of the message polynomial (52) by the polynomial $x + 1$. The relationship, of course, is that the end state of the shift register is the remainder. Note that dividing a message polynomial by $x + 1$ discloses any single bit errors. It is the same thing as a simple parity check. We can also divide the message polynomial by polynomials more complicated than $x + 1$. The machine depicted in Figure 12.19 divides a message polynomial by the polynomial

$$x^r + A_{r-1}x^{r-1} + \cdots + A_2x^2 + A_1x + 1 \tag{53}$$

The remainder is given by the contents of stages $r - 1, r - 2, \ldots, 2, 1, 0$.

Problem

Think about what is involved in division of modulo 2 polynomials modulo another mod 2 polynomial and explain, in your own words, why the circuit of Figure 12.19 performs division.

Let us look at a simple example. Suppose we wish to divide a message polynomial by $x^3 + x + 1$. Figure 12.20 shows the division circuit. Let us examine the division of the message 01101000. Table 4 gives the state of the stages of the division shift register which is initialized to all zeros at $t = 0$. We see that the remainder is 011. (The contents of stages 2, 1, and 0, respectively.)

The remainder can be used as a check for errors induced by the channel. When used for this purpose it is often called the Cyclic Redundancy Check or CRC and is emplaced in the Error Check (EC) field of the generic data

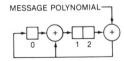

Figure 12.20. Sequential machine that divides the message polynomial by $x^3 + x + 1$.

Table 4. State of the Shift Register Stages
Over Time

Time	Message input	Stages 0	1	2
0	—	0	0	0
1	0	0	0	0
2	1	1	1	0
3	1	1	0	1
4	0	1	0	0
5	1	1	0	0
6	0	0	1	0
7	0	0	0	1
8	0	1	1	0

frame depicted in Figure 12.13. Our original message followed by the CRC based on the dividing polynomial $x^3 + x + 1$ is then 01101000011. If at the receiver we pass the received message plus its CRC through a similar division circuit we end up with all zeros in the shift register as shown in Table 5. Checking by direct division, using as a shorthand only the coefficients of the polynomials in question:

```
          1111001
1011|01101000011
     1011
      1100
      1011
       1110
       1011
        1010
        1011
         1011
         1011   Division is exact—no remainder
```

The theory of error *detecting* codes can be used to choose suitable division polynomials to *detect* a large class of errors. A properly chosen polynomial of degree d will detect all error bursts of length d or less. Detection, of course, is accomplished by examining the contents of the division shift register after the message and CRC have been divided. An error is declared detected if and only if the contents of the shift register are not *all* zeros.

Patel (1971) has worked out a method of implementing the CRC division in parallel steps versus serial shifts. This contribution is valuable for those circuits passing data frames at extremely high speed. Patel's

Table 5. State of the CRC Decoder Shift Register Stages

Time	Augmented message input	Stages 0 1 2
0	—	0 0 0
1	0	0 0 0
2	1	1 1 0
3	1	1 0 1
4	0	1 0 0
5	1	1 0 0
6	0	0 1 0
7	0	0 0 1
8	0	1 1 0
9	0	0 1 1
10	1	0 0 1
11	1	0 0 0

construction is straightforward and, through its use of matrices, leads to a natural schematic for the circuit implementer.

We write the division polynomial as

$$A(x) = x^r + A_{r-1}x^{r-1} + \cdots + A_1 x + A_0 \qquad (54)$$

We let the contents of the shift register at time t be denoted by the vector

$$\mathbf{X}_t = [x_0(t) \quad x_1(t) \quad \cdots \quad x_{r-1}(t)] \qquad (55)$$

We now construct the following companion matrix, T, to the division polynomial

$$T = \left(\begin{array}{c|c} 0 & I_{r-1} \\ \hline A_0 \quad A_1 & \cdots \quad A_{r-1} \end{array} \right] \qquad (56)$$

We let m_t represent the message bit at time t. For the division shift register, we can immediately write

$$\mathbf{X}_{t+1}^T = \mathbf{X}_t^T T + m_t \mathbf{A}^T \qquad (57)$$

where $\mathbf{A}^T$ is defined to be the vector $[A_0, A_1, \ldots, A_{r-1}]$ and $+$ is taken here to be the component-by-component modulo 2 sum.

Problem

Show that after P shift times the contents of the CRC shift register are given by

$$\mathbf{X}_{t+p}^T = \mathbf{X}_t^T T^p + m_t \mathbf{A}^T T^{p-1} + m_{t+1} \mathbf{A}^T T^{p-2} + \cdots + m_{t+p-1} \qquad (58)$$

where T^r is the rth power of T.

Now let us represent a segment of our message input by

$$\mathbf{M}_t^T = [m_{t+p-1} \quad m_{t+p-2} \quad \cdots \quad m_{t+1} \quad m_t] \tag{59}$$

and form the partitioned matrix D according to

$$D = \begin{bmatrix} \mathbf{A} \\ \hline \mathbf{A}T \\ \hline \mathbf{A}T^2 \\ \hline \mathbf{A}T^2 \\ \hline \vdots \\ \hline \mathbf{A}T^{p-1} \end{bmatrix} \tag{60}$$

Problem

Show that (58) can be written

$$\mathbf{X}_{t+p}^T = \mathbf{X}_t^T T^p + \mathbf{M}_t^T D \tag{61}$$

What Equation (61) allows is the divisional computation, in one parallel processing step, of p message bits. It is equivalent to serially shifting in p bits.

Problems

1. Show that

$$T^p = \left[\begin{array}{c|c} 0 & I_{r-p} \\ \hline & D \end{array}\right] \tag{62}$$

2. Find an efficient way to generate the entries of the partitioned matrix (60).

Now we are ready to design our parallel CRC circuit. We partition the vector $\mathbf{X}_t$ as follows:

$$\mathbf{X}_t^T = [X_t^T(1) \vdots X_t^T(2)] \tag{63}$$

where

$$\begin{aligned} X_t^T(1) &= [x_0(t) \quad x_1(t) \quad \cdots \quad x_{r-p-1}(t)] \\ X_t^T(2) &= [x_{r-p}(t) \quad \cdots \quad x_{r-1}(t)] \end{aligned} \tag{64}$$

Problem

Use (62) and (63) to show that (61) can be rewritten

$$\mathbf{X}_{t+p}^T = \mathbf{X}_t^T(1)[0 \mid I_{r-p}] + (\mathbf{X}_t^T(2) + \mathbf{M}_t^T)D \tag{65}$$

Equation (65) serves as a "schematic" for the CRC division in parallel.

As a simple example, let us consider our previous example run at twice the serial rate. For this case then, $p = 2$, $r = 3$, and

$$\begin{aligned} \mathbf{X}_t(1) &= x_0(t) \\ \mathbf{X}_t^T(2) &= [x_1(t) \quad x_2(t)] \end{aligned} \tag{66}$$

and

$$\mathbf{M}_t^T = [m_{t+1}, m_t] \tag{67}$$

and

$$\mathbf{A}^T = [1 \quad 1 \quad 0] \tag{68}$$

and

$$\mathbf{A}^T T = [1 \quad 1 \quad 0] \begin{bmatrix} 0 & 1 & 0 \\ 0 & 0 & 1 \\ 1 & 1 & 0 \end{bmatrix} = [0 \quad 1 \quad 1] \tag{69}$$

and then

$$D = \begin{bmatrix} \mathbf{A} \\ \hline \mathbf{A}T \end{bmatrix} = \begin{bmatrix} 1 & 1 & 0 \\ 0 & 1 & 1 \end{bmatrix} \tag{70}$$

Using (65) we find that

$$\mathbf{X}_{t+2}^T = [x_0(t)][0 \quad 0 \quad 1] + ([x_1(t) \quad x_2(t)] + [m_{t+1} \quad m_t]) \begin{bmatrix} 1 & 1 & 0 \\ 0 & 1 & 1 \end{bmatrix}$$

$$= [0 \quad 0 \quad x_0(t)] + [x_1(t) + m_{t+1} \quad x_2(t) + m_t] \begin{bmatrix} 1 & 1 & 0 \\ 0 & 1 & 1 \end{bmatrix}$$

$$= [0 \quad 0 \quad x_0(t)] + [x_1(t) + m_{t+1} \quad x_1(t) + x_2(t) + m_t + m_{t+1} \quad x_2(t) + m_t]$$

$$= [x_1(t) + m_{t+1} \quad x_1(t) + x_2(t) + m_t + m_{t+1} \quad x_0(t) + x_2(t) + m_t] \tag{71}$$

Figure 12.21 depicts the parallel circuit specified by (71).

Problems

1. Show that the circuit shown in Figure 12.21 does indeed perform the CRC division by stepping through it with the previously used example traffic

2. Generate a circuit according to (65) for processing traffic a byte at a time using the SDLC protocol CRC division polynomial $x^{16} + x^{12} + x^5 + 1$. Do the same for the (now defunct) Autodin II system polynomial $x^{32} + x^{26} + x^{23} + x^{22} + x^{16} + x^{12} + x^{11} + x^{10} + x^8 + x^7 + x^5 + x^4 + x^2 + x + 1$.

12.7. A Curious Problem

An interesting and potentially serious problem that can occur is the creation, by a single error, of false frames. But, you might immediately object, how can this happen? Aren't we guaranteed that the CRC will catch all single bit errors? Indeed, the CRC will do so if, and here's the rub, we know where the CRC is. Many modern data packet networks allow the frames to be of almost arbitrary length. We find the CRC by noting the flag delimiting the end of the frame. But, if a false flag should be created by an error, then the wrong bits will be interpreted as the CRC bits, and if perchance they should be the correct CRC bits for whatever preceded them, then the bogus frame will be accepted. Funk (1982) has raised this problem and provided a lively discussion and ideas for strengthening the data architectures against this rather unexpected problem. We will follow his development.

For motivation, assume we have transmitted the following octet 01011110 and that bit number three is received in error resulting in 01111110. This, of course, is the bit pattern we use for our flag. What is the probability of this happening in a BSC with error probability p? Funk proceeds by considering the six sequences that can serve as "seeds" for flags created by a single error. These sequences are listed in Table 6 along with their respective probabilities. The zero that needs to be inverted to produce a bogus flag is starred.

The probabilities are computed based on the bit stuffing of traffic modeled as coming from a balanced Bernoulli source. For clarity, let us

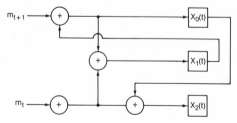

Figure 12.21. Parallel implementation of (71).

Table 6. The Six False Flag "Seeds"

Sequence	Probability of occurrence (for a balanced Bernoulli source)
0 0 * 1 1 1 1 1 0	1/126
0 1 0 * 1 1 1 1 0	1/252
0 1 1 0 * 1 1 1 0	1/252
0 1 1 1 0 * 1 1 0	1/252
0 1 1 1 1 0 * 1 0	1/252
0 1 1 1 1 1 0 * 0	1/126

calculate the probability of occurrence, at a randomly chosen position, of the second seed 01011110. The probability of the first zero is $\frac{32}{63}$. (Why?) The probability of the next seven bits being 1011110 is 2^{-7}. Thus, the probability that eight contiguous bits are 01011110 is $\frac{32}{63} \cdot 2^{-7} = \frac{1}{252}$.

Problem

Show that the probability that one of the eight bit patterns of Table 6 will occur at a randomly chosen starting point is $\frac{2}{63}$.

The probability that we will create a false flag is then, at least,

$$\frac{2}{63} \cdot p(1-p)^7 \approx 0.0317p \quad \text{for } p \ll \tfrac{1}{2} \tag{72}$$

Now if the r bits preceding the false flag happen to be the correct CRC for the bits before them, then we have the creation of a false frame. The probability that the bits integrated to be the CRC bits will be a correct CRC is 2^{-r}. Thus, the probability that a false frame, P_{FF}, will be created within a frame of n bits is

$$P_{FF} = 2^{-r}(1 - (1 - \tfrac{2}{63}p)^{n-16}) \approx \frac{(n-32)p}{2^{r-1} \cdot 63} \tag{73}$$

(We have restricted the false frame to be at least 32 bits long.)

Problems

1. How many N-bit frames would have to be passed through a BSC with parameter p before the probability of a false frame's occurrence would be greater than one in a million?

2. In what other ways might false frames be created?

12.8. Estimation of Channel Parameters

A low capacity backward or reverse channel can be used for a number of purposes; see Hershey et al. (1984). One important use is for error control. Frames in error can be identified and retransmission requested. Forward channel error estimation is also possible.

Consider that we have a high data rate channel in one direction, say from a ground station to a satellite and a low capacity channel from the satellite to the ground station as depicted in Figure 12.22. The ground station to satellite channel is assumed to be a binary symmetric channel (BSC) with error probability p. The reverse, low capacity, channel is assumed to be error-free. Let us assume that the data to be transmitted from the ground station to the satellite need not be flawless but should be kept at as high a quality as reasonably possible; for this reason an error correction code is applied. The selection of the particular code depends on the BSC parameter p; once p has been estimated, $\hat{p}$, $p = \hat{p}$ is selected. But suppose $\hat{p}$ is actually a sampling of a slowly varying random process. When $\hat{p}$ becomes very small, as it may during nighttime periods when there are fewer noise sources within the satellite's view, it might behoove us to reduce the error correction overhead. Alternately, when $\hat{p}$ is significantly larger it is imperative that we strengthen the error correction capability. Recalling that the capacity of a BSC depends solely on p, what we are really doing is adapting our information rate to the changing channel capacity.

Our method of estimating p is through results of a computation performed on successive "stretches" of the high-rate traffic. A "stretch" of

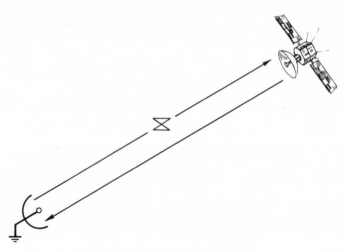

Figure 12.22. An asymmetric channel.

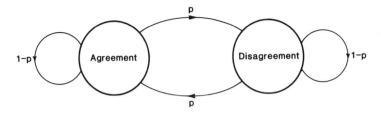

Figure 12.23. The Markov model.

traffic, as opposed to the more restrictive term "frame," means simply a time-continuous series of n bits.

We consider a stretch of traffic. We randomly select exactly m, $m \leq n$, bits from the stretch and compute their sum modulo 2. This single bit sum is sent via the low-speed, errorless, feedback channel and compared with the sum performed on the same m bits of data prior to its transmission over the high-speed channel. If we perform this experiment S times and observe d differences upon comparison of the sums then we can easily determine $\hat{p}$.

We begin with the simple Markov model depicted in Figure 12.23. An odd number of errors (one, three, five, etc.) will result in a disagreement between the two modulo 2 sums upon comparison.

Problem

Show that the probability, P, that we observe d differences in S trials is

$$P = \binom{S}{d}[1 - (1 - 2p)^m]^d[1 + (1 - 2p)^m]^{S-d} \qquad (74)$$

We find $p = \hat{p}$ in order to maximize the likelihood of P; that is, we solve

$$\left.\frac{dP}{dp}\right|_{p=\hat{p}} = 0 \qquad (75)$$

and find the following MLE for p:

$$\hat{p} = \frac{1}{2}\left[1 - \left(1 - 2\frac{d}{S}\right)^{1/m}\right] \qquad (76)$$

The form (76) is exact. Consider the expression

$$y = 1 - (1 - x)^{1/m} \tag{77}$$

where $|x| < 1$ and $m \gg 1$. We rewrite (77) and expand as follows

$$y = 1 - (1 - x)^{1/m} = 1 - e^{(1/m)\ln(1-x)}$$

$$= 1 - \exp\left(\frac{1}{m} \sum_{i=1}^{\infty} \frac{x^i}{i}\right)$$

$$= 1 - \prod_{i=1}^{\infty} e^{-x^i/mi}$$

$$= 1 - \prod_{i=1}^{\infty} \left[1 + \sum_{j=1}^{\infty} \left(-\frac{x^i}{mi}\right)^j \frac{1}{j!}\right] \tag{78}$$

Since $m \gg 1$,

$$y \approx 1 - \prod_{i=1}^{\infty} \left(1 - \frac{x^i}{mi}\right)$$

$$\approx \frac{1}{m} \sum_{i=1}^{\infty} \frac{x^i}{i} = \frac{1}{m} \ln \frac{1}{1 - x} \tag{79}$$

Thus, for large m, we may successfully approximate $\hat{p}$ as

$$\hat{p} = \frac{1}{2m} \ln \frac{1}{1 - 2(d/S)} \tag{80}$$

From (84) we note that $\hat{p}$ has desirable scaling properties. First $\hat{p}$ depends only on m and the ratio of d to S. Second, $\hat{p}$ is easily scaled for different $m(m \gg 1)$.

Problems

1. Attempt to calculate the bias in the estimator given in (76).
2. Suppose we are using a $(7, 4, 3)$ Hamming code over a BSC with parameter p and we note that code words are corrected on reception with probability P. Derive an estimator $\hat{p}$ for p. Attempt to determine its bias.

References

Brayer, K. and O. Cardinale (1967), Evaluation of Error Correction Block Encoding for High-Speed HF Data, *IEEE Transactions on Communications Technology*, Vol. 15, pp. 371–382.

Funk, G. (1982), Message Error Detecting Properties of HDLC Protocols, *IEEE Transactions on Communications*, Vol. 30, pp. 252–257.

Hershey, J., H. Gates, and R. Yarlagadda (1984), The Asymmetric Capacity Communications Channel and Divestiture, in Proceedings of the 1984 Military Communications Conference (MILCOM).

Hamming, R. (1950), Error Detecting and Error Correcting Codes, *The Bell System Technical Journal*, Vol. 29, pp. 147–160.

Mela, D. (1961), Information Theory and Search Theory as Special Cases of Decision Theory, *Operations Research*, Vol. 9, pp. 907–909.

Patel, A. (1971), A Multi-channel CRC Register, *Spring Joint Computer Conference*, Vol. 38, pp. 11–14.

13

Space Division Connecting Networks

13.1. Introduction

If nothing ever changed, this would be a simpler world; not a particularly interesting world but a simpler one. Consider that there are n nodes. These nodes may represent people, computers, ports, or anything that wishes to send and receive information. If node i always wants to communicate with node j, node k with l, and so on, then our data engineering task is simple. We connect i with j, k with l, and so on for the other pairs. But suppose i tired of j and k with l and that i wished to discourse with k and j with l. If so, we would have to disconnect these nodes and then reconnect them. This is motivation for considering the network shown in Figure 13.1.

The general connecting network allows any idle pair of nodes to be connected together; and, as its name implies, it is quite general. We study a slightly less general class of networks, one that can be used as a building block of the Figure 13.1 network. This restricted connecting network is shown in Figure 13.2. This connecting network is termed a permutation network and it effects a permutation of inputs to outputs. Such networks have been designed with various techniques and technologies. Some modern

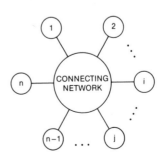

Figure 13.1. The general connecting network.

437

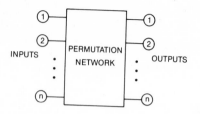

Figure 13.2. A permutation network.

electronic switches accomplish the task by frequency of time division multiplex busses; others by purely combinatorial or "space division" methods. It is the purely combinatorial switches that we discuss. These are the switches that are needed for the very highest data rate switching.

13.2. Complete Permutation Networks

A complete permutation network must possess an awesome power. It must be capable of realizing any one of $n!$ permutations. As we have seen, $n!$ grows extremely rapidly and it should therefore come as no surprise that a permutation network for even moderately sized n is quite complex.

13.2.1. The Crossbar

The simplest, essentially brute force approach to building a permutation network is a crossbar switch shown in Figure 13.3. As an example, consider that we wish to effect the permutation $\left(\begin{smallmatrix} 1 & 2 & 3 & 4 & 5 & 6 & 7 & 8 \\ 6 & 1 & 3 & 7 & 4 & 8 & 2 & 5 \end{smallmatrix}\right)$. We would set our crossbar as shown in Figure 13.4.

The problem with crossbars is that the number of switches required is n^2. For example, if $n = 10,000$, a full crossbar would require one hundred million switches. The reliability, maintenance, cost, and space problems of such a network would indeed be intimidating and thus we are encouraged to begin the search for other, more equipment efficient networks to perform permutations.

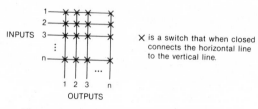

Figure 13.3. The General $n \times n$ crossbar switch.

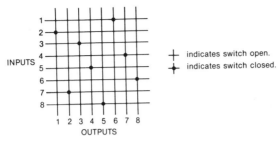

INPUTS

┼ indicates switch open.
╈ indicates switch closed.

Figure 13.4. The crossbar set to implement a particular permutation.

13.3. The Clos Network

Clos (1953) presented a connection network that is capable of performing any permutation of inputs to outputs. Although these networks are not generally practical, they are of interest for two reasons. First, historically, they showed that what required n^2 switches in the n-line-input/n-line-output square crossbar could, for sufficiently large n, be done with fewer switching elements. Second, it is easy to prove that a Clos network can perform all permutations and, although this book is not theory oriented, it is instructive to see an occasional proof especially when a fairly powerful result obtains from a few, easily followed, machinations.

We need a bit of notation before we construct our Clos network. Figure 13.5 shows a generalized crossbar (often referred to as a rectangular network) and its symbol. The $N \times M$ rectangular network has N inputs and M outputs. There are three possible cases: (1) $N > M$ in which the M output lines may be any ordering of M of the N input lines—this function is often called "concentration," (2) $N = M$ in which the M output lines are any permutation of the N input lines—this is the square crossbar, introduced previously, and (3) $N < M$ in which the N inputs may be permuted to any N-member subset of the M output lines.

We are now ready to consider the three-stage Clos network shown in Figure 13.6. The network shown in Figure 13.6 has RN inputs and RN outputs. It is capable of performing all $(RN)!$ permutations without rearranging any paths in use if $M \geq 2N - 1$. To see this we follow Clos' or

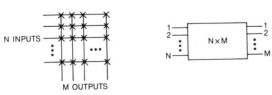

Figure 13.5. The $N \times M$ rectangular network and its symbol.

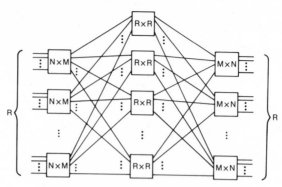

Figure 13.6. A three-stage Clos network.

Marcus' (1977) argument. Assume that we wish to assign one of the unassigned inputs to an unassigned output, that is, connect an idle input to an idle output. Let the idle input be on the rth rectangular input network. The rth input network (as are all the R input networks) has a single selectable output to each of the M $R \times R$ square crossbars in the network's middle. How many of the M $R \times R$ crossbars can not be reached by the idle input? The "worst" case occurs when the idle input is the only idle input to its $N \times M$ input network; for then $N - 1$ of the M $R \times R$ crossbars will be unreachable. By a symmetric argument, the idle output may be denied connection to at most $N - 1$ of the $R \times R$ crossbars. Thus, in order for there to be at least one middle crossbar that both the idle input *and* the idle output can reach, we must have $2(N - 1) < M$ or

$$M \geq 2N - 1 \tag{1}$$

As NR becomes increasingly large, the center crossbars of the Clos networks can be recursively replaced by three-stage Clos networks and, for sufficiently large NR, this recursion is efficient because it reduces the number of switches required by the total network. For small NR, the Clos network requires more switches than an $NR \times NR$ crossbar; however, as NR increases, the Clos network eventually exhibits a savings in number of switches over the full crossbar.

As an example of a three-stage Clos network that requires less switches than the square crossbars, we consider $NR = 100$. Keister (1967) has determined that the Clos network for $NR = 100$, shown in Figure 13.7, is optimum. A crossbar would require $100^2 = 10,000$ switches to do the same job as the Clos network of Figure 13.7. Let us now count the number of switches used by the Clos network of our example. First, there are 20 5×9 and 20 9×5 rectangular networks. Each requires 45 switches for a total of 1800 switches. Second, there are 9 middle 20×20 crossbars, each requiring

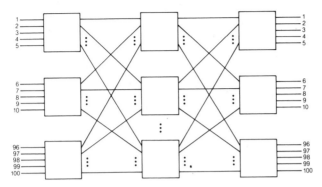

Figure 13.7. The optimum three-stage strictly nonblocking Clos network for $N = 100$.

400 switches for a total of 3600. Adding, we find that the Clos network in question requires $1800 + 3600 = 5400$ switches or only 54% of the number required by a 100×100 crossbar.

As previously mentioned, the Clos network can be recursively extended to beyond three stages. Through an exhaustive search, Keister (1967) has derived the architecture for optimum Clos networks for inputs numbering up to 80,080. As an example of an optimum Clos network that possesses more than three stages, consider Keister's find that for $N = 500$, the five-stage Clos network shown in Figure 13.8 is optimum. This particular network uses 53,200 switches or only 21.28% of the number of switches which would be required by a single 500×500 crossbar. To see how the network of Figure 13.8 uses only 53,200 switches, we count as follows:

1. There are 50 10×19 switches in stage 1. This requires 9500 switches.
2. There are similarly 50 19×10 switches in stage 2. This also requires 9500 switches.

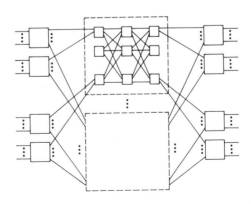

Figure 13.8. The optimum Clos network for $N = 500$.

3. There are 19 center switches. Each is a three-stage Clos network. Each of these networks has 10 5 × 9 rectangular input switches for a total of 450 switches. Each center network similarly requires 10 9 × 5 rectangular output switches for a total of 450 switches. Each center network has 9 10 × 10 square crossbars at *its* center for another 900 switches. Thus, each center three-stage Clos network requires 450 + 900 + 450 or 1800 switches. The 19 center networks thus require a total of 19 × 1800 = 34,200 switches.

Adding the total switches in 1, 2, and 3 we obtain 9500 + 9500 + 34,200 = 53,200 switches.

Problem

Consider the Clos network shown in Figure 13.9.

(a) What are the values for *N*, *M*, and *R* of the problem network?

(b) We have shown that as long as $M \geq 2N - 1$ the Clos network will be able to effect all possible permutations or connections of input lines to output lines. What we showed precisely was that if $M \geq 2N - 1$, the Clos network is strictly nonblocking and by this we mean that any series of permutations may be effected without rerouting those connections that support fixed points of the permutations. To see this, set up the problem network to realize the permutation $\left(\begin{smallmatrix} 1 & 2 & 3 & 4 & 5 & 6 \\ 4 & 2 & 3 & 6 & 1 & 5 \end{smallmatrix}\right)$. Having done this, set up the permutation $\left(\begin{smallmatrix} 1 & 2 & 3 & 4 & 5 & 6 \\ 5 & 2 & 4 & 6 & 1 & 3 \end{smallmatrix}\right)$ *without disturbing* the previously established paths $2 \to 2$, $4 \to 6$, and $5 \to 1$. In telephony theory, these fixed points of the two permutations represent "calls in progress." The strictly nonblocking Clos network $(M \geq 2N - 1)$ does not require the rerouting of calls in progress.

Keister's search has further shown that between 2736 and 52,560 inputs, the optimum Clos network comprises seven stages. Table 1, due to Keister and published by Kappel (1967), lists near optimum switch sizes for Clos networks for inputs in this range.

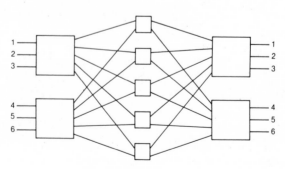

Figure 13.9. A Clos network.

Table 1. Optimum Structure of Selected Clos Networks Due to Keister

N	Rectangular switches required for stages				Approximate total of millions of switches required
	1 and 7	2 and 6	3 and 5	4	
10,000	20×39	10×19	5×9	10×10	2.86
20,020	22×43	13×25	5×9	14×14	6.93
30,030	26×51	11×21	7×13	15×15	11.6
40,222	26×51	13×25	7×13	17×17	16.8
49,980	30×59	14×27	7×13	17×17	22.1

13.4. A Rearrangeable Connecting Network and Paull's Algorithm

In the previous section, we learned that the three-stage Clos network of Figure 13.6 was strictly nonblocking if $M \geqslant 2N - 1$. The number of crosspoint switches required by this network is

$$2MNR + MR^2 \tag{2}$$

If we pick $M = 2N - 1$, we use

$$R(4N^2 + 2NR - 2N - R) \tag{3}$$

crosspoint switches. We are now going to look at a Clos-like network that is rearrangeable, that is, it will be capable of performing all the $(RN)!$ permutations doable by the network of Figure 13.6 but differs from a strictly nonblocking network. The difference appears when we require a new permutation to be implemented. The difference affects the routing of the fixed points of the permutations, those elements of the two permutations that are the same. For example, consider the following two permutations, A and B, of four elements

$$A = \begin{pmatrix} 1 & 2 & 3 & 4 \\ 2 & 4 & 1 & 3 \end{pmatrix} \quad \text{and} \quad B = \begin{pmatrix} 1 & 2 & 3 & 4 \\ 3 & 4 & 1 & 2 \end{pmatrix} \tag{4}$$

The fixed points of the two permutations above are the mappings $2 \to 4$ and $3 \to 1$. In a strictly nonblocking network, permutation B can always be effected following the setup of permutation A without disturbing the routings set up to accomplish $2 \to 4$ and $3 \to 1$. (See the problem near the end of the previous section.) In a rearrangeable network, all permutations are still possible but the paths set up for fixed points may have to be rerouted. In telephony, one says that calls in progress may have to be rerouted.

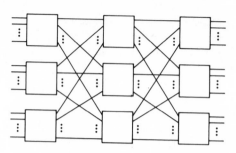

Figure 13.10. A three-stage rearrangeable network ($R \geq N$).

The rearrangeable network we consider is shown in Figure 13.10. This network is rearrangeable if and only if $R \geq N$. The proof of this statement resides in the Slepian–Duguid Theorem for which the interested reader is referred to Beneš (1962). The network of Figure 13.10 requires a total of

$$R(2N^2 + R^2) \tag{5}$$

crosspoint switches.

Paull's Algorithm (Paull, 1962; Melas, 1983) for rearranging the paths in Figure 13.10, in order to go from one permutation to another, usually requires the rearrangement of calls in progress. Paull's Algorithm proceeds by solving the problem of connecting one idle input to one idle output at a time. We develop Paull's Algorithm by example.

Assume that we have the rearrangeable network shown in Figure 13.11. The middle switches are denoted by A, B, and C. They are each, in effect, 3×3 square crossbars. Let us now assume that the example network is performing the following partial permutation, P_1,

$$P_1 = \begin{pmatrix} 1 & 2 & 3 & 4 & 5 & 6 & 7 & 8 & 9 \\ 4 & 1 & 9 & 5 & - & 8 & 7 & - & 2 \end{pmatrix} \tag{6}$$

Let us assume further that the path routing to accomplish P_1 is as shown in Figure 13.12. We note that the middle switch, A, is completely filled, the middle switch, B, has one idle pair, as does switch C.

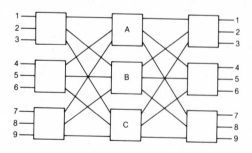

Figure 13.11. Example of a rearrangeable network with $N = 3$ and $R = 3$.

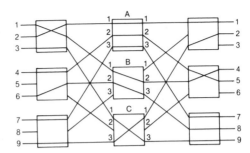

Figure 13.12. The routing used to realize P_1.

Now let us suppose we wish to add a path from input 5 to output 3, that is, we wish to perform the partial permutation

$$P_2 = \begin{pmatrix} 1 & 2 & 3 & 4 & 5 & 6 & 7 & 8 & 9 \\ 4 & 1 & 9 & 5 & 3 & 8 & 7 & - & 2 \end{pmatrix} \tag{7}$$

How can we do this? Let us first look at some constraints. Note that input 5 enters the crossbar *all* of whose output lines go to input line number 2 of the middle switches. In fact, the outputs from each input switch (or crossbar) are directed to the same input line number of the middle switches. For example, consider the inputs 4, 5, and 6 to the middle input switch. This switch can access input line 2 of middle switch A, input line 2 of middle switch B, and input line 2 of middle switch C. Similarly, all middle switch (or crossbar) outputs are directed to the same input number lines of the output switches.

Thus, to connect input 5 to output 3, it is necessary to connect input 2 to output 1 of one of the middle switches. On examining all the middle switches, we are unable to find one that possesses an uncommitted output line. We therefore must rearrange some of the existing paths or "calls in progress."

The first step of Paull's Algorithm is to pick two middle switches. Jajszczyk and Rajski (1980) have examined the problem of choosing the two middle switches. Aside from the obvious requirement that the two switches possess *between them* the idle input and output lines, the candidate switches should be those that are least used. In our case, the choice is clear. The candidate switches must be switches B and C. It helps bookkeeping if we associate with each middle switch a set of what has been called "dominoes" in the literature. A domino is a double number written as number 1/number 2 and it denotes a used path through any particular middle switch. Thus, the set of dominoes for switch A is 1/1, 2/2, 3/3. The set for switch B is 1/2, 2/3. The set for switch C is 1/3, 3/1. We call a domino a "blocking domino" if its input or output is required by the new

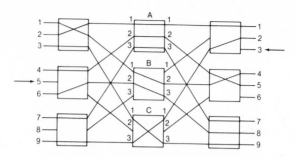

Figure 13.13. The need to connect input 5 to output 3 requires the middle switch domino 2/1.

path. The new path in our example requires the domino 2/1 and thus the blocking domino for switch B is 2/3 and the blocking domino for switch C is 3/1.

The second step in Paull's Algorithm is to move or transfer the blocking domino from one switch to another. This allows the new connection to be made but displaces a previously established connection—a "call in progress." (In equipment, a bypass line is sometimes added to each middle switch for the displaced call.) The displaced domino is then transferred to the other switch and so on. The process will eventually terminate with all paths established.

We now perform Paull's Algorithm on our example and use a separate diagram for each step for maximum clarity.

Step 0:   The Initial State of the Network (also Figure 13.12). We wish to connect input 5 to output 3. This requires the middle switch domino 2/1 as shown in Figure 13.13.

Step 1:   We choose switches B and C as discussed previously.

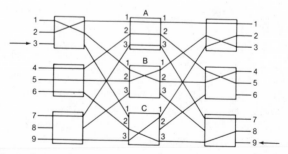

Figure 13.14. Input 5 connected to output 3 causing disconnect of path from input 3 to output 9.

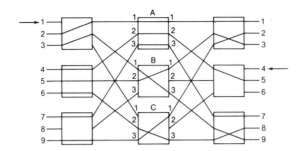

Figure 13.15. Input 3 reconnected to output 9 resulting in disconnect of path from input 1 to output 4.

Step 2:   We start (arbitrarily) with switch *B*. Switch *B*'s blocking domino is 2/3. The blocking is due to the fact that the *input* line to switch *B* is in use. We transfer the domino to switch *C* and connect our new pair through switch *B* as shown in Figure 13.14.

Note that we have disconnected the path from input 3 to output 9 (see arrows). We must now reestablish this path which requires the middle switch domino 1/3. We transfer this domino up to switch *B* as portrayed in Figure 13.15. Note that we have restored the path from input 3 to output 9 but have disconnected input 1 to output 4 (see arrows). To connect input 1 to input 4 requires the middle switch domino 1/2 which we transfer to switch *C*.

All connections required by P_2 are now established and the algorithm has terminated (see Figure 13.16). Of the seven paths originally set up for P_1, three were rearranged in forming P_2.

Problems

1. Repeat the above example but start with switch *C* in Step 2 instead of switch *B*.

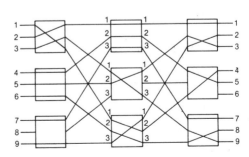

Figure 13.16. Termination of Paull's Algorithm.

2. Connect input line 1 to output line 8 in the network of Figure 13.17 Paull's Algorithm. How many paths had to be rearranged? See if you can develop a notational shorthand so that you do not have to draw each step of the process.

An exciting and very useful field of study is devoted to control of rearrangeable networks by analytic methods such as matrix decomposition. A simple example of such methods is due to Jajszczyk (1985). Jajszczyk's method is easy to program or execute in hardware and its behavior is fairly efficient.

The method of "call rearrangement" which we have studied is well suited to telephone or terminal switching. As Jajszczyk points out, however, it may be desirable to develop a method to perform an entire permutation at once for such applications as interconnecting microprocessors.

Recall that the outer switches of a rearrangeable network are of dimension $N \times R$ and the inner switches of dimension $R \times R$ as depicted in Figure 13.10. Jajszczyk's method derives the inner switch settings given for the general permutation

$$P = \begin{pmatrix} 1 & 2 & 3 & \cdots & NR \\ i_1 & i_2 & i_3 & \cdots & i_{NR} \end{pmatrix}$$

Jajszczyk's Algorithm is a five-step procedure which is executed R times. Using Jajszczyk's notation, we define the matrix $H_R = (h_{ij})$ where h_{ij} is the number of inputs to outer switch i that are output by outer switch j. (Prove that $\sum_{i=1}^{R} h_{ij} = \sum_{j=1}^{R} h_{ij} = N$.)

1. The *first* step is to write the number of zeros in each row to the side of the rows and to write the number of zeros in each column under each column.

2. The *second* step is to find the row or column which has the maximum number of zeros and "mark" any nonzero element in the row or

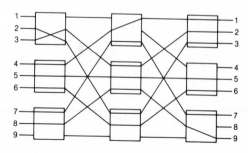

Figure 13.17. The problem is to establish a path from input 1 to output 8.

column selected. Let this element be at the intersection of rows k and column l.

3. The *third* step is to cross out all entries in row k and column l.
4. The *fourth* step is to repeat the three steps above on the remaining (non-crossed-out) elements of the matrix.
5. The *fifth* step is to form a permutation matrix $E = (e_{ij})$ defined from H_R by the rule: $e_{ij} = 1$ if h_{ij} is "marked;" $e_{ij} = 0$ otherwise.

Once a permutation matrix has been derived, it is subtracted from the H_R matrix: $H_R \leftarrow H_R - E$. The resultant H_R matrix is repeatedly subjected to the five steps of the algorithm until R permutations have been derived. These permutation matrices specify the settings of the R inner switches.

As an example we choose the rearrangeable network of Figure 13.11 for which $N = R = 3$. We wish to realize $P = \binom{1\ 2\ 3\ 4\ 5\ 6\ 7\ 8\ 9}{2\ 1\ 3\ 4\ 8\ 9\ 7\ 6\ 5}$. We find that for this example

$$H_3 = \begin{bmatrix} 3 & 0 & 0 \\ 0 & 1 & 2 \\ 0 & 2 & 1 \end{bmatrix}$$

Our first pass yields the matrix of Figure 13.18a, where the numbers along the matrix's right-hand edge and along the bottom are the numbers of zeros in the rows and columns. We see that row one or column one has the maximum number of zeros in any row or column. We are forced to pick and mark (with a circumscribing circle) element h_{11} as the only nonnegative entry in the row. We then cross out all the elements in row one and column one. Our second pass yields the matrix of Figure 13.18b. We see that we can pick any row or column for our next choice. We arbitrarily pick row two and mark element h_{22}. The remaining pass yields the matrix of Figure 13.18c. The first permutation matrix formed is thus

$$\begin{pmatrix} 1 & 0 & 0 \\ 0 & 1 & 0 \\ 0 & 0 & 1 \end{pmatrix}$$

$$\text{a}\ \begin{pmatrix} \textcircled{3} & 0 & 0 \\ 0 & 1 & 2 \\ 0 & 2 & 1 \end{pmatrix} \begin{matrix} 2 \\ 1 \\ 1 \end{matrix} \qquad \text{b}\ \begin{pmatrix} \textcircled{3} & 0 & 0 \\ 0 & \textcircled{1} & 2 \\ 0 & 2 & 1 \end{pmatrix} \begin{matrix} \\ 0 \\ 0 \end{matrix} \qquad \text{c}\ \begin{pmatrix} \textcircled{3} & 0 & 0 \\ 0 & \textcircled{1} & 2 \\ 0 & 2 & \textcircled{1} \end{pmatrix}$$

$$\begin{matrix} 2 & 1 & 1 \end{matrix} \qquad\qquad \begin{matrix} 0 & 0 \end{matrix}$$

Figure 13.18. (a) The first pass on H_3; (b) the second pass on H_3; (c) the third pass on H_3.

$$
a \begin{pmatrix} ② & 0 & 0 \\ 0 & 0 & 2 \\ 0 & 2 & 0 \end{pmatrix} \begin{matrix} 2 \\ 2 \\ 2 \end{matrix} \qquad b \begin{pmatrix} ② & 0 & 0 \\ 0 & 0 & ② \\ 0 & 2 & 0 \end{pmatrix} \begin{matrix} 1 \\ 1 \end{matrix} \qquad c \begin{pmatrix} ② & 0 & 0 \\ 0 & 0 & ② \\ 0 & ② & 0 \end{pmatrix}
$$

$$
\begin{matrix} & 2 & 2 & 2 \end{matrix} \qquad\qquad\qquad \begin{matrix} 1 & 1 \end{matrix}
$$

Figure 13.19. (a) The first pass on the new H_3; (b) the second pass on the new H_3; (c) the third pass on the new H_3.

Subtracting the permutation matrix from H_3 we find that the new

$$
H_3 = \begin{pmatrix} 2 & 0 & 0 \\ 0 & 0 & 2 \\ 0 & 2 & 0 \end{pmatrix}
$$

Going through the algorithm again, with some arbitrary selections as before, we have as the first pass the matrix of Figure 13.19a. The second pass yields the matrix of Figure 13.19b and the third pass gives the matrix of Figure 13.19c. We find that the second permutation matrix is

$$
\begin{pmatrix} 1 & 0 & 0 \\ 0 & 0 & 1 \\ 0 & 1 & 0 \end{pmatrix}
$$

Subtracting the permutation matrix from H_3 we find that the new

$$
H_3 = \begin{pmatrix} 1 & 0 & 0 \\ 0 & 0 & 1 \\ 0 & 1 & 0 \end{pmatrix}
$$

which is already a permutation matrix. Using the three permutation matrices as the inner switch settings we have the connections as shown in Figure13.20.

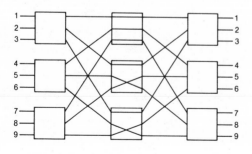

Figure 13.20. The settings for the inner switches.

Complete the connecting network's setup by setting the outer switches. How can you get their settings quickly in general? Is there a necessary order in which the permutation matrices must be assigned to the middle switches? Why or why not?

13.5. The Perfect Shuffle and the Omega Network

Imagine that we have a deck of N cards. Imagine further that N is a power of 2, that is, $N = 2^n$ and that the cards are labeled from 0 to $N - 1$ and in order with card 0 at the top of the deck and card $N - 1$ at the bottom. We now perform what is called a "perfect shuffle" (Stone, 1971; Lawrie, 1975). To perform the perfect shuffle we first divide the deck precisely in half; the first $N/2$ cards are placed in one hand, the equal sized remainder in the other. We then riffle the two half decks together such that they meld by precisely interleaving. We start the interleaving so that card 0 falls at the top of the shuffled deck. Figure 13.21 shows the resulting permutation of the card numbers.

If we examine Figure 13.21 we can convince ourselves that card i, which also starts in position i, ends up in position $i' = 2i$ if $0 \leq i \leq N/2 - 1$ and in position $i' = 2i + 1 - N$ if $N/2 \leq i \leq N - 1$. Now for something truly marvelous! Let us express card i's original position in normal binary

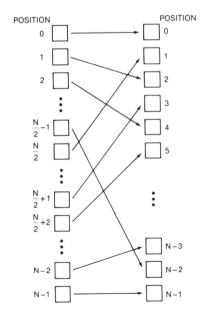

Figure 13.21. The perfect shuffle.

as

$$i = i_{n-1} \cdot 2^{n-1} + i_{n-2} \cdot 2^{n-2} + \cdots + i_1 \cdot 2 + i_0 \qquad (8)$$

1. If i lies in the range $0 \leqslant i \leqslant N/2 - 1$ then $i_{n-1} = 0$ and by our rule above $i' = 2i = i_{n-2} \cdot 2^{n-1} + i_{n-3} \cdot 2^{n-2} + \cdots + i_1 \cdot 2^2 + i_0 \cdot 2$. Because $i_{n-1} = 0$ we could as well write

$$i' = i_{n-2} \cdot 2^{n-1} + i_{n-3} \cdot 2^{n-2} + \cdots + i_1 \cdot 2^2 + i_0 \cdot 2 + i_{n-1} \qquad (9)$$

2. If i lies in the range $N/2 \leqslant i \leqslant N - 1$ then $i_{n-1} = 1$ and by our rule above $i' = 2i + 1 - N = i_{n-2} \cdot 2^{n-1} + i_{n-3} \cdot 2^{n-2} + \cdots + i_1 \cdot 2^2 + i_0 \cdot 2 + 1$. Because $i_{n-1} = 1$ we could as well write

$$i' = i_{n-2} \cdot 2^{n-1} + i_{n-3} \cdot 2^{n-2} + \cdots + i_1 \cdot 2^2 + i_0 \cdot 2 + i_{n-1} \qquad (10)$$

Note that (9) and (10) are the same! Thus, in a perfect shuffle the element occupying position i is shuffled to position i' where i' has the normal binary representation obtained by end-around-left-shifting the normal binary representation of i. We now know two additional things. First, after n perfect shuffles, all cards will be back in their orginal order. (Why?) Second, the perfect shuffle of 2^n cards is intuitively related to the binary necklace problem in combinatorics and thus the permutation induced by the perfect shuffle on 2^n cards is known to induce $(1/n) \sum_{d|n} \phi(d) 2^{n/d}$ cycles. (See the section on comma-free codes.)

Let us look at an example. Consider $n = 3$ and that we perfect shuffle the $2^3 = 8$ cards three times. This is depicted in Figure 13.22, which illustrates the two properties that we mentioned. The necklace theorem predicts $\frac{1}{3}\sum_{d|3} \phi(d) 2^{3/d} = \frac{1}{3}(1 \cdot 2^3 + 2 \cdot 2) = 4$ cycles. They are demonstrated by the orbits of position numbers occupied by the different cards:

$$0 \to 0$$

$$7 \to 7$$

$$1 \to 2 \to 4 \to 1$$

$$3 \to 6 \to 5 \to 3$$

POSITION	CARD		POSITION		POSITION		POSITION
(000)	0 ——→ 0		(000) ——→ 0		(000) ——→ 0		(000)
(001)	1	4	(001)	2	(001)	1	(001)
(010)	2	1	(010)	4	(010)	2	(010)
(011)	3	5	(011)	6	(011)	3	(011)
(100)	4	2	(100)	1	(100)	4	(100)
(101)	5	6	(101)	3	(101)	5	(101)
(110)	6	3	(110)	5	(110)	6	(110)
(111)	7 ——→ 7		(111) ——→ 7		(111) ——→ 7		(111)

Figure 13.22. Three perfect shuffles of 8 cards.

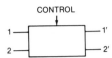

Figure 13.23. The elementary cell for the omega network.

Our next step is to introduce a switching element and then integrate it with the perfect shuffle to build what is termed an "omega network." This switching element is termed an "elementary cell" or a "β element" and is illustrated in Figure 13.23. The elementary cell may be in one of two states determined by the control bit. These states are defined in Figure 13.24.

An omega network for 2^n inputs consists of n stages each having 2^{n-1} elementary cells with the cells connected together by a perfect shuffle wiring. Figure 13.25 is an example of an omega network for $2^3 = 8$ inputs.

Now we consider how to program the omega network to perform permutations. As we can see, the 8-input omega network above requires us to set 12 switches, or control bits; in general, for an omega network for 2^n inputs, we must specify $n \cdot 2^{n-1}$ control bits. One fact should be immediately apparent: because an omega network for 2^n inputs has only $n \cdot 2^{n-1}$ states, it can not, in general, perform all possible permutations since $(2^n)!$ quickly exceeds $n \cdot 2^{n-1}$ as we increase n. So we really have to answer two questions. How do we program the omega network, and, what permutations, or class of permutations, can the omega network perform?

Let us examine the omega network in a microscopic fashion. Each elementary cell has two inputs and two outputs that are fed into a perfect shuffle network or final output stage. The positions of the two input elements, in binary notation, are $p_{n-1}p_{n-2} \cdots p_2 0$ and $p_{n-1}p_{n-2} \cdots p_2 1$, where p_{n-1} is the most significant bit. For the 8-input omega network of Figure 13.25, for example, the first elementary cell in any of the three stages accepts elements whose positions are 000 or 001. This particular cell passes inputs 000 and 001 into 000 and 001 or into 001 and 000, respectively. The important thing to note is that each elementary cell either retains the positions of the elements passed into it or inverts the least significant bit of both positions. Now we recall the amazing fact that a perfect shuffle shuffles or passes the element in position i to position i' where the normal binary representation of i' is the one bit-end-around-left-shifted normal binary representation of i. Let us again consider the 8-input omega network of Figure 13.25 and let us look at the progress of the element from position $i = b_2, b_1, b_0$. The first

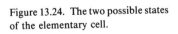

Figure 13.24. The two possible states of the elementary cell.

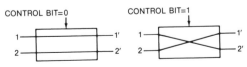

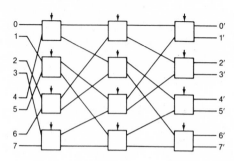

Figure 13.25. The omega network for 8 inputs.

thing element i does is to pass through a perfect shuffle. It thus encounters an elementary cell in position b_1, b_0, b_2. If this first encountered elementary cell's control is c_2, then the element is output in position $b_1, b_0, b_2 + c_2$. This element passes through a perfect shuffle and is presented to the next (the middle, in our example) elementary cell in position $b_0, b_2 + c_2, b_1$. If the middle elementary cell's control is c_1, the element is presented to the final perfect shuffle in position $b_0, b_2 + c_2, b_1 + c_1$. This final perfect shuffle takes the element to position $b_2 + c_2, b_1 + c_1, b_0$. Finally, the last elementary cell presents the original input on output line $b_2 + c_2, b_1 + c_1, b_0 + c_0$.

We now have a very simple algorithm for setting the controls in an omega network. If we wish the input element in input position $i = b_{n-1}b_{n-2} \cdots b_1 b_0$ to be sent to output position $o = d_{n-1}d_{n-2} \cdots d_1 d_0$ then the controls of the elementary cells through which the element passes should be set to $c_{n-1} = b_{n-1} + d_{n-1}$, $c_{n-2} = b_{n-2} + d_{n-2}, \ldots, c_1 = b_1 + d_1$, $c_0 = b_0 + d_0$. Again, using the 8-input omega network, let us send the element in input position $5 = 101$ to position $6 = 110$. According to the above, our controls should be 011. Figure 13.26 depicts the flow.

The omega network is capable of routing any single input line to any output line, however, it may or may not be capable of effecting a permutation of all the input lines. This limitation results from the fact that some permutations will require control conflicts. For example, if in an 8-input network

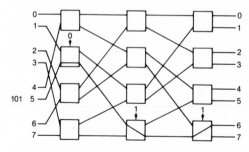

Figure 13.26. The omega network permuting input position 5 to output position 6.

we asked not only to send input line 5 to output line 6 but also input line 1 to output line 5, then we would require that the first elementary cell, which is used for both permutations, have different controls which is an impossibility.

Although the omega network for 2^n input elements can not effect all $(2^n)!$ permutations, it can perform an important subset of these permutations. One class of doable permutations is represented by

$$\{x\} \to \{ax + b\} \bmod (2^n), \quad a \text{ odd}; \quad 0 \leqslant x < 2^n \tag{11}$$

The linear permutation polynomial (11) gives rise to a rich set of permutations.

Problems

1. Show that the 2^n input omega network is capable of performing the identity permutation:

$$\begin{pmatrix} 0 & 1 & 2 & \cdots & 2^n - 1 \\ 0 & 1 & 2 & \cdots & 2^n - 1 \end{pmatrix}$$

2. Show that the 2^n input omega network is capable of performing the important reverse permutation:

$$\begin{pmatrix} 0 & 1 & \cdots & 2^n - 2 & 2^n - 1 \\ 2^n - 1 & 2^n - 2 & \cdots & 1 & 0 \end{pmatrix}$$

What are the values of a and b in (11) for this permutation? (This particular permutation is important in some implementations of the Fast Fourier Transform.)

3. Use Stirling's approximation to show that the 2^n input omega network is capable of performing what approximate fraction of all possible permutations of 2^n elements? Evaluate this fraction for $2^n = 2, 4, 8, 16, 32$.

One additional important aspect of the omega network is the following: if a 2^n input omega network is capable of performing a particular permutation, P, then an omega network of 2^{n+1} inputs is capable of the same permutation, P. (This permutation would, of course, be a subset of a permutation of 2^{n+1} elements in the larger network.)

13.6. The Beneš–Waksman Permutation Network

We now examine a network capable of realizing any permutation of 2^n input lines to output lines. The Beneš–Waksman network (Beneš, 1965;

Waksman, 1968) uses β elements and is optimized, in the limit as $n \to \infty$, in its use of these elements.

The proof that the Beneš–Waksman network is capable of performing any permutation is a constructive proof and attributed by Waksman to H. Stone.

We start with a discussion of the network depicted in Figure 13.27. The figure shows the input lines labeled from i_0 to i_{2^n-1} and the output lines labeled from o_0 and o_{2^n-1}. Note that only $2^n - 1$ β elements are used as outer switches to the networks P_1 and P_2; output lines o_0 and o_1 are not switched through a β element.

We represent P_1 and P_2 as crossbars in the following example but we should realize that P_1 and P_2 may each be replaced by a smaller network of the form of Figure 13.27; and these smaller networks by even smaller networks and so on. (We demonstrate this recursive construction later on.) We have obtained the Beneš–Waksman network when no further recursions or "decompositions" are possible.

To prove that the network of Figure 13.27 is indeed a full permutation network we proceed, via Stone, as follows. We start with output line zero, o_0, and link it to its assigned input, i_j, through P_1. If j is odd, the control line to the β element is set to 1; otherwise 0. We then connect the pair member (input line number $j + 1$ if j is even; $j - 1$ if j is odd) through P_2. (The pair member has been sent to P_2 by virtue of its pair, i_j, going through P_1.) The connection from P_2 proceeds to the appropriate output β element whose control line is then appropriately set. We continue this "looping" or "threading" process by taking the pair element of the most recently set β element and sending it through P_1 to its corresponding input line. If we should complete a cycle or "orbit" within the permutation before all input and output lines have been taken care of, we need merely start a new threading with any unlinked input or output line.

Recalling that P_1 and P_2 are full permutation networks and that both P_1 and P_2 have exactly 2^{n-1} distinct inputs and 2^{n-1} distinct outputs, the

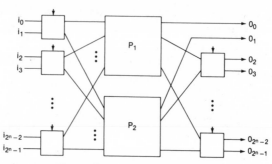

Figure 13.27. A permutation network.

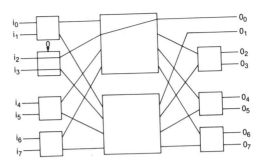

Figure 13.28. Output line o_0 linked to input line i_2.

pigeonhole principle guarantees that the network of Figure 13.27 is capable of performing any of the $(2^n)!$ possible permutations.

We now show, through the use of eight consecutive figures, how Stone's "threading algorithm" works. Suppose we wish to effect the permutation

$$P = \begin{pmatrix} 0 & 1 & 2 & 3 & 4 & 5 & 6 & 7 \\ 3 & 4 & 0 & 5 & 1 & 2 & 7 & 6 \end{pmatrix}$$

The steps are as shown in Figures 13.28–13.35.

Problem

Draw the diagram of Figure 13.27 and fill in the connections in P_1 and P_2 and set the β elements appropriately to realize the permutation

$$P = \begin{pmatrix} 0 & 1 & 2 & 3 & 4 & 5 & 6 & 7 \\ 7 & 6 & 5 & 4 & 3 & 2 & 1 & 0 \end{pmatrix}$$

Let us now consider P_1 and P_2 of Figure 13.27. Both of these permutation networks can be recursively replaced by a network of the form of Figure 13.27. The network shown in Figure 13.36 is the result of doing this.

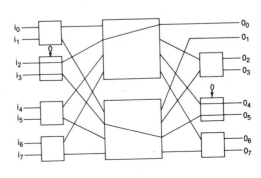

Figure 13.29. Input line i_3 linked to output line o_5.

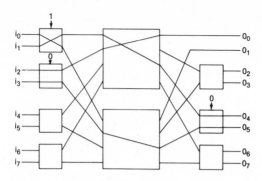

Figure 13.30. Output line o_4 linked to input line o_3.

Using the network of Figure 13.36 in place of the networks P_1 and P_2 of Figure 13.27 we have the network of Figure 13.37. Because further recursion is not possible, the network of Figure 13.37 is a Beneš–Waksman network. We have further denoted the β elements in the Beneš–Waksman network by index numbers 1 through 17 placed directly below them. The β elements included in P_1 are numbers 5, 6, 9, 10, 13 and the β elements included in P_2 are numbers 7, 8, 11, 12, 14.

The Beneš–Waksman network for 2^n inputs requires only $S(n) = 2^n(n-1) + 1$ switches. Each switch requires one bit to set it. The Beneš–Waksman network thus requires $S(n)$ bits to set. A permutation of 2^n inputs requires $F(n) = \log_2(2^n)!$ bits to specify.

Problem

1. Show that $S(n)$ is a given above.

2. Show that $\lim_{n \to \infty} [S(n)/F(n)] = 1$, that is, the Beneš–Waksman network makes as efficient use as possible (in the limit) of its switches.

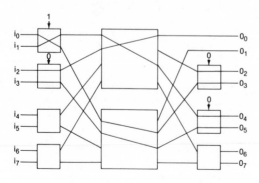

Figure 13.31. Input line i_0 linked to output line o_3.

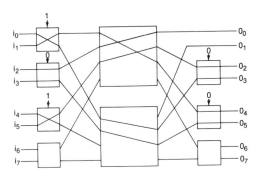

Figure 13.32. Output line o_2 linked to input line i_5.

The remaining problem is to set the switches of the Beneš–Waksman network, that is, determine the control states of the β elements. This problem area is an exciting field of study and a series of important papers have been written on this question alone. We do not consider a particular algorithm but rather attempt to show what is behind some of the algorithms. We use a linear or single processor approach. At the outset, however, let it be known that there exist parallel algorithms (Nassimi and Sahni, 1980), exhibiting concomitant time savings, for setting up the Beneš–Waksman network. These algorithms will not be investigated, but, should the reader consider implementing a switch setting algorithm, time constraints may require the employment of parallel processing. The appropriate papers are cited at the end of the chapter.

We proceed to determine all the switch settings for the example considered previously for the permutation

$$P = \begin{pmatrix} 0 & 1 & 2 & 3 & 4 & 5 & 6 & 7 \\ 3 & 4 & 0 & 5 & 1 & 2 & 7 & 6 \end{pmatrix} \tag{12}$$

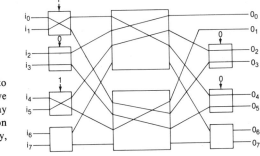

Figure 13.33. Input line i_4 linked to output line o_1. Note that we have completed or "threaded" our way through a cycle of the permutation P. Let us start again with, arbitrarily, output line o_6.

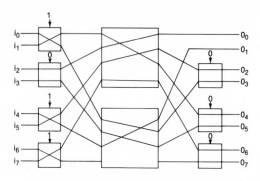

Figure 13.34. Output line o_6 linked to input line i_7.

We refer to the switch numbering scheme given in Figure 13.37, the Beneš-Waksman network for 8 inputs. The reader will recall that at the conclusion of Figure 13.35 we had accomplished two things. We had found the settings for switches 1, 2, 3, 4, 15, 16, and 17, and we had determined the permutations required of P_1 and P_2:

$$P_1 = \begin{pmatrix} 0 & 1 & 2 & 3 \\ 2 & 0 & 1 & 3 \end{pmatrix} \quad \text{and} \quad P_2 = \begin{pmatrix} 0 & 1 & 2 & 3 \\ 1 & 2 & 0 & 3 \end{pmatrix}$$

Let us examine P_1 in the form of Figure 13.27. This is shown in Figure 13.38. Using Stone's Algorithm we quickly find that the above network requires the settings as shown in Figure 13.39. Note that we now have the settings for switches 5, 6, and 13. But what are P_{11} and P_{12}? They are, of course, the single β elements 9 and 10, respectively. Their settings are immediately obvious.

In like manner we derive the subnetwork for doing

$$P_2 = \begin{pmatrix} 0 & 1 & 2 & 3 \\ 1 & 2 & 0 & 3 \end{pmatrix}$$

(See Figure 13.40.)

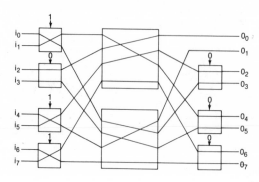

Figure 13.35. Input line i_6 linked to output line o_7. We are finished.

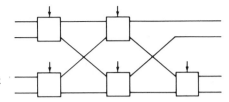

Figure 13.36. A permutation network equivalent to P_1 or P_2.

Combining all of the above, we have the Beneš-Waksman network for the permutation (12).

Problem

Draw, and set the switches for, the Beneš-Waksman network which implements the permutation

$$P = \begin{pmatrix} 0 & 1 & 2 & 3 & 4 & 5 & 6 & 7 \\ 7 & 6 & 5 & 4 & 3 & 2 & 1 & 0 \end{pmatrix}$$

of a previous problem (see Figure 13.41).

13.7. The Perfect Shuffle Network Revisited

Recall that a perfect shuffle stage can be fitted with β elements. Figure 13.42 depicts one such stage for $N = 2^3 = 8$ inputs. We recall that we built an omega network out of $\log_2 N$ stages of the form of Figure 13.42. The omega network was not capable of effecting all possible permutations and we moved on to networks that were capable of all possible permutations such as the Beneš-Waksman network. Not only is the Beneš-Waksman network capable of performing all possible permutations, but it also uses the fewest possible number of β elements as $N \to \infty$. The Beneš-Waksman

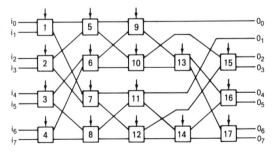

Figure 13.37. The Beneš-Waksman network for 8 inputs.

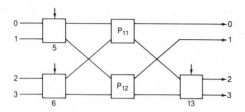

Figure 13.38. The subnetwork for doing $P_1 = \begin{pmatrix} 0 & 1 & 2 & 3 \\ 2 & 0 & 1 & 3 \end{pmatrix}$.

network, however, is not regular in the sense that every stage is not identical in its connections to every other stage. As we continue to establish ever greater densities in VLSI and as we continue to interconnect and cascade VLSI components, networks that do exhibit regularity or "modularity" in construction become of increasing interest. In this vein we want to consider a contribution due to Wu and Feng (1981). They showed that a network consisting of $3 \log_2 N - 1$ perfect shuffle stages with β elements can effect any permutation of the $N = 2^n$ inputs. The controls are derivable from the Beneš–Waksman controls.

We recite the method given by Wu and Feng to derive the controls for the β elements of the perfect shuffle network stages. We use, as an example, the permutation effected by the Beneš–Waksman network of Figure 13.41.

The first step is to perform a bit reversal permutation on P given in (12). Consider the entries in the top row of P as expressed in normal binary and reverse the order of the bits. After doing this we obtain P' given in (13):

$$P' = \begin{pmatrix} 0 & 4 & 2 & 6 & 1 & 5 & 3 & 7 \\ 3 & 4 & 0 & 5 & 1 & 2 & 7 & 6 \end{pmatrix} \tag{13}$$

Reformatting P' to the usual form for a permutation we have (14):

$$P' = \begin{pmatrix} 0 & 1 & 2 & 3 & 4 & 5 & 6 & 7 \\ 3 & 1 & 0 & 7 & 4 & 2 & 5 & 6 \end{pmatrix} \tag{14}$$

The second step is to determine the controls for the Beneš–Waksman network that implements (14). The matrix of controls is easily found and

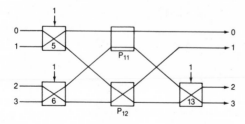

Figure 13.39. The controls for the subnetwork of Figure 13.38.

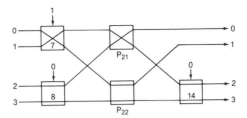

Figure 13.40. The subnetwork for doing
$P_2 = \begin{pmatrix} 0 & 1 & 2 & 3 \\ 0 & 2 & 0 & 3 \end{pmatrix}$.

is displayed in (15):

$$
M = \begin{bmatrix}
0 & 1 & 0 & \text{``0''} & \text{``0''} \\
0 & 0 & 0 & 0 & 1 \\
0 & 0 & 0 & \text{``0''} & 0 \\
1 & 1 & 1 & 0 & 0
\end{bmatrix}
\tag{15}
$$

(The "0"s are the controls for β elements which would be superfluous but which, if included, would fall spatially as indicated.) The matrix M is in general of dimension $2^{n-1} \times (2n - 1)$.

The third step is to modify M and partition it into two parts, matrices A and C. The first part is a matrix of dimension $2^{n-1} \times n$; the second is $2^{n-1} \times (n - 1)$. The entries in the new matrix M' are formed according to the following rules:

1. Let $b_{n-1}b_{n-2}\cdots b_1$ be the normal binary word designating a β element in stage i where $0 \leq i \leq 2n - 1$.
2. For the first n columns of M', that is, for $0 \leq i \leq n - 1$, the β element control bit in row $b_{n-1}b_{n-2}\cdots b_1$ of stage i is the element in row

$$
b_i \cdots b_1 b_{i+1} \cdots b_{n-1}
\tag{16}
$$

of M.

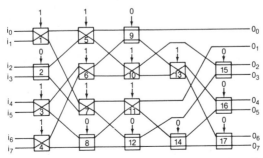

Figure 13.41. The Beneš–Waksman network which implements $P = \begin{pmatrix} 0 & 1 & 2 & 3 & 4 & 5 & 6 & 7 \\ 3 & 4 & 0 & 5 & 1 & 2 & 7 & 6 \end{pmatrix}$.

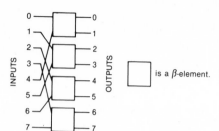

Figure 13.42. An 8-input perfect shuffle stage with elements.

3. For the last $n - 1$ columns of M', for $0 < i' \leq n - 1$, $i \equiv i' + n$, the β element control bit in row $b_{n-1}b_{n-2} \cdots b_1$ of stage i is the element in row

$$b_{i'+1} \cdots b_{n-1}b_{i'} \cdots b_1 \qquad (17)$$

of M.

For our example with M as given in (15), we see that

$$M' = [A \,|\, C] = \begin{bmatrix} 0 & 1 & 0 & \vline & 0 & 0 \\ 0 & 0 & 0 & \vline & 0 & 1 \\ 0 & 0 & 0 & \vline & 0 & 0 \\ 1 & 1 & 1 & \vline & 0 & 0 \end{bmatrix} \qquad (18)$$

The fourth step is to create a third matrix, matrix B. This matrix will control an omega network to be inserted between matrices A and C. Matrix B will be of dimension $2^{n-1} \times n$ and constructed by a series of column vector matrix partitions. The columns will either be all zeros, denoted by 0, or be 2^{n-1} long segments of the Rademacher sequences which are those sequences of zeros and ones extracted from the contents of a normal binary counter started at zero and incrementing by one. (Here we let r_i be the ith Rademacher sequence where i denotes the ith stage of the binary counter. Stage one is the stage containing the least significant bit. Thus, $r_1 = 01010101\ldots$, $r_2 = 001100110\ldots$, $r_3 = 000011110\ldots$, and so on.) The matrix B is created by the following rules:

$$B = (r_2^T | r_4^T | \cdots | r_{n-1}^T | r_1^T | r_3^T | \cdots r_{n-2}^T | 0) \qquad (n \text{ odd})$$
$$B = (r_2^T | r_4^T | \cdots | r_{n-2}^T | 0 | r_2^T | r_4^T | \cdots | r_{n-2}^T | 0) \qquad (n \text{ even})$$

For our example

$$B = \begin{bmatrix} 0 & 0 & 0 \\ 0 & 1 & 0 \\ 1 & 0 & 0 \\ 1 & 1 & 0 \end{bmatrix} \qquad (19)$$

The fifth step is to modify the A matrix in (18). We represent A in general by the partitioned matrix of columns

$$A = [a_0 \quad a_1 \quad \cdots \quad a_{\left[\frac{n}{2}\right]} \quad a_{\left[\frac{n}{2}\right]+1} \quad \cdots \quad a_{n-1}] \qquad (20)$$

We now form another matrix $K = [k_0 \ \ k_1 \ \ k_2 \ \ \cdots \ \ k_{\text{last}}]$ where

$$K = [r_1^T \ \ r_3^T \ \ r_5^T \ \ \cdots] \qquad (n \text{ odd})$$

$$K = [r_2^T \ \ r_4^T \ \ r_6^T \ \ \cdots] \qquad (n \text{ even})$$

The matrix K is of dimension $2^{n-1} \times (n - [n/2] - 1)$. We replace A with the matrix F where

$$F = [a_0 \ \ a_1 \ \ \cdots \ \ a_{\left[\frac{n}{2}\right]-1} \ \ (a_{\left[\frac{n}{2}\right]} + k_0)$$

$$(a_{\left[\frac{n}{2}\right]+1} + k_1) \ \ \cdots \ \ (a_{n-2} + k_{\text{last}}) a_{n-1}] \tag{21}$$

(The additions of the a and k vectors are done bit-by-bit modulo 2.) For our example then, (21) yields

$$F = \begin{bmatrix} 0 & 1 & 0 \\ 0 & 1 & 0 \\ 0 & 0 & 0 \\ 1 & 0 & 1 \end{bmatrix} \tag{22}$$

But we are not yet finished. We must now operate on (21) with yet another permutation operation, E. This new operation operates only on columns in *the range* $[n/2] + 1$ to $n - 1$. The operation on column $n - i - 1$ is as follows: the bit in position $p_{n-1} p_{n-2} \cdots p_1$ is permuted to position

$$b_{n-1} b_{n-2} \cdots b_{\left[\frac{n}{2}\right]-i+1} (b_{\left[\frac{n}{2}\right]-i} + b_{n-\left[\frac{n}{2}\right]-i})$$

$$(b_{\left[\frac{n}{2}\right]-i-1} + b_{n-\left[\frac{n}{2}\right]-i+1}) \ \cdots \ (b_1 + b_{n-2i-1}) \tag{23}$$

where the additions are performed modulo 2.

Carrying out this final operation on our example, (22) becomes

$$E(F) = \begin{bmatrix} 0 & 1 & 0 \\ 0 & 1 & 0 \\ 0 & 0 & 1 \\ 1 & 0 & 0 \end{bmatrix} \tag{24}$$

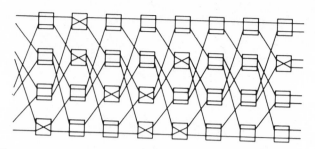

Figure 13.43. The perfect shuffle network controlled to implement (12).

The control matrix for the perfect shuffle network is then

$$[E(F) \quad B \quad C]$$

(25)

which for our example is

$$
\begin{bmatrix}
0 & 1 & 0 & 0 & 0 & 0 & 0 & 0 \\
0 & 1 & 0 & 0 & 1 & 0 & 0 & 1 \\
0 & 0 & 1 & 1 & 0 & 0 & 0 & 0 \\
1 & 0 & 0 & 1 & 1 & 0 & 0 & 0
\end{bmatrix}
$$

(26)

Figure 13.43 shows the perfect shuffle network for our example set to the proper control matrix (26). The reader should trace the inputs to outputs to verify that the method does indeed implement (12).

Problems

1. Use the Wu–Feng method to implement the permutation

$$
P = \begin{pmatrix}
0 & 1 & 2 & 3 & 4 & 5 & 6 & 7 & 8 & 9 & 10 & 11 & 12 & 13 & 14 & 15 \\
14 & 12 & 5 & 7 & 15 & 8 & 9 & 13 & 4 & 3 & 10 & 6 & 1 & 0 & 2 & 11
\end{pmatrix}
$$

2. (O'Donnell and Smith, 1982) Consider the single stage permutation network shown in Figure 13.44. The network allows for interchange of any two adjacent inputs. The network is controlled by an $n-1$ bit control

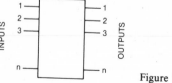

Figure 13.44.

vector. If bit i, $1 \leq i \leq n - 1$, is a 0, then inputs i and $i + 1$ are sent to outputs i and $i + 1$, respectively. If i is a 1, then inputs i and $i + 1$ are sent to outputs $I = 1$ and i, respectively. There is a restriction on the control vector and that is that the vector may not have two adjacent ones. (Why?) Show that each admissible control vector performs a unique permutation and then count the total number of permutations that the network is capable of performing. [Hint: Develop an appropriate linear, homogeneous difference equation, derive the necessary boundary conditions, and solve for general n.]

References

Beneš, V. (1962), On Rearrangeable Three-Stage Connecting Networks, *The Bell System Technical Journal*, Vol. 41, pp. 1481–1492.

Beneš, V. (1965), *Mathematical Theory of Connecting Networks*, Academic Press, New York.

Clos, C. (1953), A Study of Nonblocking Switching Networks, *The Bell System Technical Journal*, Vol. 32, 406–424.

Jajszczyk, A. and J. Rajski (1980), Effects of Choosing the Switches for Rearrangements in Switching Networks, *IEEE Transactions on Communications*, Vol. 28, pp. 1832–1834.

Jajszczyk, A. (1985), A Simple Algorithm for the Control of Rearrangeable Switching Networks, *IEEE Transactions on Communications*, Vol. 33, pp. 169–171.

Kappel, J. (1967), Nonblocking and Nearly Nonblocking Multistage Switching Arrays, in *Proceedings of the 5th International Teletraffic Congress*, pp. 238–241.

Keister, W. (1967), Unpublished work.

Lawrie, D. (1975), Access and Alignment of Data in an Array Processor, *IEEE Transactions on Computers*, Vol. 24, pp. 1145–1155.

Marcus, M. (1977), The Theory of Connecting Networks and Their Complexity: A Review, *Proceedings of the IEEE*, pp. 1263–1271.

Melas, C. (1983), Path Rearranging in a Data Switching Network, *IEEE Transactions on Communications*, Vol. 31, pp. 155–157.

Nassimi, D. and S. Sahni, Parallel Algorithms to Set-Up the Beneš Permutation Network, in *Proceedings of the Workshop on Interconnection Networks for Parallel and Distributed Processing*, pp. 70–71, IEEE Pub. No. 80CH1560-2. Conference date: April 1980.

O'Donnell, M. and C. Smith (1982), A Combinatorial Problem Concerning Processor Interconnection Networks, *IEEE Transactions on Computers*, Vol. 31, pp. 163–164.

Paull, M. (1962), Reswitching of Connection Networks, *The Bell System Technical Journal*, Vol. 41, pp. 833–855.

Stone, H. (1971), Parallel Processing with the Perfect Shuffle, *IEEE Transactions on Computers*, Vol. 20, pp. 153–161.

Waksman, A. (1968), A Permutation Network, *Journal of the Association for Computing Machinery*, Vol. 15, pp. 159–163.

Wu, C-L. and T-Y. Feng (1981), The Universality of the Shuffle-Exchange Network, *IEEE Transactions on Computers*, Vol. 30, pp. 324–332.

14

Network Reliability and Survivability

The ability of a network to withstand component failure, subversion, or openly hostile actions is inextricably linked to its topology as well as its individual component survivabilities. The most basic of the topological questions is node "connectivity." Node connectivity means simply the linking together of centers that can originate or receive messages by means of pathways or, as they are more appropriately termed, "links." We will at first assume only two-way or bidirectional links which are appropriate models, for example, for microwave links that use the same physical antenna for transmission and reception through polarization or frequency diversity. We further restrict our attention, for now, to those cases that allow a maximum of one link to be placed between any two nodes.

If we have N nodes we may have up to $\frac{1}{2}N(N-1)$ links. We define N nodes to constitute an N-node network if there exists a path or series of links between any two nodes. It is now clear that a network of N nodes will have between $N-1$ and $\frac{1}{2}N(N-1)$ links.

In Chapter 4, the connectivity matrix, $C = [c_{ij}]$, is given by the rule:

$$c_{ij} = \begin{cases} 0, & \text{if there is no link between nodes } i \text{ and } j \\ 1, & \text{if there is a link between nodes } i \text{ and } j \text{ or if } i = j \end{cases} \tag{1}$$

Note that C is a symmetric matrix for a network with bidirectional links. For an example, consider the network of Figure 14.1. The nodes are represented by circles. The arbitrarily assigned node number is emplaced within the circle. The links are indicated by lines. The connectivity matrix

Figure 14.1. Example network. ①—②—③—④

469

describing the network in Figure 14.1 is

$$C = \begin{bmatrix} 1 & 1 & 0 & 0 \\ 1 & 1 & 1 & 0 \\ 0 & 1 & 1 & 1 \\ 0 & 0 & 1 & 1 \end{bmatrix} \tag{2}$$

We now define connectivity matrix multiplication (CMM) as follows. If A and B are two connectivity matrices, both of dimension $N \times N$, then $Z = A \cdot B$ is defined as follows:

$$z_{ij} = \begin{cases} 0, & \text{if } \sum_{k=1}^{N} a_{ik}b_{kj} = 0 \\ 1, & \text{if } \sum_{k=1}^{N} a_{ik}b_{kj} > 0 \end{cases} \tag{3}$$

Note that the rule in (3) is equivalent to performing the INCLUSIVE-OR on the boolean dot product between the vectors $[a_{i1} \quad a_{i2} \quad \cdots \quad a_{iN}]$ and $[b_{1j} \quad b_{2j} \quad \cdots \quad b_{Nj}]$. Let us consider

$$C^2 = C \cdot C \tag{4}$$

Using (3) we quickly determine that

$$C^2 = \begin{bmatrix} 1 & 1 & 1 & 0 \\ 1 & 1 & 1 & 1 \\ 1 & 1 & 1 & 1 \\ 0 & 1 & 1 & 1 \end{bmatrix} \tag{5}$$

The interpretation of C^2 is as follows. If c_{ij}^2 is a 0, then there is no path between nodes i and j that uses zero, one, or two links or, equivalently, starting at node i, one can not "pass through the network" to node j by traversing only zero, one, or two links. If c_{ij}^2 is a 1, then either $i = j$ or there is a path from node i to node j that uses only one or two links.

Consider now

$$C^3 = \begin{bmatrix} 1 & 1 & 1 & 1 \\ 1 & 1 & 1 & 1 \\ 1 & 1 & 1 & 1 \\ 1 & 1 & 1 & 1 \end{bmatrix} \tag{6}$$

What Equation (6) demonstrates is that there is a path between all nodes that uses three or less links.

It is very easy to show that, in general, C^k is to be interpreted as

$$c_{ij}^k = \begin{cases} 0 \rightarrow \text{there is no path between nodes } i \text{ and } j \text{ that traverses} \\ \quad\quad k \text{ or less links} \\ 1 \rightarrow \text{there is a path between nodes } i \text{ and } j \text{ that traverses} \\ \quad\quad k \text{ or less links} \end{cases}$$

For a network there will exist a $k \le N - 1$ such that $C_k = 1_N$ where 1_N is the all ones matrix of dimension $N \times N$.

We now introduce a difference matrix defined as follows:

$$D(l) = C^l - C^{l-1} \tag{7}$$

(We define $C^0 = I_N$ where I_N is the identity matrix of dimension $N \times N$.) The difference matrix $D(l), l \ge 1$, is interpreted as follows: $d_{ij} = 1$ if and only if the *shortest* path from node i to node j is *exactly* l links. We now define a counting function $P(l)$ as follows:

$$P(l) = \tfrac{1}{2} \sum_{i=1}^{N} \sum_{j=1}^{N} D_{ij}(l) \tag{8}$$

We interpret $P(l)$ as the number of shortest (symmetric) paths in the network that use exactly l links.

Problems

1. Show that $P(1)$ is the number of links in the network.
2. Show that

$$\sum_{l=1}^{N-1} P(l) = \tfrac{1}{2} N(N - 1). \tag{9}$$

As an example of what we have so far developed, consider the network of Figure 14.2. The connectivity matrix and its powers for this network are

$$C = \begin{bmatrix} 1 & 1 & 0 & 0 & 0 \\ 1 & 1 & 1 & 0 & 0 \\ 0 & 1 & 1 & 1 & 1 \\ 0 & 0 & 1 & 1 & 1 \\ 0 & 0 & 1 & 1 & 1 \end{bmatrix} \tag{10}$$

Figure 14.2. Example network.

$$C^2 = \begin{bmatrix} 1 & 1 & 1 & 0 & 0 \\ 1 & 1 & 1 & 1 & 1 \\ 1 & 1 & 1 & 1 & 1 \\ 0 & 1 & 1 & 1 & 1 \\ 0 & 1 & 1 & 1 & 1 \end{bmatrix} \tag{11}$$

$$C^3 = \begin{bmatrix} 1 & 1 & 1 & 1 & 1 \\ 1 & 1 & 1 & 1 & 1 \\ 1 & 1 & 1 & 1 & 1 \\ 1 & 1 & 1 & 1 & 1 \\ 1 & 1 & 1 & 1 & 1 \end{bmatrix} \tag{12}$$

We now compute the difference matrices

$$D(1) = \begin{bmatrix} 0 & 1 & 0 & 0 & 0 \\ 1 & 0 & 1 & 0 & 0 \\ 0 & 1 & 0 & 1 & 1 \\ 0 & 0 & 1 & 0 & 1 \\ 0 & 0 & 1 & 1 & 0 \end{bmatrix} \tag{13}$$

$$D(2) = \begin{bmatrix} 0 & 0 & 1 & 0 & 0 \\ 0 & 0 & 0 & 1 & 1 \\ 1 & 0 & 0 & 0 & 0 \\ 0 & 1 & 0 & 0 & 0 \\ 0 & 1 & 0 & 0 & 0 \end{bmatrix} \tag{14}$$

$$D(3) = \begin{bmatrix} 0 & 0 & 0 & 1 & 1 \\ 0 & 0 & 0 & 0 & 0 \\ 0 & 0 & 0 & 0 & 0 \\ 1 & 0 & 0 & 0 & 0 \\ 1 & 0 & 0 & 0 & 0 \end{bmatrix} \tag{15}$$

The counting functions are now computed:

$$p(1) = 5 \tag{16}$$

$$p(2) = 3 \tag{17}$$

$$p(3) = 2 \tag{18}$$

The counting function tells us that there are five symmetric shortest paths of one link. From Figure 14.2 we see that they are between the node pairs $(1, 2), (2, 3), (3, 4), (3, 5)$ and $(4, 5)$. Note that $p(1)$ is, in general, the number of links in the network. The counting function tells us that there are three symmetric shortest paths of two links. From the figure we see that they are between node pairs $(1, 3), (2, 4)$, and $(2, 5)$. Finally, the counting function tells us that there are two symmetric shortest paths of three links. They are, from the figure, between node pairs $(1, 4)$ and $(1, 5)$. We note also that

$$p(1) + p(2) + p(3) = \tfrac{1}{2} \cdot 5 \cdot 4 = 10 \qquad (19)$$

as expected by (9).

While enlightening and imparting a basic feel for the subject, exponentiating the connectivity matrix to investigate connectivity is not a very efficient modus operandi. A much more efficient and useful algorithm is provided by Frank (1974). The network to be analyzed has the general parameters of L links and N nodes. We seek to determine if there exist paths that interconnect all nodes. If there do not, we wish to create a table listing those sets of nodes for which there are paths linking them. Each set of linked nodes is termed a "component of the network." A network for which there exists a path of links between each pair of nodes is said to exhibit a single component.

The following list of structures and conventions are required by the algorithm. We assign a component label, $S(i)$, to each node and we let $n(i)$ denote the number of nodes in the component i. We keep the nodes in each component in a list and let the *first* node of component list k be denoted by $s(k)$. We can move through a list of component nodes by using the strings $\{l_k(j)\}$; $l_k(j)$ yields the next node in component k. The convention will be to set $l_k(j)$ to zero if j is the *last* node in component k. We additionally retain the last node in component k by the function $f(k)$. Frank's Algorithm is applied to a list of links and proceeds as diagramed in Figure 14.3.

Let us study Frank's Algorithm as applied to the network of Figure 14.4. By inspection of the figure it is clear that we have a network of two components. Often, however, we do not have the luxury of seeing the network; rather we must work solely from a "link list." The link list is just a list of node pairs that are joined together by single links. It is a serialization, if you will, of the connectivity matrix. The link list for the network of Figure 14.4 might be

$$(2, 5)$$
$$(1, 2)$$
$$(3, 4)$$
$$(1, 5)$$

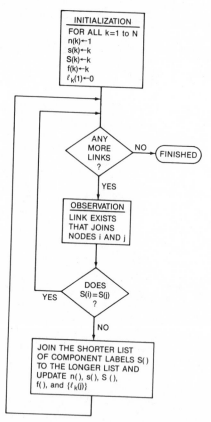

Figure 14.3. Frank's Algorithm for analysis of connectivity.

There are four links in our link list so Frank's Algorithm undergoes four iterations. These iterations are presented along with comments.

Initialization:

$$n(1) = 1 \quad n(2) = 2 \quad n(3) = 1 \quad n(4) = 1 \quad n(5) = 1$$
$$s(1) = 1 \quad s(2) = 2 \quad s(3) = 3 \quad s(4) = 4 \quad s(5) = 5$$
$$S(1) = 1 \quad S(2) = 2 \quad S(3) = 3 \quad S(4) = 4 \quad S(5) = 5$$
$$f(1) = 1 \quad f(2) = 2 \quad f(3) = 3 \quad f(4) = 4 \quad f(5) = 5$$
$$l_1(1) = 0 \quad l_2(1) = 0 \quad l_3(1) = 0 \quad l_4(1) = 0 \quad l_5(1) = 0$$

Figure 14.4. Example network having two components.

Iteration 1:   The first link examined joins nodes 2 and 5 and these nodes must then belong to the same component. We then check to see if $S(2) = S(5)$. We find that the nodes have different component labels and so we join the lists of nodes under one label. To keep our work to a minimum we will want to relabel the smallest number of nodes so we examine $n(2)$ and $n(5)$. We find that $n(2) = n(5) = 1$ and thus both lists are the same length. We arbitrarily choose to relabel the nodes under component label 5. Note that the arrays $s(\ )$, $f(\ )$, and the strings $\{l_k(j)\}$ are useful in relabeling. They serve as pointers and list delimiters for index-oriented computer languages. We thus have

$$n(1) = 1 \qquad n(2) = 2 \qquad n(3) = 1 \qquad n(4) = 1 \qquad n(5) = -$$

$$s(1) = 1 \qquad s(2) = 2 \qquad s(3) = 3 \qquad s(4) = 4 \qquad s(5) = -$$

$$S(1) = 1 \qquad S(2) = 2 \qquad S(3) = 3 \qquad S(4) = 4 \qquad S(5) = 2$$

$$f(1) = 1 \qquad f(2) = 5 \qquad f(3) = 3 \qquad f(4) = 4 \qquad f(5) = -$$

$$l_1(1) = 0 \qquad l_2(1) = 5 \qquad l_3(1) = 0 \qquad l_4(1) = 0$$

$$l_2(2) = 0$$

Iteration 2:   The next link joins nodes 1 and 2. We check, as before, to see if $S(1) = S(2)$. The labels are not the same and so we must again join two lists by relabeling one of them. This time, however, we see that $n(1) = 1$ and $n(2) = 2$. The first list is shorter and therefore the nodes previously listed under component one will be relabeled to show their absorption by component two. And we have

$$n(1) = - \qquad n(2) = 3 \qquad n(3) = 1 \qquad n(4) = 1 \qquad n(5) = -$$

$$s(1) = - \qquad s(2) = 2 \qquad s(3) = 3 \qquad s(4) = 4 \qquad s(5) = -$$

$$S(1) = 2 \qquad S(2) = 2 \qquad S(3) = 3 \qquad S(4) = 4 \qquad S(5) = 2$$

$$f(1) = - \qquad f(2) = 1 \qquad f(3) = 3 \qquad f(4) = 4 \qquad f(5) = -$$

$$\qquad\qquad l_2(1) = 5 \qquad l_3(1) = 0 \qquad l_4(1) = 0$$

$$\qquad\qquad l_2(2) = 1$$

$$\qquad\qquad l_2(3) = 0$$

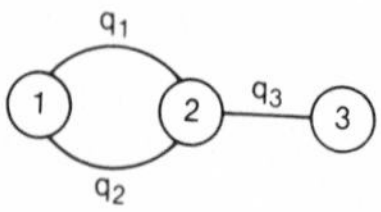

Figure 14.5. Simple network.

Iteration 3:   The next link joins nodes 3 and 4. This case is similar to Iteration 1 and we have

$$n(1) = - \quad n(2) = 3 \quad n(3) = 2 \quad n(4) = - \quad n(5) = -$$
$$s(1) = - \quad s(2) = 2 \quad s(3) = 3 \quad s(4) = - \quad s(5) = -$$
$$S(1) = 2 \quad S(2) = 2 \quad S(3) = 3 \quad S(4) = 3 \quad S(5) = 2$$
$$f(1) = - \quad f(2) = 1 \quad f(3) = 4 \quad f(4) = - \quad f(5) = -$$
$$l_2(1) = 5 \quad l_3(1) = 4$$
$$l_2(2) = 1 \quad l_3(2) = 0$$
$$l_2(3) = 0$$

Iteration 4:   The final link joins nodes 1 and 5. This time, however, $S(1)$ is the same as $S(5)$ and the algorithm terminates with the tables unchanged from the end of Iteration 3. Note that the table shows that there are two components to our network and that if we wish to see whether nodes i and j are in the same component, all we need do is test whether or not $S(i) = S(j)$.

Let us now consider cases in which the links may fail. With each link, i, we associate a probability, p_i, that the link will fail and, for convenience, the complementary probability, $q_i = 1 - p_i$, that the link is operational. We will consider only that case in which the probabilities are independent. This assumption probably belongs in reliability arguments rather than in survivability arguments since outages due to major natural disasters or extreme antisocial actions, such as bombings, tend to lead to highly correlated damages. Our model of independent outages is not bad for an introduction, however, and we proceed.

One of the most classical of the simple cases involves what is termed series/parallel reductions. Consider the network shown in Figure 14.5: it has two links from node 1 to node 2. The upper link has probability q_1 of working (or surviving) and the bottom link has probability q_2 of working. To calculate the probability that node 1 remains connected to node 2, we

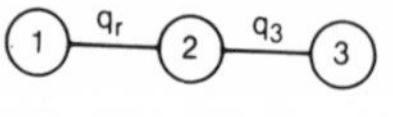

Figure 14.6. A parallel reduction of the network of Figure 14.5.

Figure 14.7. A series reduction of the network of Figure 14.6.

may replace the two links by another link whose probability of working is

$$q_r = 1 - (1 - q_1)(1 - q_2) \tag{20}$$

as shown in Figure 14.6. This type of replacement is called a parallel reduction. If we are solely interested in the probability that node 1 will be connected to node 3, then we can replace the two links between nodes 1 and 3 with an equivalent link whose probability of working is

$$q_s = q_r q_3 \tag{21}$$

as shown in Figure 14.7.

Series/parallel reductions can greatly simplify node to node pathway probability problems. But repeated series/parallel reductions alone can not simplify all cases. Consider, for example, the network of Figure 14.8. No number of series/parallel reductions will enable us to calculate the generalized probability of the existence of a path from node A to node B.

One way to analyze a network like that shown in Figure 14.8 is to consider that, as far as a path existing between nodes 1 and 2, the network can be considered a boolean function in the link variables. We say link i is operating if $x_i = 1$; if $x_i = 0$, the link is inoperable. If there are L links, the network can be thought of as a boolean function, $t(\ )$, of L variables, $t(x_1, x_2, \ldots, x_L)$; a function having 2^L entries in its truth table. The state of the boolean function is then an L-bit tuple. Now for the nice part: because each pair of L-bit tuples are mutually exclusive, the probabilities of all the "favorable cases" may be added together to calculate the probability of a "favorable event." What constitutes a "favorable event?" A favorable case is an L-tuple for which the boolean function is one; a favorable event is the presence of a path between nodes 1 and 4. In Table 1 we list all 32 entries of the truth table and even probabilities of the boolean function $t(x_1, x_2, x_3, x_4, x_5)$, representing a path between nodes 1 and 4 of Figure 14.8. The *reliability polynomial*, $R(\ )$, is the probability of there being a pathway between nodes 1 and 4. For the network of Figure 14.8 it is easy

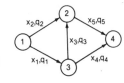

Figure 14.8. A unidirectional network with five labeled links and link operating probabilities.

Table 1. The Truth Table and Associated Probabilities for the Network of Figure 14.8

x_1	x_2	x_3	x_4	x_5	$t(x_1 - x_5)$	Probability
0	0	0	0	0	0	$p_1 p_2 p_3 p_4 p_5$
0	0	0	0	1	0	$p_1 p_2 p_3 p_4 q_5$
0	0	0	1	0	0	$p_1 p_2 p_3 q_4 p_5$
0	0	1	0	0	0	$p_1 p_2 q_3 p_4 p_5$
0	1	0	0	0	0	$p_1 q_2 p_3 p_4 p_5$
1	0	0	0	0	0	$q_1 p_2 p_3 p_4 p_5$
0	0	0	1	1	0	$p_1 p_2 p_3 q_4 q_5$
0	0	1	0	1	0	$p_1 p_2 q_3 p_4 q_5$
0	1	0	0	1	1	$p_1 q_2 p_3 p_4 q_5$
1	0	0	0	1	0	$q_1 p_2 p_3 p_4 q_5$
0	0	1	1	0	0	$p_1 p_2 q_3 q_4 p_5$
0	1	0	1	0	0	$p_1 q_2 p_3 q_4 p_5$
1	0	0	1	0	1	$q_1 p_2 p_3 q_4 p_5$
0	1	1	0	0	0	$p_1 q_2 q_3 p_4 p_5$
1	0	1	0	0	0	$q_1 p_2 q_3 p_4 p_5$
1	1	0	0	0	0	$q_1 q_2 p_3 p_4 p_5$
0	0	1	1	1	0	$p_1 p_2 q_3 q_4 q_5$
0	1	0	1	1	1	$p_1 q_2 p_3 q_4 q_5$
1	0	0	1	1	1	$q_1 p_2 p_3 q_4 q_5$
0	1	1	0	1	1	$p_1 q_2 q_3 p_4 q_5$
1	0	1	0	1	1	$q_1 p_2 q_3 p_4 q_5$
1	1	0	0	1	1	$q_1 q_2 p_3 p_4 q_5$
0	1	1	1	0	0	$p_1 q_2 q_3 q_4 p_5$
1	0	1	1	0	1	$q_1 p_2 q_3 q_4 p_5$
1	1	0	1	0	1	$q_1 q_2 p_3 q_4 p_5$
1	1	1	0	0	0	$q_1 q_2 q_3 p_4 p_5$
0	1	1	1	1	1	$p_1 q_2 q_3 q_4 q_5$
1	0	1	1	1	1	$q_1 p_2 q_3 q_4 q_5$
1	1	0	1	1	1	$q_1 q_2 p_3 q_4 q_5$
1	1	1	0	1	1	$q_1 q_2 q_3 p_4 q_5$
1	1	1	1	0	1	$q_1 q_2 q_3 q_4 p_5$
1	1	1	1	1	1	$q_1 q_2 q_3 q_4 q_5$

to see, using Table 1, that $R(p, q)$ is a 15-term sum:

$$R(p, q) = p_1 q_2 p_3 p_4 q_5 + q_1 p_2 p_3 q_4 p_5 + \cdots + q_1 q_2 q_3 q_4 q_5 \qquad (22)$$

Problem

Note that we prepared the truth table of Table 1 in a very particular way. We grouped together (by dashed lines separating the groups) all the tuples of the same weight or number of ones. Let $w(i)$ be a function which gives the number of tuples of weight i for which the function $t(\ )$ is a one. If all the links have the same probability of operating, $q_1 = q_2 = \cdots = q_L = q$, show that the reliability polynomial can be written

$$R(p, q) = \sum_{i=0}^{L} w(i)q^i p^{n-i} \qquad (23)$$

Verify that, for the truth table of Table 1, $w(0) = 0$, $w(1) = 0$, $w(2) = 2$, $w(3) = 7$, $w(4) = 5$, $w(5) = 1$ and that $R(p, q)$ for the network of Figure 14.8 is

$$R(p, q) = 2q^2p^3 + 7q^3p^2 + 5q^4p + q^5 \qquad (24)$$

Graph (24) for $0 \le q \le 1$.

Now repeat the problem for a "fully connected" network. Such a network has a link between *each* pair of nodes.

Enumeration by truth table is a general, but time costly, method of determining the probability that two nodes will be connected by a pathway of links. The following method, introduced by Wing and Demetriou (1964) is a nice, orderly way of simultaneously computing the probabilities that any node pair will be so connected.

The method steps through all 2^L possible conditions of the L-links. At each step, the probability that the L-links will be in the condition specified by the L-tuple is calculated. This probability is termed the "elementary probability." We form a "terminal reliability" matrix whose ijth entry will, after termination of the algorithm, be the probability that nodes i and j will be connected by a pathway of links.

The algorithm proceeds by filling in the connectivity matrix for each of the 2^L L-tuples and then partitioning the matrix into the form

$$C = \begin{bmatrix} C_1 & 0 & 0 & \\ 0 & C_2 & 0 & \cdots \\ 0 & 0 & C_3 & \\ & & & \vdots \end{bmatrix} \qquad (25)$$

introduced in Chapter 4. The partitioning is done, as explained in Chapter 4, by suitably permuting rows and columns. Once partitioned, the elements $C_1, C_2, C_3, \ldots$ are "fleshed out" by changing all the zero entries in each C_i to ones. Finally, the terminal reliability matrix is updated by adding the elementary probability to entry ij of the terminal reliability matrix if and only if the "fleshed out" connectivity matrix indicates that there is a pathway between nodes i and j. An example should help to resolve questions. Consider that we are working with the network of Figure 14.8 but with bidirectional links. Let us assume that we observe the algorithm as it is working on the particular 5-tuple $(0, 1, 0, 0, 1)$ indicating that links 1, 3, and 4 are inoperable. The elementary probability, e, of this event is $p_1 q_2 p_3 p_4 q_5$. (See also the ninth entry of the truth table in Table 1.) The connectivity matrix is

$$
C = \begin{array}{c} \\ 1 \\ 2 \\ 3 \\ 4 \end{array}
\begin{array}{c} \begin{array}{cccc} 1 & 2 & 3 & 4 \end{array} \\ \left[\begin{array}{cccc} 1 & 1 & 0 & 0 \\ 1 & 1 & 0 & 1 \\ 0 & 0 & 1 & 0 \\ 0 & 1 & 0 & 1 \end{array} \right] \end{array} \tag{26}
$$

We have written the row and column numbers along the matrix edges. We now partition (26) to bring it to the form of (25) and we obtain

$$
C = \begin{array}{c} \\ 1 \\ 2 \\ 4 \\ 3 \end{array}
\begin{array}{c} \begin{array}{cccc} 1 & 2 & 4 & 3 \end{array} \\ \left[\begin{array}{cccc} 1 & 1 & 0 & 0 \\ 1 & 1 & 1 & 0 \\ 0 & 1 & 1 & 0 \\ 0 & 0 & 0 & 1 \end{array} \right] \end{array} \tag{27}
$$

Fleshing out (27) (which is, of course, equivalent to exponentiating it as discussed in the early portion of the chapter) we get

$$
C = \begin{array}{c} \\ 1 \\ 2 \\ 4 \\ 3 \end{array}
\begin{array}{c} \begin{array}{cccc} 1 & 2 & 4 & 3 \end{array} \\ \left[\begin{array}{cccc} 1 & 1 & 1 & 0 \\ 1 & 1 & 1 & 0 \\ 1 & 1 & 1 & 0 \\ 0 & 0 & 0 & 1 \end{array} \right] \end{array} \tag{28}
$$

If TR is the terminal reliability matrix which is accumulating the running sums, then it is updated as follows as required by (28):

$$
TR \leftarrow TR + \left[\begin{array}{cccc} e & e & 0 & e \\ e & e & 0 & e \\ 0 & 0 & e & 0 \\ e & e & 0 & e \end{array} \right] \tag{29}
$$

where e is the elementary probability.

Wing and Demetriou also suggest that, for large networks for which enumeration of all 2^L tuples is impractical, one can resort to a random trials or "Monte Carlo" technique. In this approach, the algorithm is carried out as described except that only a small subset of all the 2^L tuples are randomly generated and the subsequent computations performed. The probabilities in the terminal reliability matrix are normalized by multiplying each of them by $2^L/S$ where S is the number of tuples generated.

Problem

Investigate the Monte Carlo reliability estimator in terms of statistical bias. Work out some examples for simple networks.

A problem that is often posed is the following: "Given that I have N nodes and L links, how should the L links be used to link the N nodes together to achieve maximum reliability or survivability?" The problem with the question is the phrase "maximum reliability or survivability;" it just is not defined. The point can be brought home by analysis of a beautiful problem from Frank (1974). Consider the two $N = 8$, $L = 12$ link networks in Figure 14.9.

For this case let us assume that nodes can fail and that when a node fails, all links connected to the failed node also fail. Let us define survivability as a two-part requirement. First, after damage there must be at *least* two nodes surviving. Second, there must be a path between all surviving nodes, that is, the remaining network must exhibit only *one* component. Which network is better? A quick analysis shows that Network A will not survive if just the two nodes 1 and 2 are destroyed; but Network B will survive if any single pair of nodes is destroyed. Is Network B then a better network? Even though we have been careful to strictly define survivability, our question can still not be answered unconditionally. To see this, to prove this seemingly blatant assertion, we need only calculate the reliability polynomials $R_A(\ )$ and $R_B(\ )$ for the Networks A and B, respectively. For convenience, assume that nodes fail independently with probability p (and survive with probability $q = 1 - p$). It is easy to show that

$$R_A(p, q) = 12q^6p^6 + 36q^7p^5 + 55q^8p^4 + 50q^9p^3 + 27q^{10}p^2 + 8q^{11}p + q^{12} \tag{30a}$$

$$R_B(p, q) = 12q^6p^6 + 24q^7p^5 + 44q^8p^4 + 48q^9p^3 + 28q^{10}p^2 + 8q^{11}p + q^{12} \tag{30b}$$

Figure 14.10 shows the graph of $R_A(\ )$ and $R_B(\ )$ for the region $0.0 \leqslant p \leqslant 0.25$. Note that for some values of q, Network A is superior; but

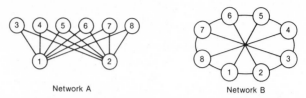

Figure 14.9. Two 8-node, 12-link networks. © 1974 IEEE.

that for other values of q, Network B is preferred. Thus, we see that for a simple case, the quest for a simple answer to the question of preferred structure may be fraught with woe.

Problem

As we have seen, the computation of node-to-node reliability in a network can easily be an immense computational task. Frank and Frisch (1971) point out that there are cases, however, in which there are symmetries which can be exploited. As an example, consider the network of Figure 14.11. We assume that the sender (S) and receiver (R) nodes of the network of Figure 14.11 are "hard" and invulnerable to destruction or failure. All other nodes may fail, independently, each with probability p. (They survive independently with probability $q = 1 - p$.) We seek the probability that, after node failures, a message can still propagate from S to R. Wei et al. (1983) point out that a situation which might be modeled by the network of Figure 14.11 is that of a broadcasting network which relays received messages. Each transmitter is only strong enough to reach the next two receivers in the direction that we wish to propagate our message. A familiar

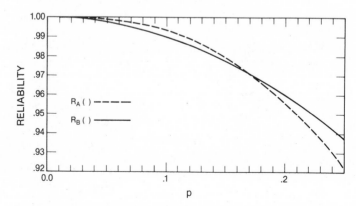

Figure 14.10. Graphs of the reliability polynomials $R_A(\ )$ and $R_B(\ )$.

Figure 14.11. Sender and receiver nodes connected by a network possessing great internal symmetry.

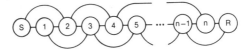

example of such a network obtains for the case of some of the microwave relay chains across the United States. In fact, it is because one transmitter's signal often reaches the receiver beyond the next relay tower that polarization diversity or "frequency frogging" is used.

Looking at Figure 14.11 we see that the n-node network that produces a pathway from S to R possesses a lot of symmetry. How do we exploit this symmetry to calculate the probability, P, that, after damage, a pathway will exist for messages to propagate from S to R.

Probably the easiest way to develop an expression for P is to note that a pathway will exist if and only if no two adjacent nodes are damaged. We can then proceed to write a difference equation for evaluating P. Do so and solve the difference equation using suitable boundary conditions.

References

Frank, H. (1974), Survivability Analysis of Command and Control Communications Networks—Part II, *IEEE Transactions on Communications*, Vol. 22, pp. 596–605.

Frank, H. and I. Frisch (1971), *Communication, Transmission, and Transportation Networks*, Addison-Wesley, Reading, MA.

Wei, V., F. Hwang, and V. Sös (1983), Optimal Sequencing in a Consecutive-2-out-of-n System, *IEEE Transactions on Reliability*, Vol. 32, pp. 30–33.

Wing, O. and P. Demetriou (1964), Analysis of Probabilistic Networks, *IEEE Transactions on Communication Technology*, Vol. 12, pp. 38–40.

Index